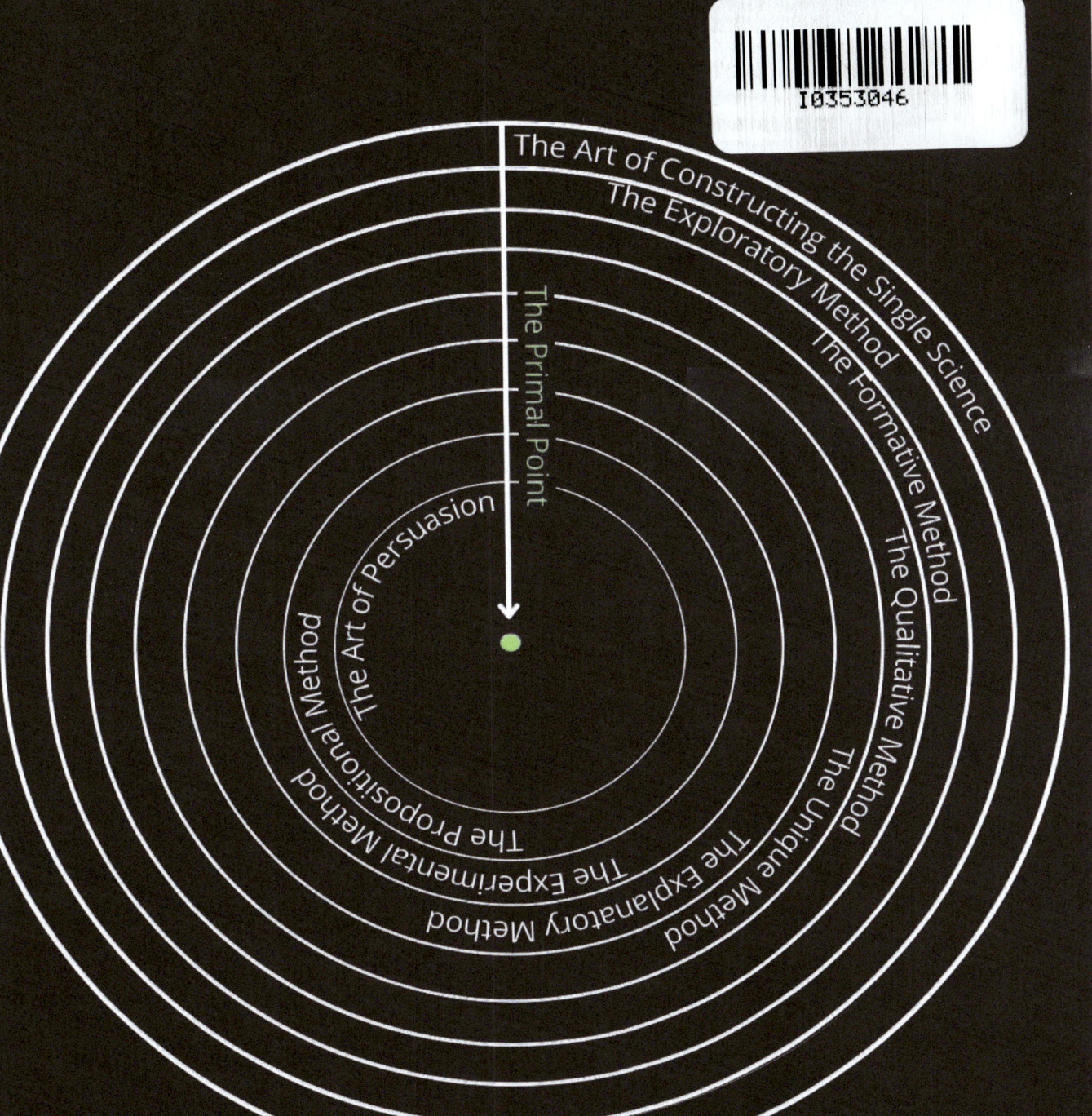

LEONARDO DUQUE

The Single Science

Second Edition

By Leonardo Duque, Marzo 2023

Table of Contents

- Table of Contents 5
- Biographic Note About the Author 11
- Preface 15
- 1 Introduction to The Single Science 19
  - 1.1 Classes and categories are the foundation of philosophy 27
  - 1.2 The methodology followed in the construction of the single science 30
- 2 The Cause of Maturity and the Exploratory Method 37
  - 2.1 Philosophical Argument 40
    - 2.1.1 Mercy Encompasses the Entire Creation 41
    - 2.1.2 Situations with challenges, problems, tests, and difficulties 43
    - 2.1.3 Challenges are blessings 44
    - 2.1.4 Intelligence 44
    - 2.1.5 Power 45
    - 2.1.6 Human powers 46
    - 2.1.7 Patience 47
    - 2.1.8 Discernment and self-control 47
    - 2.1.9 Moral principles 49
    - 2.1.10 Wisdom 50
    - 2.1.11 Sources of wisdom 50
    - 2.1.12 Tenderness, mercy, and empathy 50
    - 2.1.13 Two natures 52
    - 2.1.14 Considering the heart's role in the decision-making process 53
    - 2.1.15 Free will 53
    - 2.1.16 Human kingdom and other kingdoms in nature 54
    - 2.1.17 Stages and phases 55
    - 2.1.18 Sense of our own powerlessness 57
    - 2.1.19 Relationships 57
    - 2.1.20 Laws 57
  - 2.2 Research Steps for the Exploratory Method 58
  - 2.3 The exploratory phase of the farmers' situation 62

3	Summary of the Four Causes Taught by the Greek Philosophers	67
4	The Formative Cause and the Formative Method	69
4.1	Philosophical Argument	72
4.1.1	General properties of matter	76
4.1.2	Religious and scientific principles	76
4.1.3	Love is a Bond Inherent in the Realities of Things	78
4.1.4	The soul	79
4.1.5	Hormones and emotions	80
4.1.6	Power of imagination	80
4.1.7	Sense of religion	80
4.1.8	Sense of justice	80
4.1.9	Relationships	81
4.1.10	Parameters and indicators	81
4.2	Research Steps for the Formative Method	83
4.3	The formative phase of the farmers' situation	85
5	The Material Cause and the Qualitative Method	93
5.1	Philosophical Argument	98
5.1.1	Senses	100
5.1.2	The Spirit Pervades Creation	100
5.1.3	Sense of spirituality	103
5.1.4	Power of knowledge	103
5.1.5	The specific properties of matter	107
5.1.6	Common faculty	107
5.1.7	Supplies	108
5.1.8	Relationships	108
5.1.9	Parameters and indicators	108
5.2	Research Steps for the Qualitative Method	108
5.3	The qualitative phase of the farmers' situation	111
6	The Root Cause and the Unique Method	121
6.1	Science and destiny	129
6.2	Fate, Predestination, and Destiny	130
6.3	Philosophical Argument	132

	6.3.1	Providence:	133
	6.3.2	Fear of God	134
	6.3.3	Grace	137
	6.3.4	Our Destiny is Fixed, and the Whole Planet's too	139
	6.3.5	Are we supposed to reach a high destiny?	140
	6.3.6	The inner being:	142
	6.3.7	Transformation	145
	6.3.8	How to find the root cause of the problem?	148
	6.3.9	Examples of root causes:	153
	6.3.10	The Divine Teachers are Gardeners	160
	6.3.11	Oneness is a law	162
	6.3.12	The concept of reality	163
	6.3.13	Methods	165
6.4		Research Steps for the Unique Method	180
6.5		The Root Phase of the Farmers' Situation	180
7		The Final Cause and the Explanatory Method	189
7.1		Indeed, there is a purpose in nature	190
7.2		Philosophical Argument	194
	7.2.1	Divine Laws are the Breath of Life unto All Created Things	198
	7.2.2	Sense of obligation also called the sense of responsibility	201
	7.2.3	The senses of fear and shame	202
	7.2.4	Memory	204
	7.2.5	Relationships	205
	7.2.6	Parameters and indicators of protection	205
7.3		Research Steps for the Explanatory Method	206
7.4		The explanatory phase of the farmers' situation	210
8		The Efficient Cause and the Descriptive and Experimental Methods	217
8.1		Philosophical Argument	218
	8.1.1	The mind	219
	8.1.2	"To be" and "to do."	220
	8.1.3	The Word of God Pervades All Created Things	220
	8.1.4	Relationships	222

	8.1.5	Power of thought	223
	8.1.6	Power of prayer	224
	8.1.7	Power of reason	225
	8.1.8	Power of expression	225
	8.1.9	Parameters of efficiency	230

8.2 Research Steps for the Experimental and Descriptive Methods ... 230

8.3 The descriptive and experimental phase of the farmers' situation ... 233

9 The Cause of Maturity and the Propositional Method ... 245

9.1 Philosophical Argument ... 245

	9.1.1	Intelligence	247
	9.1.2	Discernment, self-control, and volition	247
	9.1.3	Human kingdom	247
	9.1.4	Kingdoms in nature	247
	9.1.5	Stages and stations	247
	9.1.6	Wisdom	247
	9.1.7	Mechanism of control	247
	9.1.8	Parameters and indicators of progress	247

9.2 The quintessence of knowledge ... 248

9.3 Research steps for the Propositional method ... 249

9.4 The farmers' situation and the propositional phase ... 252

	9.4.1	Increasing small farmers' income	252
	9.4.2	Enabling adolescents to become protagonists of social change	255
	9.4.3	Following up on the suggested farmer's strategy	259

10 The Art of constructing The Single Science ... 261

10.1 The Art of Constructing the Cause of Maturity ... 262

10.2 The Art of Constructing the Formative Cause ... 266

10.3 The Art of Constructing the Material Cause ... 269

10.4 The Art of Constructing the Root Cause ... 272

10.5 The Art of Constructing the Final Cause ... 274

10.6 The Art of Constructing the Efficient Cause ... 278

11 To Fly, a Bird Needs Two Wings Equally Developed ... 281

11.1 Employment and biodiversity's potential ... 287

11.2	Addressing the environmental crisis	294
11.3	Food security for all	297
11.4	Our aspiration for peace all over the world	300
11.5	The Science's Value-free Ideal, Stripped of Emotions and Feelings	310
11.6	Dealing with multiple crises	326
12	The Primal Point of Nine Concentric Circles	335
13	Works Cited	337

Biographic Note About the Author

The author worked as a professor in Colombia for FUNDAEC's University for Rural Wellbeing (Non-Profit Foundation for the Application and Teaching of Science) and Javeriana University.

His search for profound answers to the farmers' dire situation started when three indigenous leaders from Latin American tribes, among others, were invited in 1989 to a seminar about FUNDAEC's almost 80 textbooks fruit of research about agriculture, engineering, education, and agroindustry for rural communities wellbeing. After presenting the results of 15 years of research, he asked them about their opinion. There were a Mapuche from Chile, a Kariña from Venezuela, and a Quechua from Bolivia. At the end of the seminar, he asked them what they thought about FUNDAEC's work during a break. They look at each other and the oldest, Sabino Ortega the Amauta[1]Quechua, said: Leo, what you have is good, but indigenous people perceive nature differently. When Leo asked Sabino, how come? Sabino answered: If I observe a special bird flying over my home, I know visitors started walking to visit us, and they will be here in two or three days. Leo told them that he would study their concern and look for an answer. Their response transformed the course of his life. From the writings of Bahá'u'lláh, he was aware of focusing the search on the ancient Greek philosophers.

It is essential to acknowledge that Counsellor Dr. Eloy Anello, founder, and president of Nur University in Bolivia, was Leonardo's thesis advisor. Leo's thesis became the source of his first book *En Síntesis: ¡La Ciencia es Amor!*

In his first publication, Edmundo Gutiérrez, President of FUNDAEC's "University Center for Rural Well-Being," in the prologue of *En Síntesis: ¡La Ciencia es Amor! Aplicaciones Prácticas en los Organismos No Gubernamentales* (1992), wrote:

> Leonardo Duque in his book addressed to Non-Governmental Organizations which in the last two decades are showing signs of having become a powerful force for change, proposes a new way to face their challenges; his proposal is framed in a context of integration that includes the understanding of man as a multidimensional being, a close link of theory with practice and a fusion between the material and the spiritual.
>
> I can affirm that when reading it, the book forces us to think, to reflect deeply on each of its approaches and leads to the need for a deeper exploration of that fundamental and complementary border, the relationship between the material and the spiritual.

In a second publication, *En la Encrucijada: Una Perspectiva Nueva. En Honor a las Mujeres del Mundo* (1999), Dr. Jairo Roldan, Ph.D. in Philosophy of Science from the Sorbonne University and one of Colombia's foremost physicists, said in the prologue:

> Professor Leonardo Duque argues in this book, and this constitutes one of his main theses, that the complete clarification of the art of governing requires the analysis of what is understood as Science. He maintains that Science and Power are fundamentally in harmony with tenderness, love and mercy. That conviction, together with the fact that the aforementioned qualities have been predominant

---

[1] The Inca tribe is guided by 1.000 wise men, and Sabino was recognized as an experienced teacher.

distinctive attributes in women, inspires him to dedicate his book to the women of the world.

The third chapter, as mentioned above, forms the core of the conceptual framework of the book, a framework that is based on Aristotle's four causes. The author carries out a personal and original interpretation of these difficult topics in order to find the way to a synthesis of knowledge and apply it to the work of grassroots organizations. The most original approach is the attempt to link the four causes to different powers, laws, properties, principles and values, referring to God, nature, and the human being; and the effort to transform a highly abstract topic into a valuable tool to solve practical problems.

It is also in this chapter that the author presents another of his main theses according to which Science is Love. Inspired by Plato's Dialogue 'The Banquet', where the great philosopher speaks of 'the Science of the Beautiful', and in the Writings of Abdu'l-Bahá where Love is identified as the great power that holds the Universe together, Professor Duque identifies Science with Love. The question arises: Love to what? It is not difficult to imagine that one answer would be: love of Beauty. However, it is possible to attempt an identification of Truth with Beauty and, in a mystical sense, to identify true science with a search for the attributes of the Creator, including Beauty. With the Creator being One Reality, love for His beauty, for His reality, would be one thing. By having Religion as a purpose, the search for spiritual truths, and since reality is one, both Science and Religion would seek the same thing in the background. Hence the essential harmony between them, which is one of the fundamental principles accepted by Professor Duque. His thesis, as can be seen, is inspiring novel research on the nature of Science.

Finally, in the ninth chapter, the author presents a summary of the advantages that he believes will be derived from the application of his conceptual framework in the resolution of the various problems that afflict the world today.

In closing, I would like to add that the book is an honest effort to help those most in need solve their problems. The task that the author has set himself is enormous and he does so inspired by his love for humanity and his religious convictions. To what extent Professor Duque achieves his tasks is up to the readers to decide.

Dr. Farzam Arbab, in January 7, 2000 wrote the following in reference to the same book: *At the crossroads, A New Perspective. In honor of the Women of the World*, in Spanish:

Dear Leonardo,

Just a note to thank you for the copies of your book "*En la Encrucijada: Una Perspectiva Nueva.*" I have finally had a chance to look through it, especially the sections you suggested. It is really wonderful that you have been confirmed in carrying out this work. Doors are steadily opening for the Faith in Colombia in ways we could not have imagined a couple of decades ago, and the community is fortunate to have members like you who can walk through them. Be assured of my prayers in the Holy Shrines for you and your family.

In a third publication, Dr. Stephen Beebe - a senior bean researcher at the International Center for Tropical Agriculture (CIAT) in the prologue to *A New Approach to Science* (Duque 2018), said:

Biographic Note About the Author

A hallmark of the age in which we live is the freedom to investigate reality and truth. Indeed, beyond the common concept of "freedom", which implies the possibility of a "take-it-or-leave-it" attitude toward a search for truth, it is indeed the responsibility of each individual to investigate reality. Search is part of our essential being; it is not an option.

If science is the study of cause and effect in the material world, then the causes of phenomena occur at multiple levels: as the raw substance of material reality, as the form that material reality takes, as the action or influence that shapes material reality, and as the purpose or intention behind a given expression of material reality. When put in the context of human decisions, man's spiritual reality and social context, the maturity engendered in light of the Revelation of Bahá'u'lláh guides this process of knowledge toward a more just and peaceful society. The current study investigates the implications of bringing such maturity to bear on the current state of agriculture and smallholder farmers that are prominent in the so-called developing world. It is another contribution to the search after knowledge that the Bahá'í Revelation has provoked in the collective life of humanity.

Leonardo was a member of the National Spiritual Assemblies of the Bahá'ís of Costa Rica (1979-1981) and Colombia (1983-2006) and served in the Local Spiritual Assembly of Central Dekalb in Atlanta from 2007 to today.

He says that: without continuously serving God's Cause, none of his books will realize fruition. The tests and challenges endured while practicing and teaching His Cause became opportunities to advance and serve.

Preface

This book enriches the four causes of Aristotelian thought—i.e., the material, formative, efficient and final causes (Duque, 1.992) – and a previously suggested fifth cause (Duque, 1.999), the Maturity Cause, using the same methodology discovered while working with the four causes.

The evolving conceptual framework led the author to suggest in, The New Approach to Science (2018) to link the four causes taught by the Greeks to the qualitative, the formative, the experimental, and the explanatory methods. Because the author concluded that science is the attraction to the beauty of truth, he discarded the quantitative approach and suggested the Formative Method of Science. Beauty is related to the desire form, shape, or arrangement, searching for justice, unity, harmony, symmetry, and moderation. The exploratory and propositional methods emerged from the Maturity Cause. The art of inter-weaving the causes horizontally and vertically, using the grouping criteria for constructing the "single science" also appears from a very challenging question from the distinguished scholar, Ian Kluge. A total of six methods of science and one art.

Each of the causes is associated with a group of crucial notions and a set of human faculties. The coordinated interaction of those notions and faculties results in the development of each method of science.

Later, the author found a quote from Baha'u'llah in The Summons of the Lord of Hosts: "Say: We have revealed Our verses in nine different modes. Each one of them bespeaketh the sovereignty of God, the Help in Peril, the Self-Subsisting. A single one of them sufficeth for a proof unto all who are in the heavens and on the earth; yet the people, for the most part, persist in their heedlessness. Should it be Our wish, We would reveal them in countless other modes."[2]

That quote inspired the author to continue searching. In this book, the author suggests the Root Cause and the Unique Method of Science and the Art of Persuasion titled: *"To Fly, a Bird Needs Two Wings Equally Developed,"* chapters 6 and 11.

This conceptual framework is a valuable tool for solving everyday common issues and resolving significant individual and collective challenges; its framework to perceive, interpret and transform reality profoundly questions the definition of science today. "The spiritual world is like unto the phenomenal world. They are the exact counterpart of each other. Whatever objects appear in this world of existence are the outer pictures of the world of heaven".[3] What would be the synergic effect when "science and religion, the two most potent forces in human life,"[4] become reconciled?

The author dedicated his other books to the relentlessly persecuted Bahá'ís of Iran. This book is also in their honor for being so resilient and steadfast human beings.

My eternal gratitude to my wife and our dear children, their spouses, and

---

[2] Bahá'u'lláh, The Summons of the Lord of Hosts

[3] The Promulgation of Universal Peace, p. 813

[4] Shoghi Effendi in the Introduction of a book by Bahá'u'lláh, The Proclamation of Bahá'u'lláh, p. xi.

grandkids; I will never forget the joy of being with them and their support, patience, and love in accepting the long hours I dedicate to my research.

ԃԃԥ

To the readers unfamiliar with the Bahá'í Faith, its Central Figures, Institutions, and Writings the author suggests looking for further information in the Bahá'í Reference Library www.bahai.org. Besides English, on this web page, you can find information in the following languages: Arabic, Chinese, French, Hindi, Kiswahili, Persian, Portuguese, Russian, and Spanish. Because the Bahá'í Faith is a world religion, most likely, you will find a web page in your mother's tongue.

Bahá'ís do not proselytize. Individuals use their intelligence to find the truth; however, imitation is a considerable obstacle to seeing it:

> The divine Prophets have revealed and founded religion. They have laid down certain laws and heavenly principles for the guidance of mankind. They have taught and promulgated the knowledge of God, established praiseworthy ethical ideals and inculcated the highest standards of virtues in the human world. Gradually these heavenly teachings and foundations of reality have been beclouded by human interpretations and dogmatic imitations of ancestral beliefs. The essential realities, which the Prophets labored so hard to establish in human hearts and minds while undergoing ordeals and suffering tortures of persecution, have now well nigh vanished. Some of these heavenly Messengers have been killed, some imprisoned, all of Them despised and rejected while proclaiming the reality of Divinity. Soon after Their departure from this world, the essential truth of Their teachings was lost sight of and dogmatic *imitations* adhered to.
>
> Inasmuch as human interpretations and blind *imitations* differ widely, religious strife and disagreement have arisen among mankind, the light of true religion has been extinguished and the unity of the world of humanity destroyed. The Prophets of God voiced the spirit of unity and agreement. They have been the Founders of divine reality. Therefore, if the nations of the world forsake *imitations* and investigate the reality underlying the revealed Word of God, they will agree and become reconciled. For reality is one and not multiple.
>
> The nations and religions are steeped in blind and bigoted *imitations*. A man is a Jew because his father was a Jew. The Muslim follows implicitly the footsteps of his ancestors in belief and observance. The Buddhist is true to his heredity as a Buddhist. That is to say, they profess religious belief blindly and without investigation, making unity and agreement impossible. It is evident, therefore, that this condition will not be remedied without a reformation in the world of religion. In other words, the fundamental reality of the divine religions must be renewed, reformed, revoiced to mankind.[5]

There is no clergy in the Bahá'í Faith and voluntary contributions to support teaching efforts come exclusively from bahá'ís:

> The crucial point in deciding whether or not funds may be accepted from non-Bahá'í sources is the purpose for which the funds are to be used. As you know, it is absolutely forbidden in the Faith to accept from non-Bahá'ís contributions towards the work of the Cause itself. However, in addition to the work of spreading the Faith

---

[5] 'Abdu'l Bahá, The Promulgation of Universal Peace. Emphasis added.

and establishing its institutions, Spiritual Assemblies also engage in humanitarian activities, and contributions from non-Bahá'í sources may be accepted towards such activities. Indeed, although we never ask individual non-Bahá'ís for funds, it sometimes happens that a person who has a great admiration for the Faith insists on contributing. In such a case the contribution may be accepted, with the express provision that it will be used only for charitable and humanitarian purposes.

1. Introduction to The Single Science

While living in Atlanta, Georgia, let me introduce this chapter with a recent discovery when serving the Local Spiritual Assembly of Central Dekalb to establish a solid foundation for the core activities.

When reflecting on the following outcomes, coordinated sets of small groups of students and researchers should form to get used to practicing team research, which seems to be the future of interdisciplinary analysis of a complex reality. Here is an example of the challenge ahead: 'Abdu'l-Bahá said: "The more microscopic animals exist in the soil, the better the plants will grow."[6] [7]

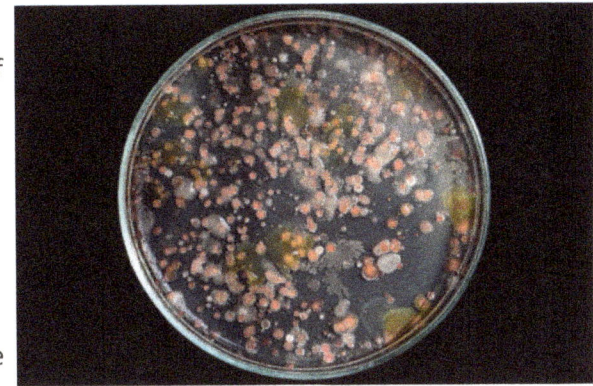

- Will the Master's quote affect the world's farmers for a least one thousand years? Is that the lasting period for the line of inquiry given to us?
- What is the situation of the soils today? (Hobson, Jeremy, 2019)[8], (Lori, Martina, et al., 2017)[9]
- Will it increase the productivity of the harvest meaningfully? ≈ -13% (Pearsons, Kirsten Ann, et al., 2022)[10], ≈ 47% and 60% (de Kroon, Hans, et al., 2011.)[11], ≈ 200% (Kevin E. Mueller, et al. 2013.)[12], ≈ 500% (Schnitzer, Stefan A., et al. 2011.)[13]. Will it improve the profit of organic farmers?
  - Will it contribute to addressing climate change by sequestering carbon in the soil? (Zomer, Robert J., et al., 2017)[14], (Thakur, Madhav Prakash, et al., 2015)[15], (Prommer, Judith, et al., 2019)[16], (Renee Cho, 2018)[17]
- If productivity increases, would it be a motif for farmers to gradually expand the

---

[6] 'Abdu'l-Bahá, Additional Tablets, Extracts and Talks.

[7] Photo 143153200 © Sarawut Chainawarat | Dreamstime.com.

[8] https://www.wbur.org/hereandnow/2019/09/20/soil-degradation-climate-change

[9] https://pubmed.ncbi.nlm.nih.gov/28700609/

[10] https://www.mdpi.com/2071-1050/14/2/631/htm

[11] https://besjournals.onlinelibrary.wiley.com/doi/10.1111/j.1365-2745.2011.01906.x

[12] https://esajournals.onlinelibrary.wiley.com/doi/10.1890/12-1399.1

[13] http://www.ncbi.nlm.nih.gov/pubmed/21618909

[14] https://www.nature.com/articles/s41598-017-15794-8

[15] https://pubmed.ncbi.nlm.nih.gov/26118993/

[16] https://onlinelibrary.wiley.com/doi/full/10.1111/gcb.14777

[17] https://news.climate.columbia.edu/2018/02/21/can-soil-help-combat-climate-change/

heterogeneity of poly-cultures, cover crops (Finney, D.M., et al., 2017)[18],(Vukicevich, Eric, et al., 2016)[19], (Kima, Nakian, et al., 2020)[20], and microorganisms in the soil? (Tilman, David, et al. 2014)[21], (Steinauer, Katja, et al. 2016)[22], (Thakur, Madhav Prakash, et al., 2015)[23], (Prommer, Judith, et al., 2019)[24], (Chen, Chen, et al. 2019)[25], (Mueller, Kevin, et al., 2013)[26]. Will it reduce the number of square miles planted with monocultures?

- If we want to increase the diversity of microorganisms in the soil, what are the sources of abundant diversity of microscopic organisms? (Noah Fierer and Robert B. Jackson, et al., 2006)[27], (Baldrian, Petr, 2016)[28], (Clemmensen, K. E. *et al.*, 2013)[29]. Will farmers gradually opt for agroforestry because the grounds where you find a greater variety of microorganisms are in the forest? (Smith, Laurence G., et al., 2022)[30].

- What lab equipment is required to test the diversity and quantity of microbes at the farmers' farms and in the soil amendment? (Laboratory Equipment, University of Maine)[31], (microBIOMETER®)[32].

- What are the methods (Pearsons, Kirsten Ann, et al., 2022)[33], (Chen, Chen, et al. 2019)[34], (Zhen, Zhen, et al., 2014)[35], and requisites (Björn Berg, 2000)[36] to be used by the storehouse to collect, reproduce, preserve, and distribute them?

---

[18] https://www.jswconline.org/content/72/4/361.short

[19] https://link.springer.com/article/10.1007/s13593-016-0385-7

[20] https://www.mssoy.org/uploads/files/kim-et-al-soil-biol-biochem-jan-2020.pdf

[21] https://www.cedarcreek.umn.edu/biblio/fulltext/Tilman-etal_annurev-ecolsys-2014.pdf

[22] https://pubmed.ncbi.nlm.nih.gov/26286355/

[23] https://pubmed.ncbi.nlm.nih.gov/26118993/

[24] https://onlinelibrary.wiley.com/doi/full/10.1111/gcb.14777

[25] https://www.nature.com/articles/s41467-019-09258-y

[26] https://esajournals.onlinelibrary.wiley.com/doi/10.1890/12-1399.1

[27] https://www.pnas.org/content/103/3/626

[28] https://academic.oup.com/femsre/article/41/2/109/2674172

[29] https://www.science.org/doi/abs/10.1126/science.1231923

[30] https://www.sciencedirect.com/science/article/abs/pii/S0308521X21003103

[31] https://umaine.edu/ednalab/equipment/

[32] https://microbiometer.com/our-story/

[33] https://www.mdpi.com/2071-1050/14/2/631/htm

[34] https://www.nature.com/articles/s41467-019-09258-y

[35] https://journals.plos.org/plosone/article?id=10.1371/journal.pone.0108555

[36] https://www.researchgate.net/publication/223880248_Litter_decomposition_and_organic_matter_turnover_in_northern_forest_soils

Introduction to The Single Science

Will it increase our willingness to recycle organic matter, food leftovers, and garden waste? activity of fungi especially in litter where their contribution to microbial transcription was > 50% (Žifčáková, Lucia, et al., 2016)[37], ≈ 32 % of the global yield gap for maize, and 60 % of the gap for wheat (Oldfield, Emily E., et al., 2019)[38], compost with manure ≈ 45% and 200% (Ouédraogo, E., et al., 2001)[39].

- If productivity increases meaningfully, would it be a motive for farmers to refrain from applying insecticides, herbicides, and fungicides? (Gunstone, Tari, et al., 2021)[40], (Barros-Rodríguez, Adoración, et al., 2021)[41], and heavy machinery? (Sekara, Udayakumar and Sandeep Kumar, 2021)[42].
- If seeds descendants are planted during several cycles in soils with a greater diversity of microorganisms and the loving care of a good gardener, what perfections may start to appear?
  - Will it generate millions of jobs? How many barriers will farmers grow, and how many durable terraces will farmers decide to build to deal with soil erosion?
  - One of the reasons for the productivity increase and host-specific *disease* decrease is increasing microbial community diversity (Schnitzer, Stefan A., et al., 2011)[43]. Will it improve the health of plants? (Garbeva, P., et al., 2005)[44], (Eric Vukicevich, et al., 2016)[45], (Mommer, Liesje, et al., 2018)[46], (Cao, Xia, et al., 2022)[47], (, et al.,). Soils in "organic systems had 32% to 84% greater microbial biomass carbon", in a meta-analysis that integrated data from 56 mainly peer-reviewed papers (Lori, Martina, et al., 2017)[48]. Organic farming brings better nutritional value (Pearsons, Kirsten Ann, et al., 2022)[49]. Yet, it seems not evident that animals' health improves (Sutherland, Mhairi A., et al., 2013)[50] (Brodziak,

---

[37] https://pubmed.ncbi.nlm.nih.gov/26286355/

[38] https://soil.copernicus.org/articles/5/15/2019/

[39] https://www.sciencedirect.com/science/article/abs/pii/S0167880900002462

[40] http://unscilenced.com/gunstone-et-al-2021/

[41] https://www.mdpi.com/2223-7747/10/11/2325

[42] https://link.springer.com/chapter/10.1007/978-981-16-0827-8_17

[43] http://www.ncbi.nlm.nih.gov/pubmed/21618909

[44] https://sfamjournals.onlinelibrary.wiley.com/doi/abs/10.1111/j.1462-2920.2005.00888.x

[45] https://link.springer.com/article/10.1007/s13593-016-0385-7

[46] https://pubmed.ncbi.nlm.nih.gov/29468690/

[47] https://link.springer.com/article/10.1007/s13580-021-00373-8

[48] https://pubmed.ncbi.nlm.nih.gov/28700609/

[49] https://www.mdpi.com/2071-1050/14/2/631/htm

[50] https://www.ncbi.nlm.nih.gov/pmc/articles/PMC4494360/

Introduction to The Single Science

Aneta, et al., 2021).[51] In humans? (Mayo Clinic Staff).[52]

Those are some of the potentialities of a meaningful quote from the Master, with only 14 words. The author believes that 'Abdu'l-Bahá was the most outstanding scientist ever!

In the Writings, it is clear that moderation is crucial to success. It is prudent to protect farmers by being cautious. Implement the guidance mentioned above gradually without putting the harvest at risk, for example, using sources of microorganisms from healthy environments, not rotten[53] trees.

It is evident that if productivity increases meaningfully, it will not guarantee food security for all.

The problem is not just productivity, in *Ending Poverty*, the UN says: According to the most recent estimates, in 2015, 10 percent of the world's population or 734 million people lived on less than $1.90 a day. One out of five children live in extreme poverty, and the negative effects of poverty and deprivation in the early years have ramifications that can last a lifetime. In 2016, 55% of the world's population –about 4 billion people – did not benefit from any form of social protection.[54] [55]

"About 3 billion people worldwide lack adequate facilities to safely wash their hands at home."[56] [57]

The author believes that Eduardo Galeano accurately described the **The Nobodies'** situation:

---

[51] https://www.mdpi.com/2076-2615/11/10/2760

[52] https://www.mayoclinic.org/healthy-lifestyle/nutrition-and-healthy-eating/in-depth/organic-food/art-20043880

[53] Putrid or bad smelling.

[54] United Nations, Ending Poverty. Web. February 2022 <https://www.un.org/en/global-issues/ending-poverty>.

[55] Photo ID 43884710 © Paop | Dreamstime.com

[56] CDC- Centers for Disease Control and Prevention, Access to Clean Water, Sanitation, and Hygiene.

[57] Photo ID 20638996 © Djembe| Dreamstime.com

Fleas dream of buying themselves a dog[58], and nobodies dream of escaping poverty: that one magical day good luck will suddenly rain down on them— will rain down in buckets. But good luck didn't rain down yesterday, today, tomorrow, or ever. Good luck doesn't even fall in a fine drizzle, no matter how hard the nobodies summon it,  even if their left hand is tickling, or if they begin the new day with their right foot, or start the new year with a change of brooms.

The nobodies: nobody's children, owners of nothing. The nobodies: the no ones, the nobodied, running like rabbits, dying through life, screwed every which way.

Who are not, but could be.

Who don't speak languages, but dialects.

Who don't have religions, but superstitions.

Who don't create art, but handicrafts.

Who don't have culture, but folklore.

Who are not human beings, but human resources.

Who do not have faces, but arms.

Who do not have names, but numbers.

Who do not appear in the history of the world, but in the police blotter of the local paper.

The nobodies, who are not worth the bullet that kills them.

Continuing the conversation about the potentialities of the quote, "The more microscopic animals exist in the soil, the better the plants will grow,"[59] this book is about empowering the nobodies populating the grassroots of society with scientific methods to become peaceful protagonists of social change. They own an untapped wealth that will benefit the whole earth. "Regard man as a mine rich in gems of inestimable value. Education can, alone, cause it to reveal its treasures, and enable mankind to benefit therefrom."[60]

☙❧

Harmony between science and religion is one of the principles of the Bahá'í Faith. For a long time, they have been considered incompatible. Jacques Monod, said:

Armed with all the powers, enjoying all the wealth they owe to science, our societies are still trying to practice and to teach systems of values already destroyed at the roots by that very

---

[58] Photo ID 76259988 / Fleas © Dimarik16 | Dreamstime.com

[59] 'Abdu'l-Bahá, Additional Tablets, Extracts and Talks.

[60] Bahá'u'lláh, Tablets of Bahá'u'lláh Revealed after the Kitáb-i-Aqdas, pp. 161–162

science. Man knows at last that he is alone in the indifferent immensity of the universe, whence which he has emerged by chance. His duty, like his fate, is written nowhere. —[61]

I am aware that prejudices and ideologies cause bias that distorts the truth, and I believe that "Truthfulness is the foundation of all human virtues. Without truthfulness, progress and success, in all the worlds of God, are impossible for any soul. When this holy attribute is established in man, all the divine qualities will also be acquired"[62], I sought to explore the harmony of science and religion by scrutinizing the four causes taught by the ancient Greek philosophers, which I explained more fully below. I did this for the below mentioned three reasons:

The first reason is outlined in the following quotes by Roger Walsh, from his "*Staying Alive: the psychology of human survival*":

> Our current crises are seen as expressions of the mistaken desires, fears, and perceptions that arise from our mistaken identity. Since all the major threats to human survival and wellbeing are human-caused, they are of course, deeply, though not exclusively, psychological in origin. The state of the world, then, is a reflection of the state of our individual and collective minds. He points out psychological, educational, and sociological strategies that may mitigate our situation.
>
> World issues proceed, as a last resort, from human minds and from the actions that these minds unleash; because of this, the efforts to subdue our fears and egoisms; to overcome people's mistaken beliefs or transcend non-solidary attitudes are keys in the planetary evolution.
>
> Inasmuch as the state of the world reflects the state of our minds, then what we have called our global "problems" are actually global "symptoms": symptoms of our individual and collective pathologies and immaturities. Therefore, truly curative responses will need to address these internal sources. We will need multifaceted interventions in which we not only try, for example, to feed the hungry and work for peace, but also attempt to address the psychological roots of these problems: the greed, hatred and delusion, and lack of Love, compassion, and wisdom, which created them in the first place. We are in a race between consciousness and catastrophe.[63]
>
> The perennial philosophy, which lies at the heart of the great religions and is increasingly said to represent their deepest thinking, suggests a very different view. It views consciousness as central and its development as the primary goal of existence. This development is said to culminate in the condition variously known in different traditions as enlightenment, liberation, salvation …
>
> The descriptions of this condition show remarkable similarities across cultures and centuries. Its essence is said to be the recognition that the distortions of our usual state of mind are such that we have been suffering from a case of mistaken identity.[64]

---

[61] https://todayinsci.com/M/Monod_Jacques/MonodJacques-Quotations.htm

[62] 'Abdu'l-Bahá qtd. by Shoghi Effendi, The Advent of Divine Justice, p. 26.

[63] Walsh, Staying Alive.

[64] Walsh, El Compromiso con el Planeta.

The second reason is that the Greek philosophers have been, throughout history, universally recognized for their great wisdom, and their books amply disseminated.

Third, and most importantly, is that Bahá'u'lláh, when referring to the Greek philosophers, said,

> Although it is recognized that the contemporary men of learning are highly qualified in philosophy, arts and crafts, yet were anyone to observe with a discriminating eye he would readily comprehend that most of this knowledge hath been acquired from the sages of the past, for it is they who have laid the foundation of philosophy, reared its structure and reinforced its pillars. Thus doth thy Lord, the Ancient of Days, inform thee. The sages aforetime acquired their knowledge from the Prophets, inasmuch as the latter were the Exponents of divine philosophy and the Revealers of heavenly mysteries.
>
> Consider Hippocrates, the physician. He was one of the eminent philosophers who believed in God and acknowledged His sovereignty. After him came Socrates who was indeed wise, accomplished and righteous. ... He is the most distinguished of all philosophers and was highly versed in wisdom.
>
> After Socrates came the divine Plato who was a pupil of the former and occupied the chair of philosophy as his successor. He acknowledged his belief in God and in His signs which pervade all that hath been and shall be. Then came Aristotle, the well-known man of knowledge. He it is who discovered the power of gaseous matter.
>
> These men who stand out as leaders of the people and are pre-eminent among them, one and all acknowledged their belief in the immortal Being Who holdeth in His grasp the reins of all sciences.[65]

The great Greek philosophers based their explanation of all phenomena on four fundamental causes, or reasons for phenomena: the Material Cause, the Formative Cause, the Efficient Cause, and the Final Cause. Today's academic world seems uninterested in the significance of the four Causes taught by the Greek philosophers.

Academics seem apprehensive and uncertain regarding the possibility of unifying knowledge. Consider, for example, Mikael Stenmark's conception of science:

> The idea seems to be that science should be free from not merely ideological or religious values but also ideological or religious beliefs. ... With this clarification in mind, I propose that we define the value-free view of science more precisely in this way: The value-free view of science is the standpoint that science should be autonomous, neutral, impartial, non-responsible, and non-normative.[66]

In supporting his perspective about an impartial science, Stenmark says: "Science should be impartial in the sense that it should not presuppose the truth of any particular political vision, religion or ideology in the validation of scientific theories."[67]

About Stenmark's statement, I would not waste my time trying to demonstrate that God's teachings are false; I would dedicate my ephemeral life to understanding their validity.

---

[65] Tablets of Bahá'u'lláh Revealed After the Kitáb-i-Aqdas, Tablet of Wisdom, pp. 146-47.

[66] Mikael Stenmark, qtd. in LeRon Shults (ed.), The Evolution of Rationality, p. 51.

[67] ibid. p. 51.

Let us go back to what Roger Walsh said:

"our current crises are seen as expressions of the mistaken desires, fears, and perceptions that arise from our mistaken identity," and that "the current threats to human survival and wellbeing *are actually symptoms*, symptoms of our individual and collective mindset. The state of the world is, therefore, a creation and expression of our own minds, and it is to our own minds that we must look for solutions."[68]

The following questions, among many others, arise from these reflections:

- Which options do I have once I reach the age of maturity?
- How do I envision myself?
- What am I?
- Am I unique?
- Why do I exist? For what do I exist?
- Who am I?

To examine the approach to answering these questions, I will start the philosophical discussion of the causes with two quotes by 'Abdu'l-Bahá that served as the foundation for this search:

Essential pre-existence is an existence which is not preceded by a cause; essential origination is preceded by a cause. Temporal pre-existence has no beginning; temporal origination has both a beginning and an end. For the existence of each and every thing depends upon four causes: the efficient cause, the material cause, the formal cause, and the final cause. So this chair has a creator who is a carpenter, a matter which is wood, a form which is that of a chair, and a purpose which is to serve as a seat. Therefore, this chair is essentially originated, for it is preceded by, and its existence is conditioned upon, a cause. This is called essential or intrinsic origination.[69]

'Abdu'l-Bahá is reported to have said:

Someone desires an explanation of the terms soul, mind and spirit. The terminology of ancient and modern philosophers differs. According to the great ancient philosophers the words soul, mind and spirit implied the underlying principles of life; the essence was expressed under different names and these three terms designated the various functions of the absolute reality, or the operations of the one single essence; for instance, when they dealt with the *sensations of emotion* they called it the *soul*; when they desired to express that *power which discovers the reality of phenomena* they gave it the appellation of *mind* and when they discussed the *consciousness* which pervades the world of creation they gave it the title of *spirit*.[70]

When I started my research, I was not aware of the epistemological discussions about understanding that would allow one to say that one owes the knowledge of something, nor had I read about the metaphysical search for the meaning of "truth." I did not receive guidance on

---

[68] Roger Walsh, Staying Alive: the psychology of human survival. Emphasis added.

[69] 'Abdu'l-Bahá, Some Answered Questions, pp.155-56.

[70] 'Abdu'l-Bahá, Divine Philosophy, p. 119. Emphasis added.

comprehending the philosophical questions related to the concepts of causes and principles. I believe that I likely would have lost the path if I had these things!

The assumption upon which I based my investigation started believing deeply in my heart that it was possible to bring about harmony between science and religion. In this book, the reader will find faith and all science disciplines fused in a common conceptual framework: a single science. The problem today is the fragmentation of knowledge into scientific specialties and the separation of science and religion in two entirely separate domains. I started the puzzle, putting a few notions next to each of the four causes that the Greek philosophers taught and then searching their writings for a meaningful connection. It was frustrating but ultimately rewarding with patience and perseverance.

Each of the six causes, the four original Causes taught by the Greek philosophers, the Maturity Cause and the Root Cause, is associated with one or two scientific methods and a set of human faculties. The Formative Cause to the formative method; the Material Cause to the qualitative method; the Final Cause, taught by the Greeks, to the explanatory method, and the Efficient Cause to the descriptive and experimental methods. The exploratory and propositional methods emerged from the Maturity Cause and the Root Cause to the Unique Method of Science.

The author will refer to Diotima's text, Socrates's teacher, the wise woman Diotima of Mantineia, in each one of the causes taught by the Greek Philosophers to support the conclusion that a single science is "Love": attraction to the beauty of truth.

1.1 Classes and categories are the foundation of philosophy

Classes: clusters, groups, categories, sets, conglomerates, and the answer to: "What type of object or being?" "What species of plants or insects?"

Socrates explains why "categories" should be included in the Efficient Cause, which we will study later on in this document:

Stranger- Then, not to exclude anyone who has ever speculated at all upon the nature of being, let us put our questions to them as well as to our former friends.

Theaetetus- What questions?

Stranger- Shall we refuse to attribute being to motion and rest, or anything to anything, and assume that they do not mingle, and are incapable of participating in one another? Or shall we gather all into one class of things communicable with one another? Or are some things communicable and others not? Which of these alternatives, Theaetetus, will they prefer?

Theaetetus- I have nothing to answer on their behalf. Suppose that you take all these hypotheses in turn, and see what are the consequences which follow from each of them. …

Stranger- And now, if we suppose that all things have the power of communion with one another -what will follow?

Theaetetus- Even I can solve that riddle. Stranger- How?

Theaetetus- Why, because motion itself would be at rest, and rest again in motion, if they could be attributed to one another.

Stranger- But this is utterly impossible.

Theaetetus- Of course.

Stranger- Then only the third hypothesis remains. Theaetetus- True.

Stranger- For, surely, either all things have communion with all; or nothing with any other thing; or some things communicate with some things and others not.

Theaetetus- Certainly.

Stranger- And two out of these three suppositions have been found to be impossible.

Theaetetus- Yes.

Stranger- Every one then, who desires to answer truly, will adopt the third and remaining hypothesis of the communion of some with some.

Theaetetus- Quite true.

Stranger- This communion of some with some may be illustrated by the case of letters; for some letters do not fit each other, while others do.

Theaetetus- Of course.

Stranger- And the vowels, especially, are a sort of bond which pervades all the other letters, so that without a vowel one consonant cannot be joined to another.

Theaetetus- True.

Stranger- But does every one know what letters will unite with what? Or is art required in order to do so?

Theaetetus- What is required.

Stranger- What art?

Theaetetus- The art of grammar.

Stranger- And is not this also true of sounds high and low?-Is not he who has the art to know what sounds mingle, a musician, and he who is ignorant, not a musician?

Theaetetus- Yes.

Stranger- And we shall find this to be generally true of art or the absence of art.

Theaetetus- Of course.

Stranger- And as classes are admitted by us in like manner to be some of them capable and others incapable of intermixture, must not he who would rightly show what kinds will unite and what will not, proceed by the help of science in the path of argument? And will he not ask if the connecting links are universal, and so capable of intermixture with all things; and again, in divisions, whether there are not other universal classes, which make them possible?

Theaetetus- To be sure he will require science, and, if I am not mistaken, the very greatest of all sciences.

Stranger- How are we to call it? By Zeus, have we not lighted unwittingly upon our free and noble science, and in looking for the Sophist have we not entertained the philosopher unawares?

Theaetetus- What do you mean?

Stranger- Should we not say that the division according to classes, which neither makes the same other, nor makes other the same, is the business of the dialectical science?

Theaetetus- That is what we should say.

Stranger- Then, surely, he who can divide rightly is able to see clearly one form pervading a scattered multitude, and many different forms contained under one higher form; and again, one form knit together into a single whole and pervading many such wholes, and many forms, existing only in separation and isolation. This is the knowledge of classes which determines where they can have communion with one another and where not.

Theaetetus- Quite true.[71]

Including "categories" as part of this proposal allows others to bring about higher levels of excellence. The author tried to delineate the grouping criterion used for each of the causes, carefully setting the group's identity. For example, some of the notions closely associated with the Material Cause imply to develop a spiritual foundation and to acquire valuable knowledge, as wellsprings of material wellbeing in determining prosperity in the quality of life. An honest individual, a faithful husband, a clean person will have a much better quality of life than the contrary. Be mindful that all the notions closely associated with the other causes are complementary sets of the one considered. For example, collective responsibility, unity of thought, unity of vision, and collective volition are part of the ideal to reach a better quality of life for all and are mentioned in other Causes, but not in the Material Cause. The reader will find that all causes are intertwined.

In a search for a harmonious relationship between science and religion, the task implied exploring and linking concepts in both realms: the Book of Creation and the Book of Revelation. As the reader can perceive, to suggest a new idea of science, the author grouped a carefully chosen set of notions to each cause. The novelty of the way of the adopted grouping is that each "cause" has a conceptual framework that includes:

- a principle that encompasses all creation,
- concepts closely related to that principle,
- a structure of relationships,
- certain general laws,
- a method,
- a set of human faculties,
- some spiritual values and key notions,
- and the associated parameters and indicators.

'Abdu'l-Bahá, in *Some Answered Questions*, taught:

> That is why man is said to be the greatest sign of God—that is, he is the Book of Creation—for all the mysteries of the universe are found in him. Should he come under the shadow of the true Educator and be rightly trained, he becomes the gem of gems, the light of lights, and the spirit of spirits; he becomes the focal center of divine blessings, the wellspring of spiritual attributes, the dawning-place of heavenly lights, and the recipient of divine inspirations. Should he, however, be deprived of this education, he becomes the embodiment of satanic attributes, the epitome of animal vices, and the source of all that is oppressive and dark.[72]

---

[71] Plato, Sophist.

[72] 'Abdu'l-Bahá, Some Answered Questions, p. 236.

1.2 The methodology followed in the construction of the single science

Scientific research can be a journey with different stages, as shown in the graphic above. We can examine a problem using all the methods or just utilizing some of them. Both ways are valid. Individuals or groups can begin exploring a problem and, after some reflection, propose a solution.

The single science I have been looking for, requires the capacity to classify into one of the causes, any notion relevant to the problem. Then, in a systematic search for the "beauty of truth, proceed through the phases of the cycle of scientific research."

Please refer to the table below, showing the methodology followed in constructing the "Single Science" by inter-weaving the causes horizontally and vertically.

Ian Kluge, a bahá'í scholar philosopher, revised the first draft of my previous book, The New Approach to Science". He started asking some questions, and one of them was very profound. It took me several weeks to answer it because the methodology gradually emerged when applying the scientific methods proposed in my previous books for decades.

When Ian received the below-mentioned chart, he said:

> "I think the idea underlying the chart is excellent and draws attention to correspondences within reality as analyzed from the perspective of Aristotle's four-fold causality. Contemporary world-views are very weak vis-à-vis correspondences in reality because they lack the vision of reality as made up of different ontological levels. This literally blinds them to some of the relationships that your chart makes apparent."

Introduction to The Single Science

	FINAL CAUSE	MATERIAL CAUSE	FORMATIVE CAUSE	EFFICIENT CAUSE	MATURITY CAUSE	
1	**Question:** Why? For what?	**Question:** With what?	**Question:** How is organized?	**Questions:** With what being? With whom? With what type?	**Question:** With which option? When addressing, issues, problems, needs, and difficulties	CAUSE OF MATURITY
2	**Essence:** the power of law itself, i.e., that which is a condition to a particular aspect of life or nature, and explains its purpose, its mission	**Essence:** The Spirit. **Elements, substances, and raw materials.** **Specific properties of matter,** peculiarities, and things' features. **Virtues** also called spiritual values. **Knowledge, social values and beliefs**	**Essence:** Love. **The general properties of matter:** form, size, mass, temperature, movement, and position in time and space **Hormones, feelings, and emotions**	**Essence:** The Word of God. **To be:** animated and unanimated beings. **Forces:** the power of the mind, energy, capacities, potential, talents, vocations, arguments, concepts. **Categories:** genres, clusters, groups. Interacting entities within a system.	**Essence:** Mercy. **Sources of Wisdom:** What history or experience, the diverse disciplines of science and the world religions say about the addressed issue **Moral principles** **Kingdoms, Cycles, and phases**	MATERIAL CAUSE
3	Social order, laws, and norms, natural laws, Religious laws, ordinances, pacts, and agreements. Individual and collective commitment to ideal **Goal:** purpose, and mission	Spiritual laws, the standards that regulate quality, and laws of possession **Goals:** objectives	**Laws** related to unity, harmony, justice, and the law of gravity. **Goal:** Aspirations and desires to reach the expected vision or design	**Laws** of thermodynamic, labor law, and tools regulations. Grammar rules. The Word of God is Law **Fruits of labor:** results, harvest, services, products, and leftover materials or efforts	**Laws:** related to the inherent responsibilities of the assumed decision taking into account the effects in the other Kingdoms and the Human Kingdom. **Goal:** Stewardship in the management of the trust	FINAL CAUSE

Introduction to The Single Science

EFFICIENT CAUSE

Powers: Memory and senses of responsibility, fear, and shame **Advice from those with experience.** The consequences of obeying and disobeying.	**Powers:** conscience, knowledge, the common faculty, senses of spirituality, and appreciation **Specific lines of action or policies** to consolidate ethics, knowledge, quality, and efficacy	**Powers:** imagination and senses of justice and religion **The practical application of the principles related to the general properties of the matter** concerning the desired design, organization, and arrangement	**Powers:** thought, reason, and expression **To do:** movements: methods, processes, activities, abilities, skills, arts, technologies, mechanisms.	**Powers:** Intelligence, patience, self-control, discernment, compassion, empathy, volition **Decisions made at the individual and collective level:** Governmental Institutions, private enterprises, NGOs, families, or individuals. The human choices made concerning the stages of a plan or phases of a cycle **The spontaneous differential response to the stimuli received** by the different kingdoms **The effects of the decisions made by humans** in all kingdoms **The developmental stage** of the individual, plant or animal **Monitoring and evaluating** the strategy and its stages and the cycle's phases

4

Introduction to The Single Science

#						
5	**Situations** of risk and danger. **Relationships** of gratitude, loyalty, reciprocity, fear, protection, guilt, repentance, punishment, and reward.	**Relationships** of belonging, possession, and distinction	**Religious and Scientific principles** related to the general properties of matter	**Relations of:** Cause and effect, logic, and reason.	**Challenging Situations** in which the human kingdom chooses how to respond or, in the case of the mineral, plant, and animal kingdoms, the spontaneous differential response to the stimuli received. **Relationships:** are those that interconnect the sources of wisdom	FORMATIVE CAUSE
6	**Parameters and indicators** of protection, equality, prevention and security in the fulfillment of the mission	**Parameters and indicators** of efficacy and quality in attaining the goals and objectives	**Parameters and indicators** of unity, beauty, symmetry, harmony, reciprocity, and justice in reaching the desired vision or design	**Parameters and indicators** of efficiency and productivity in achieving the results	**Parameters and indicators** of progress: Such as level of respect for the individual to exercise his (her) free will, level of commitment, number of pledges, and number of recommendations or referrals	CAUSE OF MATURITY

CEO of a factory will get a lot of referrals if she delivers a product that is safe for the customers, with high quality, a comfortable and beautiful design, at a reasonable cost because profit sharing, and that is recyclable after a lasting life.

8 An educational program will become highly recommended if it teaches adolescents to be
- individually and collectively responsible,
- nurture their spiritual endowment,
- moderate their impulses and aim towards praiseworthy aspirations,
- able to describe social and natural phenomena,
- acquire the capacity to make sound choices in their lives,
- and develop the skills and capabilities to become the protagonist of constructive social change, aiming to serve humanity

Rows 7 and 8 suggest possible outcomes when applying the conceptual framework.

The suggested conceptual framework's desired result is to deepen the bonds between man's inner nature and his material surroundings. Similarly, with clear transparency, the theoretical framework expounds on the importance of education as the foundation for the development of the peoples of the world, empowering them with the methods of science learning in the context of agriculture's biodiversity as the basis of order and system in society.

I was confronted with the disturbing reality of the rural areas while working at FUNDAEC's University for Rural Well-being[73]. I fell in Love with FUNDAEC's scientific approach to addressing the rural areas' needs, and I am still pleased and committed to addressing the farmers' dire situation.

The reader can observe in the above table that to each essence, the author associated a well-articulated set of fundamental notions:

- In the Cause of Maturity, Mercy was the reason that helped the author bring into existence the set of notions of the Exploratory and Propositional Methods of Science.
- Love as the Formative Cause showed the author how to justify the origination of the set of science's Formative Method notions.
- The Spirit as the Material Cause was vital to conceiving the fitting of the set of notions fused to the Qualitative Method.
- As the Final Cause, the Laws of God helped the author be on the right track to articulate the set of notions of the Explanatory Method of Science.
- As the Efficient Cause, the Word of God was the key used by the author to generate the set of notions of the Experimental and Descriptive Methods of Science.

Before we proceed, it is of extreme importance to mention what Ian Kluge said in *"Reason and the Bahá'í Writings"* about God's Essence:

> The ontological difference between God and humankind is intrinsic and cannot be overcome. 'Abdu'l-Bahá states categorically that "it is absolutely impossible to ascend to that plane." This impossibility forbids all claims to know "the reality" or Essence of God and rejects all claims to having attained and experienced ontological unity with God, even if only in a subjective, emotional or 'mystic' state. This impossibility is "absolute" and, therefore, falsifies any claim to have attained such union from any perspective.

> However, our understanding of this ontological difference must be fine-tuned for, as 'Abdu'l-Bahá says, "The existence of the Divine Being hath been clearly established, on the basis of logical proofs, but the reality of the Godhead is beyond the grasp of the mind." In other words, we may know by logical proofs that God exists but not what God is, i.e. we may know about His existence which can be logically demonstrated, but we cannot know His Essence. In a similar vein, Adib Taherzadeh writes: It is essential to differentiate between the 'Essence of God' which Shoghi Effendi describes as the 'innermost Spirit of Spirits' or 'Eternal Essence of Essences,' and 'God revealed' to humanity. The former is unknowable, while the latter is comprehensible to man.

>> The "Essence of God" is unknowable but "God revealed' to humanity" i.e. God as revealed in phenomenal creation — can be known. He is known to us through the revelation of the Manifestations. What the Manifestation reflects is derived from and associated with God — that is precisely what makes him a Manifestation — and what He reveals to us about God, is knowledge about God appropriate to human

---

[73] Non-Profit Foundation for the Application and Teaching of Science.

understanding.[74]

Also, the author proposes a set of human faculties for each of the seven causes. Because we cause many problems, it is imperative to understand what faculty (faculties) need further development. Recognizing such faculties is significant in determining a more efficient way to potentiate them individually or collectively during the pedagogical process and in the psychological approach to human behaviors.

The conceptual framework will be demonstrated with two examples within each method: one about a chair and another about today's farmers' situation.

The suggested conceptual framework is valuable for solving everyday issues and resolving complex individual and collective challenges.

Because the exploratory phase is considered the starting point of the research's cycle and the propositional phase, the ending point, we will start with an in-depth discussion of the Maturity Cause.

In Chapter 10, you will find examples illustrating the art of construction of The Single Science for each of the causes.

---

[74] Ian Kluge, Reason and the Bahá'í Writings

2 The Cause of Maturity and the Exploratory Method

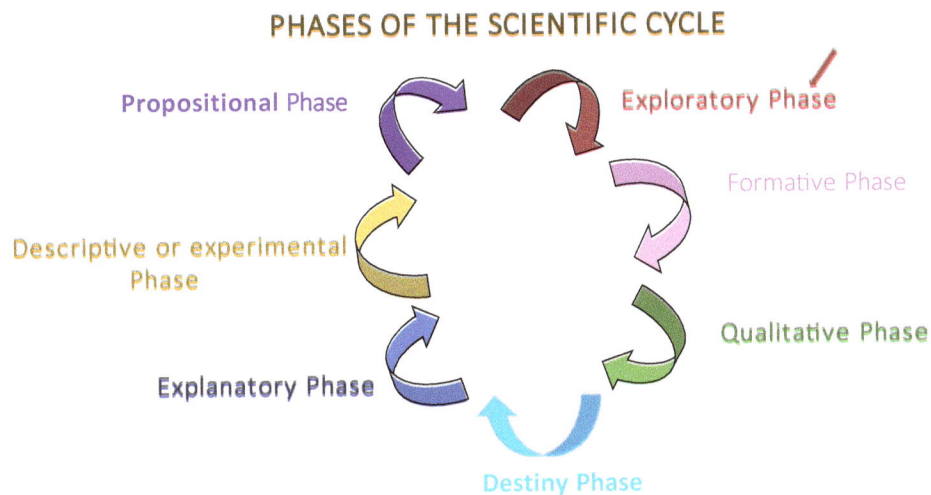

Let me introduce this chapter with a reflection on Evolution. Creation and evolution have been contentious issues between religion and science[75].

To address the considerable differences among the contenders, I would like to suggest a path that may open a possible common understanding. In the context of the notions relevant to the Cause of Maturity, I would like for the reader to reflect on the involuntary differential responses of every living organism during its maturating process.

First, let us try to understand the following quote by 'Abdu'l-Bahá: "Essential pre-existence is an existence which is not preceded by a cause; essential origination is preceded by a cause. Temporal pre-existence has no beginning; temporal origination has both a beginning and an end."[76]

Ian Kluge has some sage insights:

---

[75] ID 66307888 © Adonis1969 | Dreamstime.com

[76] 'Abdu'l-Bahá, Some Answered Questions, pp.155 -56. Emphasis added.

Because Aristotle and the Writings do not recognize a hard and fast distinction between physics and metaphysics and/or theology – a fact of enormous significance in our consideration of the unity of science and religion – the Divine is an inevitable part of any discussion of the universe's physical constitution. Not only do both see God as the "Prime Mover" but they also regard God as utterly self-sufficient, meaning, philosophically speaking, as not preceded by a cause or, as the Bahá'í Writings say, "Self-Subsisting" and, therefore, independent of all other existing things. According to Aristotle, God is also the First Mover Who is Himself unmoved or unchanged. This is because the Unmoved Mover is pure actuality, that is, has no potentials, and is, therefore, beyond all change because there are no potentials left to actualize. One might also express this by saying that God has no privations, no lacks or deficiencies requiring fulfillment. Moreover, the Divine is one and eternal that is, undivided and beyond time, characteristics which also suggest that God is not in space among other phenomenal beings. God is not limited by the normal attributes of all phenomenal, material beings. God is also alive conscious and thinking.[77]

In *Some Answered Questions*, 'Abdu'l-Bahá tells us that "all things have emanated from God; that is, it is through God that all things have been realized, and through Him that the contingent world has come to exist"[78], in other words, the potentials of things become real, are realized or brought into material existence by God's action. This explains why 'Abdu'l-Bahá considered the movement *from existence into existence* as a *degree* of change, even though Aristotle thought of specifically differentiated *types* of change: for Baha'is, "*generation*", that is, the movement from *non-existence to existence* is simply the change from potentiality to actuality, which is Aristotle's original and fundamental definition of movement. "Alternation" is a change from a something to something else, and in the Baha'i view, the movement from *non-existence*, that is, potential existence, *to existence* is simply the actualization of an already existing potentiality.[79]

Abdu'l-Bahá, The Master, said:

> All beings, whether large or small, were created perfect and complete from the first, but their perfections appear in them by degrees. The organization of God is one; the evolution of existence is one; the divine system is one. Whether they be small or great beings, all are subject to one law and system. Each seed has in it from the first all the vegetable perfections. For example, in the seed all the vegetable perfections exist from the beginning, but not visibly; afterward little by little they appear. So it is first the shoot which appears from the seed, then the branches, leaves, blossoms and fruits; but from the beginning of its existence all these things are in the seed, potentially, though not apparently.
>
> In the same way, the embryo possesses from the first all perfections, such as the spirit, the mind, the sight, the smell, the taste -- in one word, all the powers -- but they are not visible and become so only by degrees.
>
> Similarly, the terrestrial globe from the beginning was created with all its elements, substances, minerals, atoms and organisms; but these only appeared by degrees: first the mineral, then the plant, afterward the animal, and finally man. But from the first these kinds

---

[77] Kluge, Ian. The Aristotelian Substratum of the Baha'i Writings.

[78] 'Abdu'l-Bahá, Some Answered Questions, p. 203.

[79] Kluge, Ian. The Aristotelian Substratum of the Baha'i Writings. Emphasis Added.

and species existed, but were undeveloped in the terrestrial globe, and then appeared only gradually. For the supreme organization of God, and the universal natural system, surround all beings, and all are subject to this rule. When you consider this universal system, you see that there is not one of the beings which at its coming into existence has reached the limit of perfection. No, they gradually grow and develop, and then attain the degree of perfection.[80]

If we accept, that God has created every being with the potential to express its perfections gradually. Then we could imagine that through the passing of many natural cycles and the corresponding tests, the opportunity for an organism to manifest developing potentialities begins to be plausible. Because of: variations in the soil, or the atmosphere composition, or the microclimate changed, or enough biodiversity of the surrounding manifested; or its organs were able to establish ideal symbiotic relations with a virus, fungus, or bacteria. Then believing in "creation" and "evolution" seems to be reasonable if we accept that the "survival of the fittest" is just the consequence of the changes in the conditions of the environment that created the possibility for those features of perfection to appear gradually. Such differential responses happen involuntarily. Those changes evolved sequentially in each species over many generations.

I have the following questions in a racial discrimination context: Are homogeneous characteristics within one species the natural tendency of evolution? Is it a law of evolution that the organisms' trend, within one species, is toward diversity of features? What about uniqueness in the DNA?

But human beings can bring about desirable perfections and even have a lasting effect because they become heritable. The potential perfections become actualities through domestication of plants and animals, crossing, pruning, and grafting. Do DNA start to be heritable? In, *Inter-Species Grafting Caused Extensive and Heritable Alterations of DNA Methylation in Solanaceae Plants*, we read:

> Grafting has been extensively used to enhance the performance of horticultural crops. Since Charles Darwin coined the term "graft hybrid" meaning that asexual combination of different plant species may generate products that are genetically distinct, highly discrepant opinions exist supporting or against the concept. Recent studies have documented that grafting enables exchanges of both RNA and DNA molecules between the grafting partners, thus providing a molecular basis for grafting-induced genetic variation. DNA methylation is known as prone to alterations as a result of perturbation of internal and external conditions. Given characteristics of grafting, it is interesting to test whether the process may cause an alteration of this epigenetic marker in the grafted organismal products.[81]

---

[80] 'Abdu'l-Bahá, Some Answered Questions, p. 199.

[81] Rui Wu, et al., Inter-Species Grafting Caused Extensive and Heritable Alterations of DNA Methylation in Solanaceae Plants

The Cause of Maturity and the Exploratory Method

2.1 Philosophical Argument

Concerning the Maturity Cause and its relationship to the Exploratory method, consider the example of making chairs.

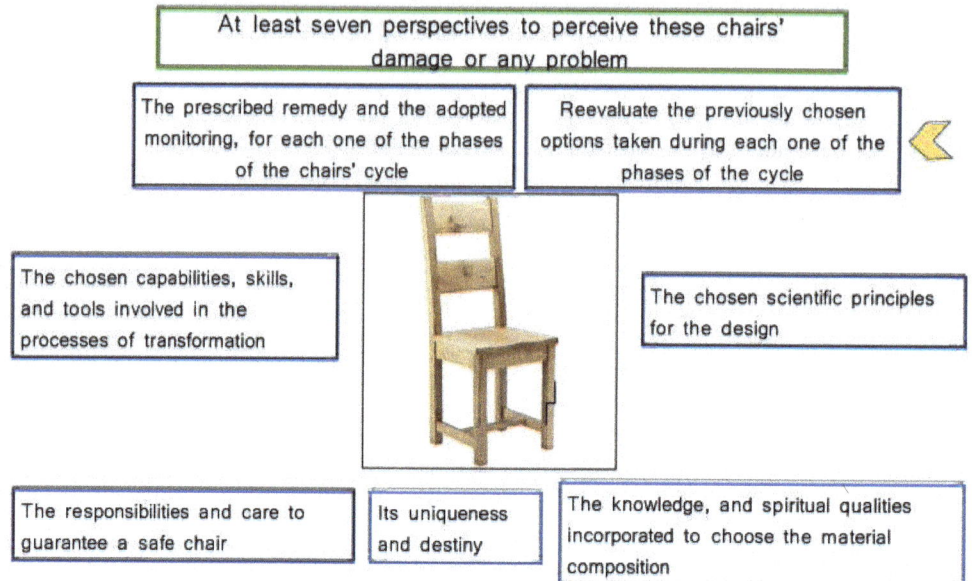

What if several chairs broke, and the question is to understand what happened, so we do not make the same mistakes when building new ones? Then, we should look at the decisions made during each phase of the cycle, possibly related to the chairs' specific damage.

To understand, "Which options did the maker of the chair have?" one must think about:

- the options and the decisions made in each of the phases of the design process;
- the examined alternatives and decisions made in selecting the materials for the chair;
- the choices and decisions made in each of the stages of the actual construction of the chair,
- including making it sturdy and safe

Let us then address the anxiety and fears related to our choices that arise from our mistaken identity. Mercy, as the Cause of Maturity, was the reason that helped the author bring into existence the set of notions of the Exploratory and Propositional Methods of Science. It will help us answer the question, "Which options do I have once I reach the age of maturity?" by understanding the role of each of the components within the following philosophical argument.

Note: the author will emphasize with *italics* the words: *kingdom*, *journey*, *cycle*, *station*, and *stage*; to help the reader connect fundamental notions to themes developed later on in this chapter.

2.1.1 Mercy Encompasses the Entire Creation

The author chose Mercy as the cause of governance and management because in the Kitáb-i-Aqdas "… He formally ordains the institution of the "House of Justice", defines its functions, fixes its revenues, and designates its members as the "Men of Justice", the "Deputies of God", the "Trustees of the All-Merciful."[82]

The Maturity Cause can be related to questions such as: In which phase or stage did the problem originate? Which options do I have? To address the situation, which is the best option? What is the strategy or master plan?

We begin our search by exploring the meaning of Mercy at a more fundamental level. Mercy's synonyms are pity, compassion, kindness, clemency, commiseration, humanitarianism, generosity, and indulgence.

O ye beloved of the Lord! The *Kingdom* of God is founded upon equity and justice, and also upon Mercy, compassion, and kindness to every living soul. Strive ye then with all your heart to treat compassionately all humankind -- except for those who have some selfish, private motive, or some disease of the soul. Kindness cannot be shown the tyrant, the deceiver, or the thief, because, far from awakening them to the error of their ways, it maketh them to continue in their perversity as before. No matter how much kindliness ye may expend upon the liar, he will but lie the more, for he believeth you to be deceived, while ye understand him but too well, and only remain silent out of your extreme compassion.[83]

Concerning the capacity of Mercy to comprehend all of creation, Bahá'u'lláh says: "I swear by the beauty of the Well-Beloved! *This is the Mercy that hath encompassed the entire creation*, the Day whereon the grace of God hath permeated and pervaded all things."[84]

---

[82] Bahá'u'lláh, The Kitab-i-Aqdas, p. 13.

[83] 'Abdu'l-Bahá, Selections from the Writings of 'Abdu'l-Bahá, p. 158. Emphasis Added.

[84] Bahá'u'lláh, Gleanings, p. 33. Emphasis Added.

The Cause of Maturity and the Exploratory Method

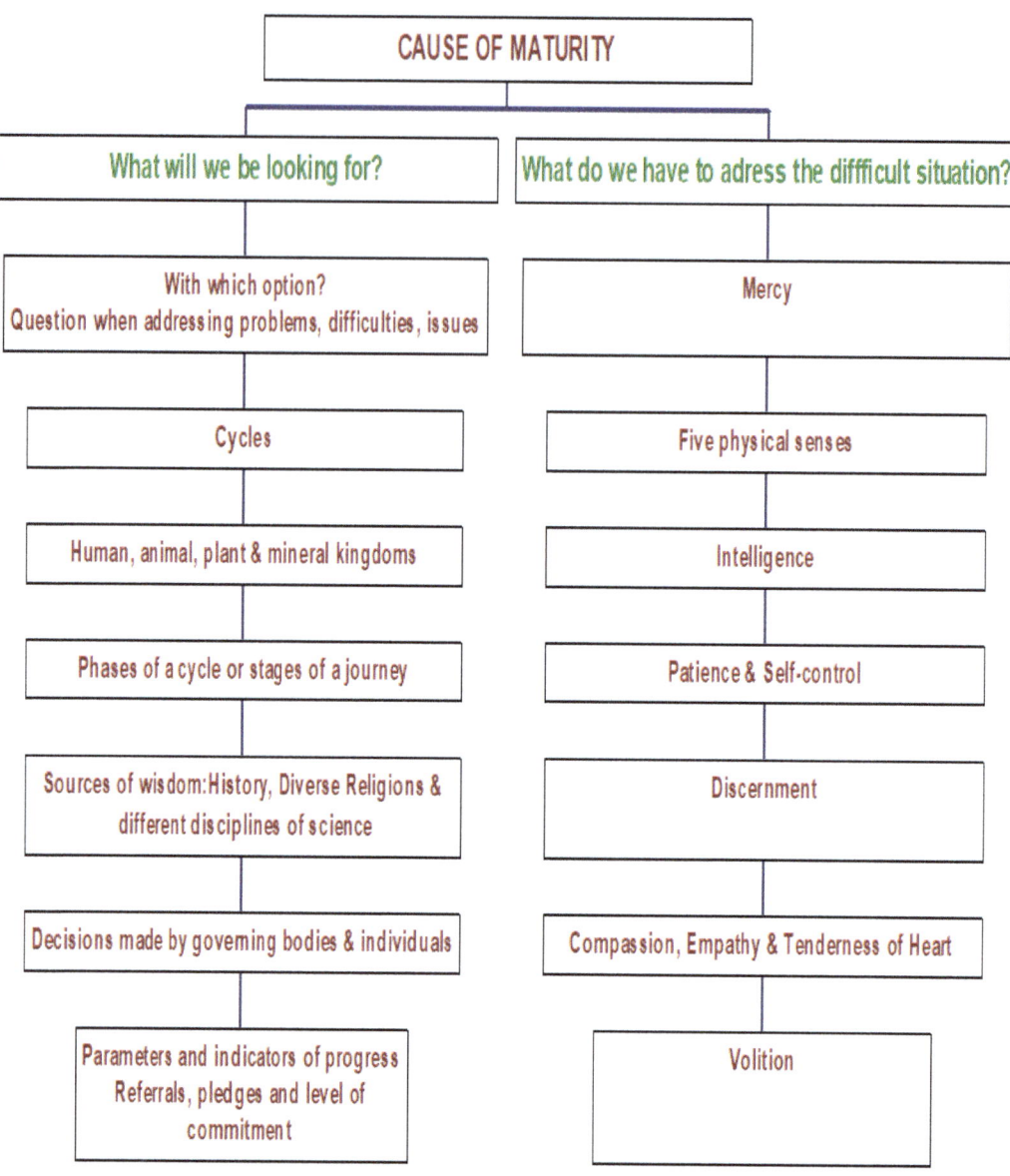

All praise to the unity of God, and all honor to Him, the sovereign Lord, the incomparable and all-glorious Ruler of the universe, Who, out of utter nothingness, hath created the reality of all things, Who, from naught, hath brought into being the most refined and subtle elements of His creation, and Who, rescuing His creatures from the abasement of remoteness and the perils of ultimate extinction, hath received them into His *kingdom* of incorruptible glory. Nothing short of His all-encompassing grace, *His all-pervading Mercy*, could have possibly achieved it. How could it, otherwise, have been possible for sheer nothingness to have acquired by itself the worthiness and capacity to emerge from its state of non-existence into the realm of being?[85]

---

[85] Bahá'u'lláh, Gleanings, pp. 64-65. Emphasis Added.

By complementing the above quote with the following quotes, which are only a small sample of the significance of Mercy, we can begin to grasp its importance when opting for alternatives:

> Thus have We recounted unto you the tales of the one true God, and sent down unto you the things He had preordained, that haply ye may ask forgiveness of Him, may return unto Him, may truly repent, may realize your misdeeds, may shake off your slumber, may be roused from your heedlessness, may atone for the things that have escaped you, and be of them that do good. Let him who will, acknowledge the truth of My words; and as to him that willeth not, let him turn aside. My sole duty is to remind you of your failure in duty towards the Cause of God, if perchance ye may be of them that heed My warning.
>
> Wherefore, hearken ye unto My speech, and return ye to God and repent, that He, through His grace, may have Mercy upon you, may wash away your sins, and forgive your trespasses. *The greatness of His Mercy surpasseth the fury of His wrath*, and His grace encompasseth all who have been called into being and been clothed with the robe of life, be they of the past or of the future.[86]

> "O son of man! *If thine eyes be turned towards Mercy, forsake the things that profit thee, and cleave unto that which will profit mankind*. And if thine eyes be turned towards justice, choose thou for thy neighbor that which thou choosest for thyself."[87]

2.1.2 Situations with challenges, problems, tests, and difficulties

"There is no philosophical high-road in science, with epistemological signposts. No, we are in a jungle and find our way by trial and error, building our roads behind us as we proceed. We do not find sign-posts at cross-roads, but our own scouts erect them, to help the rest."[88]

"The stumbling way in which even the ablest of the scientists in every generation have had to fight through thickets of erroneous observations, misleading generalizations, inadequate formulations, and unconscious prejudices rarely appreciated by those who obtain their scientific knowledge from textbooks."[89]

Bahá'u'lláh says:

> The All-Knowing Physician hath His finger on the pulse of mankind. He perceiveth the *disease*, and prescribeth, in His unerring *wisdom*, the remedy. Every age hath its own *problem*, and every soul its particular aspiration. The remedy the world needeth in its present-day *afflictions* can never be the same as that which a subsequent age may require. Be anxiously concerned with the *needs* of the age ye live in, and centre your deliberations on its exigencies and requirements.[90]

---

[86] Bahá'u'lláh, Gleanings, p. 130. Emphasis Added.

[87] Bahá'u'lláh, Epistle to the Son of the Wolf, pp. 29 -30. Emphasis Added.

[88] Born, Max, Thematic Origins of Scientific Thought, p.7.

[89] Conant, James Bryant.

[90] Bahá'u'lláh, Tabernacle of Unity, p.25. Emphasis Added.

2.1.3 Challenges are blessings

The one true God well knoweth, and all the company of His trusted ones testify, that this Wronged One hath, at all times, been faced with dire peril. But for the *tribulations* that have touched Me in the path of God, life would have held no sweetness for Me, and My existence would have profited Me nothing. For them who are endued with discernment, and whose eyes are fixed upon the Sublime Vision, it is no secret that I have been, most of the days of My life, even as a slave, sitting under a sword hanging on a thread, knowing not whether it would fall soon or late upon him. And yet, notwithstanding all this We render thanks unto God, the Lord of the worlds. Mine inner tongue reciteth, in the daytime and in the night-season, this prayer: 'Glory to Thee, O my God! But for the *tribulations* which are sustained in Thy path, how could Thy true lovers be recognized; and were it not for the *trials* which are borne for Love of Thee, how could the *station* of such as yearn for Thee be revealed?'[91]

What could mean words like development, progress, cleanliness, faith, and peace without issues or problems? What do words such as science, religion, and technology mean in the absence of needs to be satisfied? We all, individually and collectively, must excel in the solution to every problem that seems necessary. We should take it as receiving a present. A critical issue is a unique and unparalleled opportunity for growth and resolution, an opportunity for change that will affect our neighborhoods, our countries, and the world for generations to come. We become historical constructors of a new civilization.

2.1.4 Intelligence

An astonishing endowment given to human beings is the intellect to make sound choices in life. While in Paris, the Master, 'Abdu'l-Bahá in the year 1911, said:

God's greatest gift to man is that of intellect, or understanding.

The understanding is the power by which man acquires his knowledge *of the several kingdoms of creation*, and of various *stages* of existence, as well as of much which is invisible.

Possessing this gift, he is, in himself, the sum of earlier creations -- he is able to get into touch with those *kingdoms*; and by this gift, he can frequently, through his scientific knowledge, reach out with prophetic vision.

Intellect is, in truth, the most precious gift bestowed upon man by the Divine Bounty. Man alone, among created beings, has this wonderful power.

All creation, preceding Man, is bound by the stern law of nature. … [92]

"Like the animal, man possesses the faculties of the senses, is subject to heat, cold, hunger, thirst, etc.; unlike the animal, man has a rational soul, the human intelligence."[93]

"The animal may develop a wonderful degree of intelligence, but it can never attain the powers of ideation and conscious reflection which belong to man."[94]

---

[91] Bahá'u'lláh, Epistle to the Son of the Wolf, p. 94. Emphasis Added.

[92] Paris Talks, p. 715. Emphasis Added.

[93] ibid. p. 746.

[94] The Promulgation of Universal Peace, p. 954.

2.1.5 Power

In reference to the concept of power, Bahá'u'lláh says:

> Thou seest, O God of *Mercy*, Thou *Whose power pervadeth all created things*, these servants of Thine, Thy thralls, who, according to the good-pleasure of Thy Will, observe in the daytime the fast prescribed by Thee, who arise, at the earliest dawn of day, to make mention of Thy Name, and to celebrate Thy praise, in the hope of obtaining their share of the goodly things that are treasured up within the treasuries of Thy grace and bounty.[95]

Abdu'l-Bahá taught: "God, because of His Mercy and His power, lets us know His teachings. 'Abdu'l-Bahá in a talk entitled *The Holy Spirit, the Intermediary Power between God and Man*, said: "Man, then, is in extreme need of the only *Power* by which he is able to receive help from the Divine Reality, that *Power* alone bringing him into contact with the Source of all life."[96]

The Divine Reality is Unthinkable, Limitless, Eternal, Immortal and Invisible. The world of creation is bound by natural law, finite and mortal.

The Infinite Reality cannot be said to ascend or descend. It is beyond the understanding of man and cannot be described in terms which apply to the phenomenal sphere of the created world.

Man, then, is in extreme need of the only Power by which he is able to receive help from the Divine Reality, that Power alone bringing him into contact with the Source of all life.

An intermediary is needed to bring two extremes into relation with each other. Riches and poverty, plenty and need: without an intermediary power there could be no relation between these pairs of opposites.

So we can say there must be a Mediator between God and Man, and this is none other than the Holy Spirit, which brings the created earth into relation with the `Unthinkable One', the Divine Reality.

The Divine Reality may be likened to the sun and the Holy Spirit to the rays of the sun. As the rays of the sun bring the light and warmth of the sun to the earth, giving life to all created beings, so do the `Manifestations' bring the power of the Holy Spirit from the Divine Sun of Reality to give light and life to the souls of men.

Behold, there is an intermediary necessary between the sun and the earth; the sun does not descend to the earth, neither does the earth ascend to the sun. This contact is made by the rays of the sun which bring light and warmth and heat.

The Holy Spirit is the Light from the Sun of Truth bringing, by its infinite power, life and illumination to all mankind, flooding all souls with Divine Radiance, conveying the blessings of God's Mercy to the whole world. ...

Likewise the Holy Spirit is the very cause of the life of man; without the Holy Spirit he would have no intellect, he would be unable to acquire his scientific knowledge by which his great influence over the rest of creation is gained. The illumination of the Holy Spirit gives to man

---

[95] Ibid. p. 299. Emphasis Added.

[96] Paris Talks, pp. 724-25. Emphasis Added.

the power of thought, and enables him to make discoveries by which he bends the laws of nature to his will.

The Holy Spirit it is which, through the mediation of the Prophets of God, teaches spiritual virtues to man and enables him to attain Eternal Life.[97]

To choose the most robust strategy, we must examine the most relevant findings and make an effort to word them with a convincing argument. Without any doubt, Bahá'í Institutions exercise authority; however, the comprehension level of individuals, to act willingly to the institution's wants, is critical for the governance of society's affairs. All of us are part of the Bahá'í Administrative Order, in the sense that when the meeting of the institution dissolves, all are mere citizens who abide by the decisions of the Institutions. We are supposed to help others to understand the wisdom of the institution's decision.

Mercy is a vital feature of the Bahá'í communities' representatives; members of the Spiritual Assemblies are the trustees of the Merciful. To obtain the best results, individuals should show mercy and help the institutions educate humanity to have a comprehensive foundation to discern, doing it with the utmost respect for the free will of others. The exercise of the authority of the Bahá'í Administrative Order is not threatening coercion or punishment. Individuals are empowered when they decide to act using their own volition! Each person has to decide and assume the consequences of their choices. If a Bahá'í persists in wrongdoing, causing damages to others, it is the role of the Bahá'í institutions to exercise justice.

Scientists can dedicate their lives to exploiting people or establishing economic justice, creating even more powerful instruments of human destruction, or committing their creativity to peaceful conflict resolution.

2.1.6 Human powers

"The outward powers are five: the power of sight, of hearing, of taste, of smell, and of touch."[98] With our outward senses we perceive the problem, the stages of the journey and the phases of the cycle.

"Man has likewise a number of spiritual powers: the power of imagination, which forms a mental image of things; thought, which reflects upon the realities of things; *comprehension, which understands these realities*; and memory, which retains whatever man has imagined, thought, and understood. The intermediary between these five outward powers and the inward powers is a common faculty, a sense which mediates between them and which conveys to the inward powers whatever the outward powers have perceived. It is termed the common faculty as it is shared in common between the outward and inward powers.

For instance, sight, which is one of the outward powers, sees and perceives this flower and conveys this perception to the inward power of the common faculty; the common faculty transmits it to the power of imagination, which in turn conceives and forms this image and transmits it to the power of thought; the power of thought reflects upon it and, having apprehended its reality, *conveys it to the power of comprehension; the comprehension, once*

---

[97] ibid. pp. 724-25.

[98] 'Abdu'l-Bahá, Some Answered Questions, p. 56.

it has understood it, delivers the image of the sensible object to the memory, and the memory preserves it in its repository.[99]

2.1.7 Patience

The author linked the Maturity Cause to the question: *With which option?* To make thoughtful choices, we have discernment, patience, and self-control.

When we consider that each stage of the process of scientific research implies search, we can start grasping the importance of learning patience since very early in life. In *The Valley of Search* which describes, in my view, the first phase of human earthly progress[100], Bahá'u'lláh says:

> The steed of this valley is patience; without patience the wayfarer on this journey will reach nowhere and attain no goal. Nor should he ever become downhearted: If he strive for a hundred thousand years and yet fail to behold the beauty of the Friend, he should not falter. For those who seek the Kaaba of "for Us" rejoice in the tidings "In Our ways shall We assuredly guide them." In their search, they have stoutly girded up the loins of service and at every moment journey from the plane of heedlessness into the realm of search. No bond shall hold them back and no counsel deter them.[101]

2.1.8 Discernment and self-control

When we are discerning, we should carefully examine how we are going to affect the human kingdom and the other kingdoms in each one of the stages or phases.

In the human kingdom, there is discernment in the decision-making process. Because Mercy is inherent to all beings in creation, it is also essential to acknowledge that the mineral, vegetable, and animal kingdoms respond differentially per the stimulus received. The environment's cycles influence those responses:

- deciduous trees *lose their leaves in the fall and grow new ones* in the spring;
- bears' spontaneous *differential response* during winter is hibernation;
- minerals' *differential response* to heating and cooling: water and stones become warmer during sunny days.

Another way to perceive a differential response is through homeostasis, which is the property of a cell or a whole organism to seek and maintain a condition of balance or equilibrium within its internal environment, even when faced with external changes.

Homeostasis is a "Feedback Cycle … in which a variable is regulated and the level of the variable impacts the direction in which the variable changes (i.e. increases or decreases), even if there is not clearly identified loop components."[102]

Homeostasis is the property of a system within the body of an organism in which a variable, such as the concentration of a substance in solution, is actively regulated to remain very

---

[99] Some Answered Questions, p. 56. Emphasis added.

[100] Children start asking questions to understand their environment.

[101] Bahá'u'lláh, The Call of the Divine Beloved. The Seven Valleys.

[102] https://courses.lumenlearning.com/ap1/chapter/homeostasis-and-feedback-loops/.

nearly constant. Examples of homeostasis include the regulation of the body temperature of an animal, the pH of its extracellular fluids, or the concentrations of sodium (Na+), potassium (K+) and calcium (Ca2+) ions as well as that of glucose in the blood plasma, despite changes in the animal's environment, or what it has eaten, or what it is doing (for example, resting or exercising).[103]

This differential response within the mineral, vegetable, and animal *kingdoms* takes place without the exercise of free will. Human beings can exercise a conscious choosing process after considering different alternatives, implying discernment and self-control. With these faculties, we can alter the other *kingdoms*' differential response, for example, when we plant a forest or when we build a dam in a river. However, not all our relationships with the other *kingdoms* require discernment and free will. When an insect bites us, our involuntary immune system responds differentially following the chemical compounds and viruses or bacteria introduced into our system. When we breathe (Human Kingdom) we inhale oxygen (Material Kingdom), previously exhaled by the Plant Kingdom during photosynthesis. Each living cell of our organism responds when oxygen is present in the bloodstream. Then we exhale carbon dioxide, which plants take in during photosynthesis. Of course, we can consciously alter our breathing when we decide to run.

Our reflexes are also an example of spontaneous differential response to the environment's stimuli without the exercise of discernment and volition.

Bahá'u'lláh revealed:

> Know thou that, according to what thy Lord, the Lord of all men, hath decreed in His Book, the favors vouchsafed by Him unto mankind have been, and will ever remain, limitless in their range. First and foremost among these favors, which the Almighty hath conferred upon man, is the gift of *understanding*. His purpose in conferring such a gift is none other except to enable His creature to know and recognize the one true God -- exalted be His glory. *This gift giveth man the power to discern the truth in all things*, leadeth him to that which is right, and helpeth him to discover the secrets of creation.[104]

Abdu'l-Bahá, the Master, mentions self-control in the following quote:

> Make ye then a mighty effort, that the purity and sanctity which, above all else, are cherished by 'Abdu'l-Bahá, shall distinguish the people of Bahá; that in every kind of excellence the people of God shall surpass all other human beings; that both outwardly and inwardly they shall prove superior to the rest; that for purity, immaculacy, refinement, and the preservation of health, they shall be leaders in the vanguard of those who know. And that by their freedom from enslavement, their knowledge, *their self-control*, they shall be first among the pure, the free and the wise.[105]

A number of sexual problems, such as homosexuality and transsexuality can well have medical aspects, and in such cases recourse should certainly be had to the best medical assistance. But it is clear from the teaching of Bahá'u'lláh that homosexuality is not a condition to which a person should be reconciled, but is a distortion of his or her nature which should be *controlled* and overcome. This may require a hard struggle, but so also can

---

[103] Wikipedia, Homeostasis.

[104] Bahá'u'lláh, Gleanings, p. 193. Emphasis Added.

[105] 'Abdu'l-Bahá, Selections, p. 150. Emphasis Added.

be the struggle of a heterosexual person to control his or her desires. The exercise of *self-control* in this, as in so very many other aspects of life, has a beneficial effect on the progress of the soul. It should, moreover, be borne in mind that although to be married is highly desirable, and Bahá'u'lláh has strongly recommended it, it is not the central purpose of life. If a person has to wait a considerable period before finding a spouse, or if ultimately, he or she must remain single, it does not mean that he or she is thereby unable to fulfil his or her life's purpose.[106]

The entire development of the intellectual capacity of an individual requires moderating his (her) impulsiveness by learning patience and self-control. At 15 years of age, the individual is supposed to reach maturity. The right to exercise his (her) free will is not that simple; an impulsive 50 years old seems to have the right to be categorized as a mature individual, but is s/he taking enough time to discern among different alternatives? Does s/he have the patience and restraint to search into various sources of wisdom before choosing an option? Is this situation similar to those sickened with addictions?

A little more to encourage the teaching of self-control:

> Gottfredson and Hirschi advanced self-control theory in 1990 as part of their general theory of crime. Self-control is defined as the ability to forego acts that provide immediate or near-term pleasures, but that also have negative consequences for the actor, and as the ability to act in favor of longer-term interests. An individual's level of self-control is influenced by family or other caregiver behavior early in life. Once established, differences in self-control affect the likelihood of delinquency in childhood and adolescence and crime in later life. Persons with relatively high levels of self-control do better in school, have stronger job prospects, establish more stable interpersonal relationships, and attain higher income and better health outcomes.[107]

2.1.9 Moral principles

The Universal House of Justice, in a message To the Bahá'ís of the World, on 1 March 2017, said:

> "Every choice a Bahá'í makes—as employee or employer, producer or consumer, borrower or lender, benefactor or beneficiary—leaves a trace, and the *moral duty* to lead a coherent life demands that one's economic decisions be in accordance with lofty ideals, that the purity of one's aims be matched by the purity of one's actions to fulfil those aims." "The golden rule as to food is, do not take too much or too little."[108] "Lay not on any soul a load which ye would not wish to be laid upon you, and desire not for anyone the things ye would not desire for yourselves. This is My best counsel unto you, did ye but observe it ."[109] "And if thine eyes be turned towards justice, choose thou for thy neighbor that which thou choosest for thyself."[110]

---

[106] The Universal House of Justice, A Chaste and Holy Life.

[107] Michael Gottfredson, Self-Control Theory and Crime.

[108] John E. Esslemont, Bahá'u'lláh and the New Era.

[109] Bahá'u'lláh: Gleanings , Page: 128

[110] Bahá'u'lláh: Epistle to the Son of the Wolf , Page: 30

2.1.10 Wisdom

We beseech the one true God, magnified be His glory, to enable us to recognize Him *Whose unerring wisdom pervadeth all things* and that we may acknowledge His truth. For once one hath recognized Him and borne witness to His Reality, one will no longer be troubled by the idle fancies and vain imaginings of men. The divine Physician hath the pulse of mankind within His almighty grasp. At one time He may well deem fit to sever certain infected limbs, that the disease may not spread to other parts of the body. This would be the very essence of Mercy and compassion, and to none is given the right to object, for He is indeed the All-Knowing, the All-Seeing.[111]

2.1.11 Sources of wisdom

To choose the best strategy to address a problem is advisable to transcend the narrow search within one area of knowledge, which has been very common with economics. The sources of wisdom are: what history, experience, world religions, and the diverse disciplines of science say about the issue to be solved. We should even study what atheists think about the problem of interest.

2.1.12 Tenderness, mercy, and empathy

To be wiser, men should subordinate their decisions to the following points of reference. While in Paris, the Master, 'Abdu'l-Bahá said,

> As regards the constitution of the House of Justice, Bahá'u'lláh addresses the men. He says: 'O ye men of the House of Justice!'
>
> But when its members are to be elected, the right which belongs to women, so far as their voting and their voice is concerned, is indisputable. When the women attain to the ultimate degree of progress, then, according to the exigency of the time and place and their great capacity, they shall obtain extraordinary privileges. Be ye confident on these accounts. His Holiness Bahá'u'lláh has greatly strengthened the cause of women, and the rights and privileges of women is one of the greatest principles of 'Abdu'l-Bahá. Rest ye assured! Ere long the days shall come when the men addressing the women, shall say: 'Blessed are ye! Blessed are ye! Verily ye are worthy of every gift. Verily ye deserve to adorn your heads with the crown of everlasting glory, because in sciences and arts, in virtues and perfections ye shall become equal to man, and as regards *tenderness of heart and the abundance of Mercy and sympathy* ye are superior'.[112]

My suggestion is: the abundance of Mercy, tenderness and sympathy demonstrated by women is the reason why they should be welcome to participate in greater number, until they reach majority, in the administrative level of all human affairs, even at the international level, except in the Universal House of Justice as ordained by Bahá'u'lláh. To clarify the reason for such exception, Shoghi Effendi, in a letter written on his behalf to an individual believer (28 July 1936), provided the following authoritative elaboration of this theme:

---

[111] Bahá'u'lláh, Tabernacle of Unity. P.44. Emphasis Added.

[112] 'Abdu'l-Bahá, Paris Talks, p. 794. Emphasis Added.

As regards your question concerning the membership of the Universal House of Justice: there is a Tablet from 'Abdu'l-Bahá in which He definitely states that the membership of the Universal House is confined to men, and that the wisdom of it will be fully revealed and appreciated in the future. In the local as well as the national Houses of Justice, however, women have the full right of membership. It is, therefore, only to the International House that they cannot be elected. The Bahá'ís should accept this statement of the Master in a spirit of deep faith, confident that there is a divine guidance and wisdom behind it which will be gradually unfolded to the eyes of the world.[113]

In a letter, The Universal House of Justice expresses the following:

With regard to the status of women, the important point for Bahá'ís to remember is that in face of the categorical pronouncements in Bahá'í Scripture establishing the equality of men and women, the ineligibility of women for membership of the Universal House of Justice does not constitute evidence of the superiority of men over women. It must also be borne in mind that women are not excluded from any other international institution of the Faith. They are found among the ranks of the Hands of the Cause. They serve as members of the International Teaching Centre and as Continental Counselors. And, there is nothing in the Text to preclude the participation of women in such future international bodies as the Supreme Tribunal.[114]

For us, males also mean to address those faculties that we have in less abundance than females: tenderness of heart, empathy, and mercy. We need to learn, early in life, to listen to those emotions when females describe the situation of a child, the environment, or the world. We have to learn to read in those feelings your interpretation of the dimensions of the problematic situation. I suggest that boys participating in the Junior Youth Groups Spiritual Empowerment Program should learn to listen to girls. If we do that, both genders will make better choices when consulting about the solution to the needs and tests.[115]

Rather than expressing mortal ideas about mercy and the exercise of authority, concerning the tenderness of heart, and empathy, let us continue reflecting upon the Word of Bahá'u'lláh Himself:

I swear by the beauty of the Well-Beloved! This is the Mercy that hath encompassed the entire creation, the Day whereon the grace of God hath permeated and pervaded all things. The living waters of My Mercy, O Ali, are fast pouring down, and Mine heart is melting with the heat of My tenderness and Love. At no time have I been able to reconcile Myself to the

---

[113] Shoghi Effendi, qtd. in Universal House of Justice, A Compilation on Women.

[114] Universal House of Justice, 31 May 1988 – The National Spiritual Assembly of the Bahá'ís of New Zealand.

[115] Photo 87637437 © Iakov Filimonov | Dreamstime.com

afflictions befalling My loved ones, or to any trouble that could becloud the joy of their hearts.

Every time My name 'the All-Merciful' was told that one of My lovers had breathed a word that runneth counter to My wish, it repaired, grief-stricken and disconsolate to its abode; and whenever My name 'the Concealer' discovered that one of My followers had inflicted any shame or humiliation on his neighbor, it, likewise, turned back chagrined and sorrowful to its retreats of glory, and there wept and mourned with a sore lamentation. And whenever My name 'the Ever-Forgiving' perceived that any one of My friends had committed any transgression, it cried out in its great distress, and, overcome with anguish, fell upon the dust, and was borne away by a company of the invisible angels to its habitation in the realms above.[116]

God shows us the immensity of His Mercy, and we must respond to His laws. Although He is the Merciful and the Concealer, we could be inclined to disobey; because man cannot stop the effects of His commandments, we should opt for what He has ordained. The same applies when choosing between what is just and what is not; between what is conducive to wealth and what is to poverty; between science that teaches to discern and the one that explains how to divide, exploit and oppress people. We should examine the options with the lamp of divine mercy, grace, and kindness in the decision-making process.

2.1.13 Two natures

Although above the author explained the virtues linked to the Maturity Cause, there are other factors to consider when asking ourselves questions like:

- Which options do I have?
- Which decisions have to change to improve the safety, the design, the materials, and the processes?
- In which phase or stage did the problem originate?
- Which is the best option?
- What is the strategy?

It is of great transcendence to acknowledge the existence of two natures, a higher and a lower one, as an essential element in the decision-making process.

"God has created all in His image and likeness. Shall we manifest hatred for His creatures and servants? This would be contrary to the will of God and according to the will of Satan, by which we mean the natural inclinations of the lower nature. This lower nature in man is symbolized as Satan -- the evil ego within us, not an evil personality outside".[117]

While in Paris, the Master, 'Abdu'l-Bahá said:

In man there are two natures; his spiritual or higher nature and his material or lower nature. In one he approaches God, in the other he lives for the world alone. Signs of both these natures are to be found in men. In his material aspect he expresses untruth, cruelty and injustice; all these are the outcome of his lower nature. The attributes of his Divine nature are shown forth in love, Mercy, kindness, truth and justice, one and all being expressions of

---

[116] Bahá'u'lláh, Gleanings, pp. 308-09. Emphasis Added.

[117] The Promulgation of Universal Peace, p. 286.

his higher nature. Every good habit, every noble quality belongs to man's spiritual nature, whereas all his imperfections and sinful actions are born of his material nature. If a man's Divine nature dominates his human nature, we have a saint.[118]

When you wish to reflect upon or consider a matter, you consult something within you. You say, shall I do it, or shall I not do it? Is it better to make this *journey* or abandon it? Whom do you consult? Who is within you deciding this question? Surely there is a distinct power, an *intelligent ego*. Were it not distinct from your *ego*, you would not be consulting it. It is greater than the faculty of thought. It is your spirit which teaches you, which advises and decides upon matters.[119]

2.1.14 Considering the heart's role in the decision-making process

Every subject presented to a thoughtful audience must be supported by rational proofs and logical arguments. Proofs are of four kinds: first, through sense-perception; second, through the reasoning faculty; third, from traditional or scriptural authority; fourth, through the medium of inspiration. That is to say, there are four criterions or standards of judgment by which the human mind reaches its conclusions.[120]

'Abdu'l-Bahá then explains each of those criteria and demonstrates why they are individually unreliable, saying:

Consequently it has become evident that the four criteria or standards of judgment by which the human mind reaches its conclusions are faulty and inaccurate. All of them are liable to mistake and error in conclusions. But a statement presented to the mind accompanied by proofs which the senses can perceive to be correct, which the faculty of reason can accept, which is in accord with traditional authority *and sanctioned by the promptings of the heart*, can be adjudged and relied upon as perfectly correct, for it has been proved and tested by all the standards of judgment and found to be complete. When we apply but one test there are possibilities of mistake. This is self-evident and manifest.[121]

2.1.15 Free will

After consulting our lower and higher natures in our hearts, we commit our free will, also called volition, to the solution that seems to be wise:

Bahá'u'lláh revealed,

And now, concerning thy question regarding the creation of man. Know thou that all men have been created in the nature made by God, the Guardian, the Self-Subsisting. Unto each one hath been prescribed a pre-ordained measure, as decreed in God's mighty and guarded

---

[118] 'Abdu'l-Bahá, Paris Talks, p. 18.

[119] The Promulgation of Universal Peace, p. 242. Emphasis Added.

[120] Foundations of World Unity, p. 86. Emphasis Added.

[121] ibid. p.87.

Tablets. All that which ye potentially possess can, however, be manifested only as a result of your *own volition*.[122]

Man alone has *freedom*, and, by his understanding or intellect, has been able to gain control of and adapt some of those natural laws to his own needs. By the power of his intellect he has discovered means by which he not only traverses great continents in express trains and crosses vast oceans in ships, but, like the fish he travels under water in submarines, and, imitating the birds, he flies through the air in airships.[123]

"All of us know that international peace is good, that it is conducive to human welfare and the glory of man but *volition* and action are necessary before it can be established."[124]

"The attainment of any object is conditioned upon knowledge, *volition* and action. Unless these three conditions are forthcoming there is no execution or accomplishment. In the erection of a house it is first necessary to know the ground and design the house suitable for it; second, to obtain the means or funds necessary for the construction; third, to actually build it."[125]

2.1.16 Human kingdom and other kingdoms in nature

The kingdoms of nature and those in authority, in any administrative order, are also important in the Maturity Cause. When exploring a problematic situation, evaluating past decisions regarding their effects in all kingdoms is critical to determine the administration's stewardship commitment. Of course, an individual's decision also should take that principle into account. About stewardship, the following paragraph is very pertinent to today's situation, which had become worst since 1985 when The Universal House of Justice released *The Promise of World Peace*:

> The time has come when those who preach the dogmas of materialism, whether of the east or the west, whether of capitalism or socialism, must give account of the *moral stewardship* they have presumed to exercise. Where is the "new world" promised by these ideologies? Where is the international peace to whose ideals they proclaim their devotion? Where are the breakthroughs into new realms of cultural achievement produced by the aggrandizement of this race, of that nation or of a particular class? Why is the vast majority of the world's peoples sinking ever deeper into hunger and wretchedness when wealth on a scale undreamed of by the Pharaohs, the Caesars, or even the imperialist powers of the nineteenth century is at the disposal of the present arbiters of human affairs?[126]

In Wikipedia we find a fascinating reflection: In "the 1985 CBC series "*A Planet for the Taking*", Dr. David Suzuki explored the Old Testament roots of anthropocentrism and how it shaped our view of non-human animals. Some Christian proponents of anthropocentrism base their belief in the Bible, such as the verse 1:26 in the Book of Genesis:

---

[122] Bahá'u'lláh, Gleanings, p. 149. Emphasis Added.

[123] 'Abdu'l-Bahá, Paris Talks, p. 716. Emphasis Added.

[124] 'Abdu'l-Bahá, , The Promulgation of Universal Peace. Emphasis Added.

[125] Idem.

[126] The Universal House of Justice, The Promise of World Peace. Emphasis added.

And God said, Let us make man in our image, after our likeness: and let them have dominion over the fish of the sea, and over the fowl of the air, and over the cattle, and over all the earth, and over every creeping thing that creepeth upon the earth.

The use of the word 'dominion' in the *Genesis* is controversial. Many Biblical scholars, especially Roman Catholic and other non-Protestant Christians, consider this to be a flawed translation of a word meaning 'stewardship', which would indicate that mankind should take care of the earth and its various forms of life, but is not inherently better than any other form of life. The current Latin Vulgate, the official Bible of the Catholic Christian church, states that God holds man responsible for the care and fate of all earthly creatures.[127]

"All created things have their degree or *stage* of maturity. The period of maturity in the life of a tree is the time of its fruit-bearing. The maturity of a plant is the time of its blossoming and flower. The animal attains a *stage* of full growth and completeness, and in the human *kingdom* man reaches his maturity when the lights of intelligence have their greatest power and development."[128]

Also the difference of conditions in the world of beings is an obstacle to comprehension. For example, this mineral belongs to the mineral kingdom; however far it may rise, it can never comprehend the power of growth. The plants, the trees, whatever progress they may make, cannot conceive of the power of sight or the powers of the other senses; and the animal cannot imagine the condition of man— that is to say, his spiritual powers. Difference of condition is an obstacle to knowledge; the inferior degree cannot comprehend the superior degree. How then can the phenomenal reality comprehend the Preexistent Reality? Knowing God, therefore, means the comprehension and the knowledge of His attributes, and not of His Reality. This knowledge of the attributes is also proportioned to the capacity and power of man; it is not absolute. Philosophy consists in comprehending the reality of things as they exist, according to the capacity and the power of man. For the phenomenal reality can comprehend the Preexistent attributes only to the extent of the human capacity. The mystery of Divinity is sanctified and purified from the comprehension of the beings, for all that comes to the imagination is that which man understands, and the power of the understanding of man does not embrace the Reality of the Divine Essence. All that man is able to understand are the attributes of Divinity, the radiance of which appears and is visible in the world and within men's souls.[129]

Then, especially in a globalized, interconnected world, we should know how our options affect the other kingdoms.

2.1.17 Stages and phases

The Maturity Cause is necessary for preparing for adulthood and life after death. As we mature, we grow understanding and discerning to explore different alternatives to solve problems. The most desirable solution compromises our will to the highest degree to reach what God has ordained for people on earth. To a point when the individual places his absolute trust in

---

[127] "anthropocentrism." Wikipedia.

[128] 'Abdu'l-Bahá, The Promulgation of Universal Peace. Emphasis Added.

[129] 'Abdu'l-Bahá, Some Answered Questions, pp. 221-22.

God's teachings because he has found immense wisdom every time he has explored any one of them.

In *The Valley of True Poverty and Absolute Nothingness*, Bahá'u'lláh says:

> This *station* is that of dying to the self and living in God, of being poor in self and rich in the Desired One. Poverty, as here referred to, signifieth being poor in that which pertaineth to the world of creation and rich in what belongeth to the realms of God. For when the true lover and devoted friend reacheth the presence of the Beloved, the radiant beauty of the Loved One and the fire of the lover's heart will kindle a blaze and burn away all veils and wrappings. Yea, all that he hath, from marrow to skin, will be set aflame, so that nothing will remain save the Friend.
>
> When once shone forth the attributes Of Him Who is the ancient King,
>
> All mention Moses burned away
>
> Of every fleeting, transient thing.[130]
>
> In this valley the wayfarer passeth beyond the *stages* of the "unity of existence" and the "unity of appearance" and reacheth a unity that is sanctified above both of these *stations*. Ecstasy alone can encompass this theme, not utterance nor argument; and whosoever hath dwelt at this *stage* of the *journey*, or caught a breath from this garden, knoweth whereof We speak.
>
> In all these *journeys* the wayfarer must stray not a hair's breadth from the Law, for this is indeed the secret of the Path and the fruit of the Tree of Truth. And in all these *stages* he must cling to the robe of obedience to all that hath been enjoined, and hold fast to the cord of shunning all that is forbidden, that he may partake of the cup of the Law and be informed of the mysteries of Truth.[131]

All problems, difficulties, and needs relate to one or more cycles or journeys. In the quotes mentioned above, I have added emphasis to the words journey, station, and stage, which suggest phases within a journey or regularity within a cycle. The phase of the cycle or the journey's stage considered critical to the problem of interest will be the closest to the problem's features.

When we study the sources of wisdom to address a problematic situation, we should carefully examine the effect of the human kingdom's decisions in the other domains in each stage or phase. Considering the developmental stages of a situation, of an illness, of an animal, of a plant, or a human being, are vital to making the right choice. The symptoms in a little one may be more challenging than in an adult. The same idea applies to the stations within a journey.

Besides the power of intelligence, mercy, compassion, tenderness of heart, empathy, and the power of volition, human beings also have some other faculties to make sound decisions in life, such as stewardship and:

---

[130] Bahá'u'lláh, The Call of the Divine Beloved. The Seven Valleys. Emphasis Added.

[131] ibid. p. 39. Emphasis Added.

2.1.18 Sense of our own powerlessness

When we realize the unfathomable depths of His Wisdom in the Books of Revelation and Creation, we ought to develop a sense of powerlessness:

"Inspire them, O my Lord, with a sense of their *own powerlessness* before Him Who is the Manifestation of Thy Self, and teach them to recognize the poverty of their own nature in the face of the manifold tokens of Thy self-sufficiency and riches, that they may gather together round Thy Cause, and cling to the hem of Thy mercy, and cleave to the cord of the good-pleasure of Thy will."[132]

2.1.19 Relationships

We can perceive different governance patterns, such as those dominating and imposing their will by force. And those in authority, when evaluating how wise their decisions are, willing to listen to the previously well educated, empowered masses and the laments of the orphan, the grief of the widows, and the "nobodies" destitute poor.

Other relationships considered in the Maturity Cause are those interconnecting the sources of wisdom.

2.1.20 Laws

In the Cause of Maturity, there are:
- Rules related to the inherent responsibilities of the assumed decision,
- Child care regulations that apply when passing from one stage of existence to the next, for example, the rights of a child to cross the street by himself (herself) and those for adolescents,
- Laws related to plants, animals, and humans when reaching the phase of extinction.
- Parameters and indicators of progress

The follow-up controls of parameters and indicators and the regulation and performance evaluation mechanisms, in the phases within the cycle or at the stages of a journey, are crucial to exploring the situation's evolution and proposing alternative solutions.

In *En Síntesis la Ciencia es Amor* (1992), the author included parameters and indicators because of his interest as a scientist for each of the four causes (Final, Material, Efficient and Formative) taught by the Greek Philosophers. Later on, while working on the Cause of Maturity (1999), the author realized that parameters and indicators were critical to choosing among alternatives as pivotal for Governance and Management.

ɷɸɷ

In conclusion, one of the phases of the cycle of scientific research is the exploratory phase. The philosophical argument and the set of human faculties articulated to the Maturity Cause are ideal for answering the question: Which options do I have once I reach the age of maturity?

---

[132] Bahá'u'lláh, Prayers and Meditations by Bahá'u'lláh, p. 47. Emphasis Added.

2.2 Research Steps for the Exploratory Method

The author made an effort to relate this method to the Valley of Search, as revealed in The Seven Valleys. These Valleys "mark the wayfarers' journey from their mortal abode to the heavenly homeland"[133]; in the Valley of Search, we find:

> One must judge of search by the standard of the Majnún of love. It is related that one day they came upon Majnún sifting the dust, his tears flowing down. They asked, "What doest thou?" He said, "I seek for Laylí." "Alas for thee!" they cried, "Laylí is of pure spirit, yet thou seekest her in the dust!" He said, "I seek her everywhere; haply somewhere I shall find her."
>
> Yea, though to the wise it be shameful to seek the Lord of Lords in the dust, yet this betokeneth intense ardour in searching. "Whoso seeketh out a thing and persisteth with zeal shall find it.[134]

The following are just general ideas that may serve as guiding steps for the Exploratory method of science. The suggested steps should identify the problem's history, describe its facts and symptoms, and connect to the faculties related to the Cause of Maturity, such as intelligence, wisdom, human power, patience, discernment, self-control, tenderness of heart, mercy, and sympathy. Also, looking at the link to God's Power and the progress parameters in exercising stewardship to manage the trust. Finally, let us be aware that laws always regulate decisions, and tests will come back as part of the evolutionary process.

From a letter dated 6 February 1973 written by the Universal House of Justice to all National Spiritual Assemblies, we read:

> In considering the effect of obedience to the laws on individual lives, one must remember that the purpose of this life is to prepare the soul for the next. Here one must learn to control and direct one's animal impulses, not to be a slave to them. Life in this world is a succession of tests and achievements, of falling short and of making new spiritual advances. Sometimes the course may seem very hard, but one can witness, again and again, that the soul who steadfastly obeys the law of Bahá'u'lláh, however hard it may seem, grows spiritually, while the one who compromises with the law for the sake of his own apparent happiness is seen to have been following a chimera: he does not attain the happiness he sought, he retards his spiritual advance and often brings new problems upon himself.[135]

Expected results of the Cause of Maturity are healing, regeneration, rehabilitation, renewal, homeostasis, resilience, and progress. Those results among humans are fruits of perseverance, tenacity, and self-determination; in nature, they are spontaneous responses to the stimuli received.

First: The exploratory step. Search for the decision(s) that caused the problem's facts and symptoms as responses to the stimuli received by members of the kingdoms on earth.

- Which is the phase of the cycle or the journey's stage, where the symptoms' emergent properties appear?
- Which evidence (facts and symptoms) support the existence of a problem in the kingdoms on earth?

---

[133] Bahá'u'lláh, The Seven Valleys, The Call of the Divine Beloved.

[134] Idem.

[135] The Universal House of Justice, A Chaste and Holy Life.

- Which is the problem?
- Which decision might be the cause of the problem's facts and symptoms?
- In which stage of the journey or phase of the cycle emerged the problem's facts and symptoms, detected in the spontaneous response of nature due to the stimulus received?
- Description of the history of the evolving approach(es) to problem-solving. Are there diverse approaches to the history of the problem for building a holistic strategy?
- Which is the strategy to learn stewardship in managing the kingdoms on earth?
- To contribute to the control of damaging insects, are there any predators that may act on harmful pests during the oviposition or larval stages?

Second: The formative step, focusing on the general properties of matter and the faculties closely related to the Formative Cause

- What results brought the monitoring and evaluation processes in improving the arrangement changes agreed to be implemented during the last cycle?
- Search for the location, arrangement, shape, and form of the problem's facts and symptoms
- On natural cycles in all kingdoms, which augmentative features of minerals, plants, animals, and humans, should we focus our attention on discovering latent endowed perfections related to the general properties of matter?

Third: The qualitative step, focusing on the specific properties of matter

- What results brought the monitoring and evaluation processes in improving the qualitative changes agreed to be implemented during the last cycle?
- Determine the characteristics of the problem's facts and symptoms
- On natural cycles, which features of minerals, plants, animals, and humans, should we focus on discovering latent specific perfections?

Fourth: Determine the possible root causes of the problem's facts and symptoms. To be considered collectively or privately by an individual. Negligence, imitation, lack of self-accountability of faults, transgressions, omissions, failing, oversight, inadvertence, or mere ignorance are examples of Root Causes. We should not be threatening or encouraging confession or finger-pointing. Instead, search for the source of lack of accompaniment, harmony, justice, unity in addressing the strategy agreed upon in the last cycle. Search for identifying the origin of any difference in opinion about the strategy previously agreed

- Search for the root causes(s) in reaching the set goals of the strategy and its relationship to our collective high destiny?
- Was there any lack of understanding about the wisdom of the agreed guidance?
- Search for which may be obstacles of our own making, becoming the root cause for reaching a higher destiny?
- What shackles did I put on myself to regularly reflect, comprehend, and deal with problems and difficulties of the collectivity? Is it a lack of self-accountability? Is it selfishness?

Fifth: The explanatory step, focusing on the cycles' laws and the Law of Conservation of

Matter. "The Law of Cycles postulates a model of the universe where processes, events, or phenomena repeat themselves in a recurring way at fixed periods of time."[136]

- What results brought the monitoring and evaluation processes in complying with the changes agreed to be implemented during the last cycle?
- What do you think about the following proposal: "Rights of nature acknowledges that nature in all its life forms has the right to exist, persist, maintain and regenerate its vital cycles."[137] What could be the impact of disturbing a natural cycle? For example, climate change has already started affecting the ocean currents' cycle. Is this right to be adopted and obeyed collectible by the world's citizens and its governments?
- Explain the risks caused by the perceived problem, facts, and symptoms related to the laws regulating the periodicity of recurrent phenomena.
- Should we focus our attention on natural cycles' in all kingdoms, observing preventive risk features of minerals, plants, animals, and humans? For example:

"Scientists have found that tiny, golden-winged warblers can detect a storm before The Weather Channel knows its coming.

In April, a massive thunderstorm unleashed a series of tornadoes that tore through the central and southern United States. The 84 twisters decimated homes and buildings, causing more than $1 billion in damage across 17 states. In the wake of the natural disaster, 35 people lost their lives.

Now, scientists say a peculiar event took place just two days before the storm: Flocks of songbirds fled the area en masse. Many golden-winged warblers[138] had just finished a 1,500-mile migration to Tennessee when they suddenly flew south on a 900-mile exodus to Florida and Cuba. At that time, the storm was somewhere between 250 and 560 miles away. The researchers said that the birds somehow knew about the impending storm.

"At the same time that meteorologists on The Weather Channel were telling us this storm was headed in our direction, the birds were apparently already packing their bags and evacuating the area," Henry Streby, a population ecologist from the University of California, Berkeley, said in a statement. He and his research team had been examining the birds' *migratory* patterns when they made their discovery.

Initially, the team was studying if warblers, which weigh the same as four dimes, could carry half-gram geo-locators over long distances. After retrieving data from five of the 20 tagged birds, the team noticed the birds were nowhere near the path they'd expected. Why, the researchers wondered, would these tiny birds travel so far from their already-grueling

---

[136] Helena Petrovna Blavatsky, Law of Cycles.

[137] The Global Alliance for the Rights of Nature, What is Rights of Nature?

[138] Photo 167122707 © Wirestock | Dreamstime.com

migratory route? Upon further inspection, the scientists found that the dates the birds broke with the pattern coincided with the beginnings of the storm. In a paper reported today in the journal *Current Biology*, the team suggests that the birds made their "evacuation *migration*" because their keen sense of hearing alerted them to the incoming natural disaster.[139]

It is essential to acknowledge that golden-winged warblers (Vermivora chrysoptera) were breeding season during their stay in Tennessee before departure. Was the next generation at risk?

Sixth: The Experimental Step

- What results brought the monitoring and evaluation of the efficiency during the last cycle?
- Give a reasonable argument of the processes that may have resulted in the problem, the facts, and the symptoms
- Which educational experiences contribute to developing further the following faculties related to the Cause of Maturity: the five senses, patience, self-control, mercy, sympathy, tenderness of heart, discernment, volition, and intelligence?

Seventh: The Propositional Step

- What can the researcher do to excel in his(her) understanding of patience, self-control, discernment, mercy, sympathy, tenderness of heart, and volition; when exercising stewardship on his(her) decisions and their effect in all kingdoms?
- Which decision(s) cause the problem, facts, and symptoms reflected in response to the stimuli received by the kingdoms?
- Which parameters or indicators of resilience, alleviation, healing, rehabilitation, regeneration do we want to discard, improve, or define and formulate anew?

---

[139] Nicholas St. Fleur, Songbirds Can Hear Tornadoes Long Before They Form.

2.3 The exploratory phase of the farmers' situation

SUMMARY OF THE EXPLORATORY & PROPOSITIONAL METHODS

Question: With which option? when addressing problems, difficulties, issues, and needs

Essence: Mercy.

Sources of Wisdom: What history or experience, the diverse disciplines of science and the world religions say about the issue of concern

Emergent properties: The solid pieces of evidence and signs gradually evolve during the different stages of an illness and its healing, or the responsive adaptation to the seasonal variations of plants and animals to reach their potential perfections. Something similar happens with processes of regeneration, rehabilitation, renewal, resurrection, homeostasis, and resilience.

Situations that challenge the human kingdom to choose but in the mineral, plant, and animal kingdoms, they respond differentially to the stimuli received.

Moral principles

Kingdoms, Cycles, and phases

Laws related to the responsibilities of the assumed decision, considering the effects in the other Kingdoms and the Human Kingdom.

Goal: Stewardship in the management of the trust

Powers: Intelligence, patience, self-control, discernment, compassion, empathy, volition

Decisions made at the individual and collective level: Governmental Institutions, Administration of private enterprises, NGOs, families, or individuals. The human choices concerning the stages of a plan or the phases of a cycle

The spontaneous differential response to the stimuli received by the different kingdoms

The effects of the decisions made by humans in all kingdoms

The developmental **stage** of the individual, plant, or animal

Monitoring and evaluating the strategy and its stages and the cycle's phases

Relationships: are those that interconnect the sources of wisdom

Situations in which the human kingdom deals with the need to choose or, in the case of the mineral, plant, and animal kingdoms, the appearance of a spontaneous differential response to the stimuli received.

Parameters and indicators of progress: Such as level of respect for the individual to exercise his (her) free will, level of commitment, number of pledges, and number of recommendations or referrals

To help the reader understand the farmers' example, the author emphasizes the discourse's fundamental notions reasonably linked to each one of the Causes.

Let us then start with our example of the farmers with the exploratory phase using the Maturity Cause and then proceed with the suggested sequence of phases of the scientific cycle.

We can find choices made by governments, farmers, and Bahá'u'lláh's instructions to His Administrative Order in the following reflections. Let us try to perceive how those *choices* have affected and will affect the *human kingdom and the plant, the animal, and mineral kingdoms*:

Assuming a rural population with little desire to continue performing agricultural labor, they are unhappy with life in the countryside. Instead, *this population wants to migrate to the city*, with all the consequences that this step implies.

The following is from the Food and Agricultural Organization of the United Nations:

> Small farmers produce much of the developing world's food. Yet, they are generally much poorer than the rest of the population in these countries and are less food secure than even the urban poor. Furthermore, although rapid urbanization is taking place in many developing countries, farming populations in 2030 will not be much smaller than they are today. For the foreseeable future, therefore, *dealing with poverty and hunger in much of the world means confronting the problems that small farmers and their families face in their daily struggle for survival*.
>
> *Investment priorities and policies* must take into account the immense diversity of opportunities and *problems facing small farmers*. The resources on which they draw, their *choice* of activities, indeed the entire structure of their lives, are linked inseparably to the biological, physical, economic, and cultural environment in which they find themselves and over which they only have *limited control*. While every farmer is unique, those who share similar conditions also often share *common problems and priorities* that transcend *administrative* or political borders.[140]

Let us imagine the possibility of *widespread hunger* in a region and the level of *respect towards the government*.

How do you explain to a 4-year-old, who has been playing outside all afternoon, that tonight he (she) will not have dinner?

The farmers need to be aware of how acutely will their *descendants* born in the misery belts of the big cities (Rio de Janeiro, Mexico City, Calcutta, Bogota) be affected in their *IQ*. The author sadly concluded that around *1.2 billion people deprived* of a significant portion of their potential *intellectual* capacities because of a lack of protein of their mothers *during pregnancy* (Duque, 1999) (Joseph Enamuthu. Iron deficiency 1.6 billion) (Reynaldo Martorell, Iodine deficiency 2.1 billion).

As an example of *governmental decisions*, the "Food Stamps" program of the United States of America, despite the abuses, is worthy of being *evaluated to address this issue*. To *explore* the causes of their situation, small farmers could examine:

- They learned the issues in each *crop's cycle phase, choosing* which options to adopt before the next planting *season*.

---

[140] Dixon, Gulliver and Gibbon, Farming Systems and Poverty: Improving Farmers' Livelihoods In A Changing World. Emphasis added.

- *The decisions* made based on the climatic cycle during the planting, growing, and harvesting phases may have affected *the mineral kingdom*, for example, the lack of enough water to help dissolve the mineral nutrients in enough quantity.
- Based on their *decisions* on what happened during the *climatic cycle* and the *planting, growing,* and *harvesting phases,* for example, an impressive growth spurt of *weeds* during the crop's *growing stage*.
- The ignorance of the consequences of *decisions* made during the *planting, growing, and harvesting phases* concerning the effects of the intensive use of herbicides, fungicides, pesticides, chemical fertilizers, and heavy machinery during many *cycles*. For example, what happened to the worms, fungus, insects, and microorganisms that used to be in abundance in the soil?

The following is a concrete example if we assume that such soils are almost sterile. In this experiment, we should also notice that the productivity increase and host-specific *disease* decreased with increasing microbial community diversity:

> Ecosystem productivity commonly increases asymptotically with plant species diversity, and determining the mechanisms responsible for this well-known pattern is essential to predict *potential changes in ecosystem productivity with ongoing species loss*. Previous studies attributed the asymptotic diversity–productivity pattern to plant competition and differential resource use (e.g., niche complementarity). Using an analytical model and a series of experiments, we demonstrate theoretically and empirically that host-specific soil microbes can be major determinants of the diversity–productivity relationship in grasslands. In the presence of soil microbes, *plant disease* decreased with increasing diversity, and productivity increased nearly 500%, primarily because of the *strong effect of density-dependent disease on productivity at low diversity*. Correspondingly, *disease was higher in plants* grown in conspecific-trained soils than heterospecific-trained soils (demonstrating host-specificity), and productivity increased and host-specific *disease* decreased with increasing community diversity, suggesting that *disease was the primary cause of reduced productivity in species poor treatments. In sterilized, microbe-free soils, the increase in productivity with increasing plant species number was markedly lower than the increase measured in the presence of soil microbes*, suggesting that niche complementarity was a weaker determinant of the diversity–productivity relationship.
>
> Our results demonstrate that soil microbes play an integral role as determinants of the diversity–productivity relationship.[141]

- Evaluate *pest management* practices and their effect on beneficial insects, spiders, birds, and bats that are predators of *damaging insects* during the *planting, growing, and harvesting phases*.
- The *historical decisions* made by those in authority concerning rural areas and agriculture. For a sample of how the exercise and imposition of governmental power, read Brocket's *Land Power and Poverty. Agrarian Transformation and Political Conflict in Central America*. He also describes South Korea's agricultural reform's positive experience.

---

[141] Schnitzer, Stefan A. et al., Soil microbes drive the classic plant diversity-productivity pattern. Emphasis added.

To come up with a strategic solution and increase their understanding, and be *empowered* to present it to those in *authority*, it is necessary to determine the implications and the *phases* and *stages* that they have to *monitor* to *follow up decisions* made to:

- decrease the use of *pesticides* because *opting* for an integrated pest *management strategy*,
- *rotating* polycultures instead of permanent monocultures,
- reducing chemical fertilizers applications and *opting* for composting and natural fertilizers
- to minimize water usage,
- prevent soil *erosion* towards a sustainable agriculture
- Besides, *evaluate* if the farmers rely more or less on income sources outside the farm and how much of the aggregated value goes back to the farmers' pockets.

Despite the *difficult situation, I choose* being inspired by Religion as a *source of wisdom*. For over two thousand years, Christians have been praying:

"'Our Father in heaven, hallowed be your name, Your *Kingdom* come, Your *Will* be done, on earth as it is in heaven. ...'"

In the Bible:

And it shall come to pass in the last days, that the mountain of the LORD's house shall be established in the top of the mountains, and shall be exalted above the hills; and all nations shall flow unto it.

And many people shall go and say, Come ye, and let us go up to the mountain of the LORD, *to the house of the God* of Jacob; and He will teach us of His ways, and we will walk in His paths: for out of Zion shall go forth the Law, and the word of the LORD from Jerusalem. And He shall judge among the nations, and shall rebuke many people: *and they shall beat their swords into plowshares,* and their *spears into pruning hooks*: nation shall not lift up sword against nation, neither shall they learn war any more.[142]

In a statement, Shoghi Effendi mentions that Bahá'u'lláh aims to reconcile *conflicting* creeds:

His mission is to proclaim that *the ages of the infancy and of the childhood* of the human race are past, that *the convulsions* associated with the present *stage of its adolescence* are slowly and *painfully* preparing it to attain the *stage of manhood*, and are heralding the approach of that *Age of Ages when swords will be beaten into plowshares, when the Kingdom* promised by Jesus Christ will have been established, and the peace of the planet definitely and permanently ensured.[143]

In the Bahá'í Writings, we find that promise ratified by Bahá'u'lláh:

Whilst in the Prison of 'Akká, We revealed in the Crimson Book that which is conducive to the advancement of mankind and to the reconstruction of the world. The utterances set forth therein by the Pen of the Lord of creation include the following which constitute the fundamental principles for the *administration* of the affairs of men:

---

[142] Isaiah, 2:2 –2:4. Emphasis Added.

[143] Summary Statement - 1947, Special UN Committee on Palestine. Emphasis Added.

> **First:** It is incumbent upon the *ministers of the House of Justice* to promote the Lesser Peace so that the people of the earth may be relieved from the burden of exorbitant expenditures. This matter is imperative and absolutely essential, inasmuch as *hostilities and conflict* lie at the root of *affliction and calamity*.
>
> **Second:** Languages must be reduced to one common language to be taught in all the schools of the world.
>
> **Third:** It behoveth man to adhere tenaciously unto that which will promote fellowship, kindliness and unity.
>
> **Fourth:** Everyone, whether man or woman, should hand over to a trusted person a portion of what he or she earneth through trade, agriculture or other occupation, for the training and education of children, to be spent for this purpose with the knowledge of the *Trustees of the House of Justice*.
>
> **Fifth:** Special regard must be paid to agriculture. Although it hath been mentioned in the fifth place, unquestionably it precedeth the others. Agriculture is highly developed in foreign lands, however in Persia it hath so far been grievously neglected. It is hoped that His *Majesty the Shah* -- may God assist him by His grace -- will turn his attention to this vital and important matter.[144]

The author thinks that the agriculture situation precedes lesser peace, and it is crucial to understand its challenge and the need to propose a solution.

In his *"Study of History,"* Arnold Toynbee thoroughly explains the relationship between the Law and the exercise of *free will* in History; and demonstrates the irrefutable submission of human affairs to the laws of nature. When explaining the collapse of the Syrian, Indic, and Hellenic civilizations, he mentions the lack of food security and the deterioration of the population's health when agriculture crashed in some critical regions. He also said in clear terms that corruption of Religion is the cause of the falling of civilizations. Are these not the exact reasons why the actual civilization is failing?

> The time has come when those who preach the dogmas of materialism, whether of the east or the west, whether of capitalism or socialism, must give account of the moral stewardship they have presumed to exercise. Where is the "new world" promised by these ideologies? Where is the international peace to whose ideals they proclaim their devotion? Where are the breakthroughs into new realms of cultural achievement produced by the aggrandizement of this race, of that nation or of a particular class? Why is the vast majority of the world's peoples sinking ever deeper into hunger and wretchedness when wealth on a scale undreamed of by the Pharaohs, the Caesars, or even the imperialist powers of the nineteenth century is at the disposal of the present arbiters of human affairs?[145]

---

[144] Bahá'u'lláh, Tablets of Bahá'u'lláh, pp. 89 – 90. Emphasis Added.

[145] The Universal House of Justice, The Promise of World Peace.

3 Summary of the Four Causes Taught by the Greek Philosophers

FORMATIVE CAUSE:	MATERIAL CAUSE:
- How is the arrangement? - **Essence:** love and soul as its power - General properties of matter: form, size, mass, temperature, movement, and position in time and space - Scientific principles related to organization, arrangement, and design - The 11 Principles of Unity in accordance with the teachings of Bahá'u'lláh - Laws related to unification and gravity - Aspirations, desires, feelings, and emotions - **Parameters and indicators** of unity, beauty, symmetry, harmony, reciprocity, and justice	- With what? - **Essence:** the Spirit - Specific properties of matter and characteristics of things - Virtues also called spiritual values - Knowledge, beliefs, and social values - Elements, substances, and raw materials - Spiritual laws, the standards that regulate quality and laws of possession - Specific lines of action or policies to consolidate ethics, knowledge, quality, and efficacy - Relationships of belonging, possession, distinction, and characterization - **Parameters and indicators** of efficacy and quality
FINAL CAUSE:	**EFFICIENT CAUSE:**
- Why? For what? - **Essence:** the power of law itself, i.e., that which is a condition to a certain aspect of life or nature, and explains its purpose, its mission - Situations of risk and danger - Pacts, agreements, and covenants - Natural laws - Religious laws and ordinances - Social order, laws, and norms - Advice from those with experience - Individual and collective commitment to ideals - **Relationships** of gratitude, loyalty, reciprocity, mutuality, of fear and protection; and also guilt, repentance, punishment, and reward - **Parameters and indicators** of protection, equality, prevention, and security	- With what being? With whom? With what type? - **Essence:** the Word of God - Categories: genres, clusters, groups - Animated and inanimated beings - Forces: the power of mind, energy, capacities, potential, talents, vocations, arguments, concepts - To do: movements: methods, processes, activities, abilities, skills, arts, technologies, mechanisms - Interacting entities within a system - Fruits of labor: services, products, and leftover materials or efforts - Laws of thermodynamics, labor law, and tools regulations. Grammar rules - **Relations** of cause and effect, logic and reason - **Parameters and indicators** of efficiency and productivity

Summary of the Four Causes Taught by the Greek Philosophers

'Abdu'l-Bahá in Paris Talks, enumerates The Eleven Principles out of the Teaching of Bahá'u' lláh: "The Search after Truth. The Unity of Mankind. Religion ought to be the Cause of Love and Affection. The Unity of Religion and Science. Abolition of Prejudices. Eliminating Extremes of Poverty and Wealth. Equality of Men before the Law. Universal Peace. Non-Interference of Religion and Politics. Equality of Sex -- Education of Women. The Power of the Holy Spirit." Let us now study the four causes taught by the Greek philosophers.

The author attached to each "cause" the notions included in the above summary, which will explain thoroughly later on.

First, we will study the Formative and Material Causes in Chapters fourth and fifth. Then, in Chapters seventh and eighth, we will explore the other two Causes taught by Aristotle and other Greek philosophers, the Final and the Efficient Causes. We will make a parenthesis in Chapter sixth to learn about the Root Cause.

In *Why Teleology Isn't Dead*, by John Farrell, we read: "(t)he Ancient Greeks ... classic four causes Aristotle listed as essential to natural philosophy: the Material cause, the Efficient cause (both still crucial to Science), and also the Formal cause, and the Final cause, both of which modern Science has essentially dismissed as irrelevant and best relegated to the debates of philosophers."[146]

---

[146] John Farrell, Why Teleology Isn't Dead

4   The Formative Cause and the Formative Method

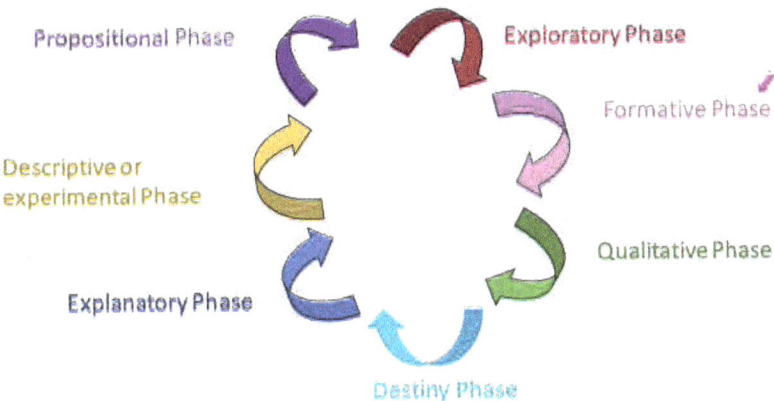

Let me introduce this chapter, briefly mentioning the forces that keep united or separated elements, compounds, tissues, organs, animals, and humans:

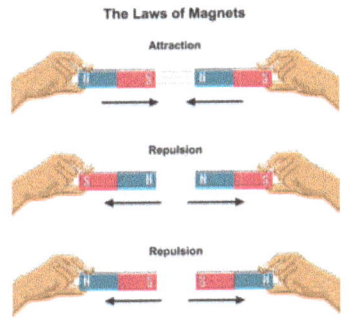

Someone said that the Master taught: "In the divine world there is love, symbolized in the material world by *magnetism*."[147] Thus, in all kingdoms, we find unifying and separating forces: "The electromagnetic force can be attractive or repulsive."[148] [149]

---

[147] 'Abdu'l-Bahá, Paris Talks, p. 120. Emphasis added.

[148] https://www.chegg.com/homework-help/definitions/electromagnetic-force-2

[149] Illustration 255765107 © Nandalal Sarkar | Dreamstime.com

The Formative Cause and the Formative Method

The Moon's mass is 7.35×10^{22} kilograms[150] [151], and the Earth's oceans' mass is 1.4×10^{21} kilograms[152] [153]. If we are on the beach on the side of the earth facing the moon, we will see a high tide.[154]

Because the moon is much closer to planet earth than the sun, the moon's mass has a more significant influence on tides.

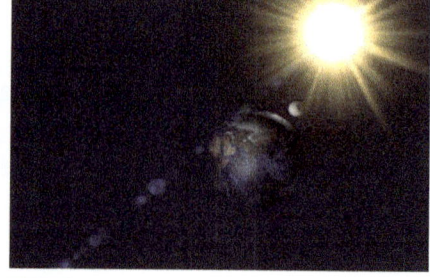

"The gravitational force is a force that attracts any two objects with mass."[155]

The picture is the hand of a gecko attached to the glass[156]; what forces keep it from falling? "Van der Waals forces are the sum of the attractive and repulsive electrical forces between atoms and molecules. These forces differ from covalent and ionic chemical bonding because they result from fluctuations in the charge density of particles. Examples of van der Waals forces include hydrogen bonding, dispersion forces, and dipole-dipole interactions."[157]

---

[150] Nasa, Moon Fact Sheet. March 2021
<https://nssdc.gsfc.nasa.gov/planetary/factsheet/moonfact.html>.

[151] Photo 11318077 / Low High Tide © Jorge Hernando | Dreamstime.com

[152] Nasa, Earth Fact Sheet. March 2021
<https://nssdc.gsfc.nasa.gov/planetary/factsheet/earthfact.html>.

[153] Photo 11317987 / High Tide © Jorge Hernando | Dreamstime.com

[154] Photo 47128749 © Jeffplay | Dreamstime.com

[155] Ohio State Test - Physical Science: Practice & Study Guide / Science Courses, Gravitational Force: Definition, Equation & Examples.

[156] Photo 45071438 © Mario Madrona Barrera | Dreamstime.com

[157] ThoughtCo. Van der Waals Forces: Properties and Components.

The Formative Cause and the Formative Method

- In minerals: chemical affinity, ligand, binding, macroaggregates.

What forces are keeping the molecules of H_2O together, forming a drop?[158]

- In plants; phototropism, gravitropism, thigmotropism, hydrotropism, thermotropism, chemotropism[159]

- "The evolution of multicellular organisms permitted specialized cells and tissues to form; a flowering plant has at least 15 cell types, and a vertebrate hundreds. In both plants and animals, cells that are specialized to carry out a particular task are found together in the tissues in which the task is performed: a xylem or meristem; a liver, a muscle, or a nerve ganglion. Different types of cells in a tissue are often arranged in precise patterns of staggering complexity. For instance, the hundreds of different types of neurons in the human brain are interconnected to one another through a network of some 10^{15} synaptic connections! The coordinated functioning of many types of cells within tissues, and of multiple specialized tissues, permits the organism as a whole to move, metabolize, reproduce, and carry out other essential activities.

    A key step in the evolution of multicellularity must have been the ability of cells to contact tightly and interact specifically with other cells. Various integral membrane proteins, collectively termed cell-adhesion molecules (CAMs), enable many animal cells to adhere tightly and specifically with cells of the same, or similar, type; these interactions allow populations of cells to segregate into distinct tissues. Following aggregation, cells elaborate specialized cell junctions that stabilize these interactions and promote local communication between adjacent cells. Animal cells also secrete a complex network of proteins and carbohydrates, the extracellular matrix (ECM), that creates a special environment in the spaces between cells. The matrix helps bind the cells in tissues together and is a reservoir for many hormones controlling cell growth and differentiation. The matrix also provides a lattice through which cells can move, particularly during the early stages of differentiation. Defects in these connections lead to cancer and developmental malformations."[160]

- In animals: Also, aggregation and alarm pheromones.
- In the human kingdom: love, prejudice, and hate

---

[158] Photo 55833435 © Peter Beattie | Dreamstime.com

[159] Regina Bailey, Understanding Plant Tropisms

[160] Harvey Lodish, et al. Molecular Cell Biology. 4th edition. New York: W. H. Freeman; 2000. Chapter 22, Integrating Cells into Tissues.

The Formative Cause and the Formative Method

4.1 Philosophical Argument

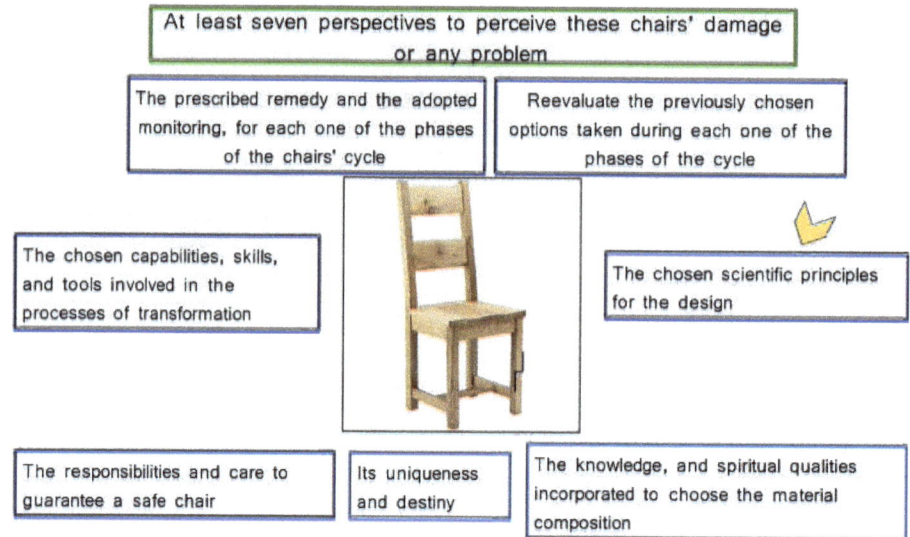

To introduce the Formative Cause, let us continue with the chair's example. To understand *"How was the chair made?"* it is necessary to consider, among others:

- The principles of anthropometry, including the average size of the person who is going to sit in it
- The principles of economics, including the proportions of the materials to be used
- The principles of engineering, estimating the maximum weight of the person who will sit in the chair and the thickness of the material that will support such weight
- The principles of ergonomics concerning the chair's backrest and arms rest are designed to eliminate discomfort and risk of injury due to work.

The Formative Cause and the Formative Method

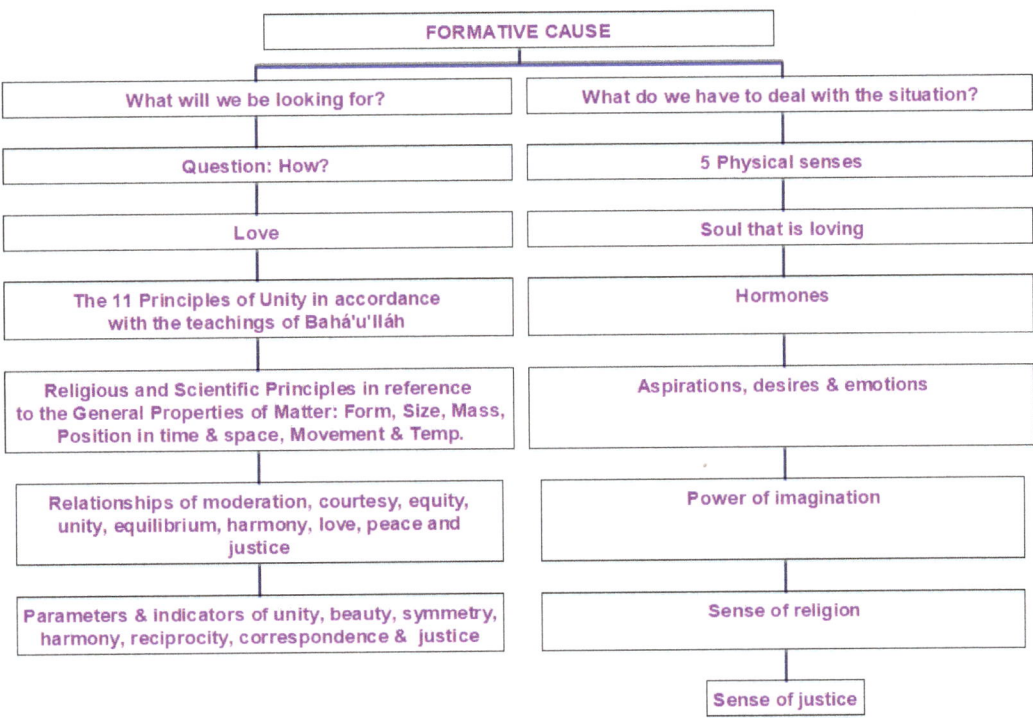

The first approach to solve the Formative Cause and each one of the other causes (Material, Efficient and Final) is to establish the connection between the two quotes by 'Abdu'l-Bahá, that serve as the foundation for this search. In the first quote, he mentions the Formal Cause:

> Essential pre-existence is an existence which is not preceded by a cause; essential origination is preceded by a cause. Temporal pre-existence has no beginning; temporal origination has both a beginning and an end. For the existence of each and every thing depends upon four causes: the efficient cause, the material cause, *the formal cause*, and the final cause. So this chair has a creator who is a carpenter, a matter which is wood, *a form which is that of a chair*, and a purpose which is to serve as a seat. Therefore, this chair is essentially originated, for it is preceded by, and its existence is conditioned upon, a cause. This is called essential or intrinsic origination.[161]

'Abdu'l-Bahá is reported to have said:

> Someone desires an explanation of the terms soul, mind and spirit. The terminology of ancient and modern philosophers differs. According to the great ancient philosophers the words soul, mind and spirit implied the underlying principles of life; the essence was expressed under different names and these three terms designated the various functions of the absolute reality, or the operations of the one single essence; for instance, when they dealt with the *sensations of emotion* they called it the *soul*; when they desired to express that power which discovers the reality of phenomena they gave it the appellation of mind and

---

[161] 'Abdu'l-Bahá, Some Answered Questions, pp.155 -56. Emphasis added.

when they discussed the consciousness which pervades the world of creation they gave it the title of spirit.[162]

The essence of Love showed the author how to justify the origination of the set of notions and the corresponding set of human faculties associated with science's formative method. It will help us answer the question "How do I envision myself?" by understanding the role of each of the components within the following philosophical argument:

After thinking, reading other quotes, the Greek Philosophers, other documents, and learning from my own mistakes, I gradually concluded that:

- The Formative Cause and the delight for the chair's design, probably associated with the sensations of emotion, which the ancient philosophers called the soul.
- The Efficient Cause possibly intertwined to that "power which discovers the reality of phenomena which they gave the appellation of mind" and the carpenter as the creator of the chair.
- The Material Cause and the matter used to make the chair which is the wood, most likely was related to the consciousness which pervades the world of creation which they gave the title of spirit.

I will make an effort to clarify each one of them. In the following reflections, you will find that I linked "form" to the arrangement of things:

Aristotle's hylomorphism concludes that matter receives and supports the form, and form determines matter. In *Human Culture,* Leon Robin says: "… what is interesting to Aristotle on the space is not the vague location of a being within that common and universal place called the world; but the *appropriate place* for that being."[163]

"Every string [of the lyre] is *set in the precisely right position* for the due production of the tones within its capacity……. The universe is good not when the individual is a stone, but when everyone throws his own voice towards a *total harmony*, singing out a life thin, harsh, imperfect though it be. The *harmony* is made of tones unequal, differing, but *together they form the perfect consonance.*"[164]

The form does not apply only to material objects. Truthfulness, justice, harmony, and unity are virtues and, as such, are not perceivable by our physical senses. When, with scientific evidence, someone demonstrates the validity of the hypothesis or is working against racial prejudices, she is acting accordingly with those virtues. Anyone can appreciate it with his (her) own physical senses.[165] [166]

---

[162] 'Abdu'l-Bahá, Divine Philosophy, p. 119. Emphasis added.

[163] El Pensamiento Griego y los Orígenes del Espíritu Científico. La Evolución de la Humanidad, p. 264. Emphasis added.

[164] Plotinus, The Six Enneads. Emphasis added.

[165] Photo 79933166 © Vampy1 | Dreamstime.com

[166] Photo 95607681 © Dmytro Zinkevych | Dreamstime.com

The Formative Cause and the Formative Method

Abdu'l-Bahá said:

> For an action to take place we need a force. A common accepted definition in Physics expresses force, as the result of multiplying mass by acceleration. "Love is the most great law that ruleth this mighty and heavenly cycle, the *unique power that bindeth together the divers elements* of this material world, the supreme *magnetic force* that directeth the movements of the spheres in the celestial realms."[167] "God has willed that love should be a vital *force* in the world, and you all know how I rejoice to speak of love."[168]

But an action or movement to take place needs to have a direction. Social and religious forces have forms, determined by the aspirations, the principles, the motives, the emotions, and the impulses involved, for example, "Let not a man glory in that he loves his country; let him rather glory in this, that he loves his kind."[169] "And if thine eyes be turned towards justice, choose thou for thy neighbor that which thou choosest for thyself."[170]

Love then is a force, but also it is a form, a spiritual quality, a law, and an option. We could assume that the intention of the action, leading to the elimination of the extremes of wealth and poverty, is the direction of the force.

When a jury identifies who is telling the truth in relation to the facts, the truth, or its opposite, is the form used by the witnesses to describe the facts of the situation.

When there is harmony and unity within a family because they consult together, consulting within the family is the form used by that existential entity to live in unity and harmony. That direction is the form adopted by the consultative force to achieve the expected results.

"Stuart Kauffman of the Santa Fe Institute (and author of Reinventing the Sacred), for an example, argues that self-organization, operating well below the level of biological organisms, is crucial to a better understanding of the origin of life".[171]

---

[167] 'Abdu'l-Bahá, Selections from the Writings of 'Abdu'l-Bahá.

[168] 'Abdu'l-Bahá, Paris Talks, p. 120. Emphasis added.

[169] Shoghi Effendi, The World Order of Bahá'u'lláh

[170] Bahá'u'lláh, Epistle to the Son of the Wolf, p. 29.

[171] Farrell, John. Why Teleology Isn't Dead

4.1.1 General properties of matter

When human beings organize the arrangement of things, we manage the general properties of matter (form, size, mass, temperature, movement, phases of matter, and position in time and space). In the case of institutions, farms, and industries, it implies the organization of human and material resources, manufacturing supplies, products, and services.

The form, arrangement, and design are much more important than the size to perceive the beauty of justice, harmony, and unity. I will call it the Formative Method of Science instead of the "quantitative method" to keep its proper perspective.

Taking agriculture as an example:

- **Form and arrangement**: Which variables best fit the plots' shape and the furrows' blueprint? How about the water drainage and the sun's trajectory?
- **Size**: Organization of human resources required in proportion to the size of the plots.
- **Mass**: Organization of the amount of seed and fertilizer required in proportion to the size of the plots.
- **Temperature**: Organization of chores in conformance with the weather forecast.
- **Position in space**: Distribution of the population and the resources, the location of infrastructure, jobs, and residences. Agricultural zones. The elevation or altitude in meters above sea level.
- **Position in time**: National, institutional and communal calendars and timetables. Dates of appraisal, project due dates, or market days. Time for planting or harvesting. Periods of high or low demand. Periods of scarcity and abundance.
- **Movement**: Utilization of systems of transportation. Attending places of education and others. Administrative activities, such as planning, monitoring, and evaluating. Household chores. Farm and industrial labor. Entertainment and resting. We also organize movements when we vary the speed—i.e., accelerating or decelerating.
- **Phases of the matter**: Activities corresponding to climatic seasons. The developmental stage of a child and the medical plan observing the actual evolutionary stage of her illness.

4.1.2 Religious and scientific principles

It is helpful to look at the following relationships to understand how a group of people organizes itself to accomplish a task or what motives are taken into account when an organization distributes its resources:

- the principles of world unity[172] taught by the Bahá'í Faith, and the organization of humankind;
- the principles of management for monitoring and evaluating its performance, and the way an institution arranges its resources;

---

[172] 'Abdu'l-Bahá in Paris Talks, when enumerating The Eleven Principles out of the Teaching of Bahá'u' lláh.

The Formative Cause and the Formative Method

- the principles of economics, education, public health and governance, and the organization of society;
- the principles of environmental preservation and the distribution of natural resources;
- the principles of engineering and the design of infrastructure, machinery, and equipment;
- the principles of decency and clothing design;
- the principles of fasting and the environmental temperature during that period[173] or where we work, and so forth, and
- the principles of consultation and teamwork and how they organize themselves at the meeting place.

Compare the room arrangement, when we see the basics of debate and when the consultation principles are applied. Compare the outcome in both.

| Politicians of each party sit in a separate section with their own agenda about their true interests[174] | Friends learn together how to search and find the truth, examining every problem from different perspectives[175] |

When searching for a solution by applying the organizing principles of Science and Religion, we should try to establish the following relationships:

- Religious and scientific principles related to form, organization, design, and trajectory
- Religious and scientific principles related to size: length, area, volume, proportion, the magnitude
- Religious and scientific principles related to mass
- Religious and scientific principles related to temperature
- Religious and scientific principles related to the position in time and space
- Religious and scientific principles related to phases of matter
- Religious and scientific principles related to movement.

---

[173] JE Esslemont, Bahá'u'lláh and the New Era, p. 189.

[174] Photo 22475686 © Ron Zmiri | Dreamstime.com

[175] Photo © Monkey Business Images | Dreamstime.com

The Formative Cause can be directly associated with the concepts of structure and model.

4.1.3 Love is a Bond Inherent in the Realities of Things

The fact that "the principles of world unity" refer to the planet's organization allowed me to understand that the generating force of the Formative Cause is Love. 'Abdu'l-Bahá is reported to have said:

> "According to the great ancient philosophers the words *soul*, mind and spirit implied the underlying principles of life; the essence was expressed under different names and these three terms designated the various functions of the absolute reality, or the operations of the one single essence; for instance, when they dealt with the *sensations of emotion* they called it the *soul* ... ."[176]

There is a relationship between love and affection that is conducive to self-sacrifice. Is it not love that drives us to increase the scope of a health project or reach out to vulnerable children? Is it not love that moves us to exert ourselves to polish a piece to reach the form and excellence desired? In truth, the power of love moves us toward unity.

The following quotes express the power of love:

> Know thou of a certainty that Love is the secret of God's holy Dispensation, the manifestation of the All-Merciful, the fountain of spiritual outpourings. Love is heaven's kindly light, the Holy Spirit's eternal breath that vivifieth the human soul. Love is the cause of God's revelation unto man, *the vital bond inherent, in accordance with the divine creation, in the realities of things.* Love is the one means that ensureth true felicity both in this world and the next. Love is the light that guideth in darkness, the living link that uniteth God with man, that assureth the progress of every illumined soul. Love is the most great law that ruleth this mighty and heavenly cycle, the unique power that bindeth together the divers elements of this material world, the supreme magnetic force that directeth the movements of the spheres in the celestial realms. Love revealeth with unfailing and limitless power the mysteries latent in the universe. Love is the spirit of life unto the adorned body of mankind, the establisher of true civilization in this mortal world, and the shedder of imperishable glory upon every high-aiming race and nation.[177]

> All things are beneficial if joined with the love of God; and without His love all things are harmful, and act as a veil between man and the Lord of the Kingdom. When His love is there, every bitterness turneth sweet, and every bounty rendereth a wholesome pleasure. For example, a melody, sweet to the ear, bringeth the very spirit of life to a heart in love with God, yet staineth with lust a soul engrossed in sensual desires. And every branch of learning, conjoined with the love of God, is approved and worthy of praise; but bereft of His love, learning is barren -- indeed, it bringeth on madness. Every kind of knowledge, every science, is as a tree: if the fruit of it be the love of God, then is it a blessed tree, but if not, that tree is but dried-up

---

[176] 'Abdu'l-Bahá, Divine Philosophy, p. 119. Emphasis added.

[177] 'Abdu'l-Bahá, Selections, p. 27. Emphasis added.

wood, and shall only feed the fire.[178]

4.1.4 The soul

The Master, 'Abdu'l-Bahá, is reported to have said, that when the great philosophers of the past "dealt with the sensations of emotion they called it the *soul*"[179], which is why the author associated it with love. "If we are caused joy or pain by a friend, if a love prove true or false, it is the *soul* that is affected. If our dear ones are far from us -- it is the *soul* that grieves, and the grief or trouble of the *soul* may react on the body. …The *soul* does not evolve from degree to degree as a law -- it only evolves nearer to God, by the Mercy and Bounty of God."[180]

Elsewhere we find:

> The human spirit, which distinguishes man from the animal, is the rational soul, and these two terms—the human spirit and the rational soul—designate one and the same thing. This spirit, which in the terminology of the philosophers is called the rational soul, encompasses all things and as far as human capacity permits, discovers their realities and becomes aware of the properties and effects, the characteristics and conditions of earthly things. But the human spirit, unless it be assisted by the spirit of faith, cannot become acquainted with the divine mysteries and the heavenly realities. It is like a mirror which, although clear, bright, and polished, is still in need of light. Not until a sunbeam falls upon it can it discover the divine mysteries.[181]

> Know, verily, that the soul is a sign of God, a heavenly gem whose reality the most learned of men hath failed to grasp, and whose mystery no mind, however acute, can ever hope to unravel. It is the first among all created things to declare the excellence of its Creator, the first to recognize His glory, to cleave to His truth, and to bow down in adoration before Him. If it be faithful to God, it will reflect His light, and will, eventually, return unto Him. If it fail, however, in its allegiance to its Creator, it will become a victim to self and passion, and will, in the end, sink in their depths.[182]

'Abdu'l-Bahá is reported to have said:

> If the soul identifies itself with the material world it remains dark, for in the natural world there is corruption, aggression, struggles for existence, greed, darkness, transgression and vice. If the soul remains in this station and moves along these paths it will be the recipient of this darkness; but if it becomes the recipient of the graces of the world of mind, its darkness will be transformed into light, its tyranny into justice, its ignorance into wisdom, its aggression into loving kindness; until it reach the apex. Then there will not remain any struggle for existence. Man will become free from egotism; he will be released from the material world; he will become the personification of justice and virtue, for a sanctified soul illumines humanity and is an honor to mankind, conferring life upon the children of men and

---

[178] ibid. p. 181.

[179] 'Abdu'l-Bahá, Divine Philosophy, p. 120.

[180] 'Abdu'l-Bahá, Paris Talks, p. 729. Emphasis added.

[181] 'Abdu'l-Bahá, Some Answered Questions, p. 55.

[182] Bahá'u'lláh, Gleanings, p. 158.

suffering all nations to attain to the station of perfect unity. Therefore, we can apply the name 'holy soul' to such a one.[183]

Besides having a soul and being able to exercise the power of love, human beings have other faculties linked to the formative cause:

4.1.5 Hormones and emotions

Our bodies respond differently to a stimulus with feelings, emotions, and their relationship with hormones and the endocrine system.

4.1.6 Power of imagination

After evaluating the past and the current arrangement or organization, we have to imagine diverse scenarios to reach the desired outcome. Imagination is a vital instrument of creativity, but it requires the proof of science.

Focusing on the power of imagination, let us consider the following quotes: "Man has likewise a number of spiritual powers: *the power of imagination, which forms a mental image of things*; thought, which reflects upon the realities of things; comprehension, which understands these realities; and memory, which retains whatever man has *imagined*, thought, and understood."[184]

If religious beliefs and opinions are found contrary to the standards of science they are mere superstitions and *imaginations*; for the antithesis of knowledge is ignorance, and the child of ignorance is superstition. Unquestionably there must be agreement between true religion and science. If a question be found contrary to reason, faith and belief in it are impossible and there is no outcome but wavering and vacillation.[185]

4.1.7 Sense of religion

In reaching the promises of a golden age for humanity, our faith grows when acknowledging the existence of the sense of religion. "Taken in general, women today have a stronger *sense of religion* than men. The woman's intuition is more correct; she is more receptive and her intelligence is quicker. The day is coming when woman will claim her superiority to man."[186]

How could we expect to attain "the principles of world unity" without enhancing our senses of religion and justice?

4.1.8 Sense of justice

O SON OF SPIRIT!

The best beloved of all things in My sight is *Justice*; turn not away therefrom if thou desirest Me, and neglect it not that I may confide in thee. By its aid thou shalt see

---

[183] 'Abdu'l-Bahá, Divine Philosophy, p. 121.

[184] Some Answered Questions, p. 56. Emphasis Added.

[185] The Promulgation of Universal Peace, p. 181. Emphasis Added.

[186] 'Abdu'l-Bahá, 'Abdu'l-Bahá in London, p. 104. Emphasis Added.

with thine own eyes and not through the eyes of others, and shalt know of thine own knowledge and not through the knowledge of thy neighbor. Ponder this in thy heart; how it behooveth thee to be. Verily justice is My gift to thee and the sign of My loving-kindness. Set it then before thine eyes.[187]

Why should man, who is endowed with the *sense of justice* and sensibilities of conscience, be willing that one of the members of the human family should be rated and considered as subordinate? Such differentiation is neither intelligent nor conscientious; therefore, the principle of religion has been revealed by Bahá'u'lláh that woman must be given the privilege of equal education with man and full right to his prerogatives.[188]

Of course, comprehending that religion is progressive, we can examine the validity of God's teachings in different epochs, for example:

On the societal level, the principle of collective security enunciated by Bahá'u'lláh (see Gleanings from the Writings of Bahá'u'lláh, CXVII) and elaborated by Shoghi Effendi (see the Guardian's letters in The World Order of Bahá'u'lláh) does not presuppose the abolition of the use of force, but prescribes 'a system in which Force is made the servant of Justice', and which provides for the existence of an international peace-keeping force that 'will safeguard the organic unity of the whole commonwealth. In the Tablet of Bishárát, Bahá'u'lláh expresses the hope that 'weapons of war throughout the world may be converted into instruments of reconstruction and that strife and conflict may be removed from the midst of men'.

In another Tablet Bahá'u'lláh stresses the importance of fellowship with the followers of all religions; He also states that 'the law of holy war hath been blotted out from the Book'.[189]

With this set of human faculties, common to all human beings, we get an idea of the arrangement's beauty.

4.1.9 Relationships

In general, the afore-mentioned religious and scientific principles frame relationships of moderation, courtesy, equity, beauty, unity, correspondence, equilibrium, harmony, love, peace, and justice. Within those relationships, it is implicit an organization of things.

The laws that are related to the Formative Cause are the law of love, the laws that regulate relationships, the law of gravity, and the law of attraction of masses. There is a relationship between love and affection that is conducive to self-sacrifice.

4.1.10 Parameters and indicators

One of the roles of governments, managers of enterprises, and mature individuals is to monitor and evaluate the efforts made to achieve their vision. We should establish parameters

---

[187] Bahá'u'lláh, The Hidden Words of Bahá'u'lláh. Emphasis Added.

[188] The Promulgation of Universal Peace, p. 106. Emphasis added.

[189] Bahá'u'lláh. The Kitáb-i-Aqdas, p. 241, Note 173.

and indicators to monitor and assess the advancement in the achievement of beauty, justice, unity, decency, and harmony in the context of the adopted organization or arrangement.

In conclusion, the cycle of scientific research also has a formative phase. The notions and the set of human faculties associated with the Formative Cause are appropriate to answer questions such as: How are we organized? How do we envision our lives?

☙❧

In his search for a single science Aristotle wrote:

> The minute accuracy of *mathematics* is not to be demanded in all cases, but only in the case of things which have no matter. Hence method is not that of natural science; for presumably the whole of nature has matter. Hence we must inquire first what nature is: for thus we shall also see what *natural science* treats of (and whether it *belongs to one science or to more to investigate the causes and the principles* of things).[190]

Further down he wrote:

> ... some do not think there is anything substantial besides sensible things, but others think there *are eternal substances* which are more in number and more real; e.g.

> Plato posited two kinds of substance-the *Forms and objects of mathematics*-as well as a third kind, viz. *the substance of sensible bodies*. And Speusippus made still more kinds of substance, beginning with the One, and assuming principles for each kind of substance, one for *numbers*, another for *spatial magnitudes*, and then another for the *soul*; and by going on in this way he multiplies the kinds of substance. And some say *Forms and numbers* have the same nature, and the other things come after them- *lines and planes*-until we come to the substance of the material universe and to *sensible bodies*.[191]

Let us read again what the wise Diotima taught Socrates about the meaning of love in connection to the words where the author added emphasis:

> For he who would proceed aright in this matter should begin in youth to visit *beautiful forms*; and first, if he be guided by his instructor aright*, to love one such form* only-out of that he should create *fair* thoughts; and soon he will of himself perceive that *the beauty* of one *form* is akin to the *beauty* of another; and then *if beauty of form* in general is his pursuit, how foolish would he be not to recognize that the *beauty in every form* is and the same! And when he perceives this he will abate his violent *love of the one*, which he will despise and deem a small thing, and will become *a lover of all beautiful forms* ... until he is compelled to contemplate and see the *beauty* of institutions and laws, and to understand that the *beauty* of them all is of one family, and that personal *beauty* is a trifle; and after laws and institutions he will go on to the sciences, that he may see their *beauty*, being not like a servant in *love with the beauty* of one youth or man or institution, himself a slave mean and narrow-minded, but drawing towards and contemplating the vast sea of beauty, he will create many *fair* and noble thoughts and notions in *boundless love of* wisdom; until on that shore he grows and waxes strong, and at last the *vision* is revealed to him of a *single science*, which is the science of *beauty everywhere*.[192]

---

[190] The Metaphysics. Emphasis added.

[191] Ibid.

[192] Plato, Symposium. Emphasis added.

4.2 Research Steps for the Formative Method

If we follow the art of constructing the Single Science, we conclude that each method of science is intertwined with the other methods.

The author made an effort to relate this method to the Valley of Love, as revealed in The Seven Valleys. These Valleys "mark the wayfarers' journey from their mortal abode to the heavenly homeland"[193]; in the Valley of Love, we find: "The steed of this valley is *pain*, and if there be no *pain* this journey will *never* end. In this plane the *lover* hath no thought save the Beloved, and seeketh no refuge save the Friend. At every moment he offereth a *hundred* lives in the path of the Loved One, at every step he throweth a *thousand* heads at His feet."[194]

The following are just general ideas that may serve as guiding steps for the formative method of science. The suggested steps should link the problem, its facts, and symptoms to the harmony, equity, and beauty in the design or arrangement and the other general properties of matter; and the faculties related to the Formative Cause.

Let us always remember that forces keep things together: gravity, friction, magnetism, Van der Waals forces, chemical affinity, ligand, binding, macroaggregates, keeping together tissues and organs in plants, highly complex structures of the senses in animals; and love among human beings.

First: The exploratory step. Search for the arrangement, form, or design modifications that caused the problem's facts and symptoms. Search for what may be the impact of the problem's facts and signs in the harmony, equity, and beauty in the design or arrangement:

- Which decision caused the problem's facts and symptoms' impact on the relationships among the diverse participants in the arrangement?
- Which general property (properties) of matter should be modified to improve our arrangement's harmony, equity, and beauty?
- Which is the strategy to develop the capacity for applying "The best beloved of all things in My sight is Justice."[195]?
- How can a change in the sequence of stages contribute to increasing the beauty of the results, for example, in prioritizing girls' education when there are not enough resources to educate both genders?

Second: The formative step, focusing on the general properties of matter and the faculties closely related to the Formative Cause

- How are we proceeding to modify the general properties of matter to reaching our aspiration by increasing harmony, equity, and beauty in our arrangement?
- Will the solution to the problem's facts and symptoms in the design or arrangement contribute to reaching the desired harmony, equity, and beauty?
- Will the new design or arrangement contribute to repel herbivorous insects and attract beneficial ones?
- What level of unity is within our reach as our short-term goal towards the

---

[193] Bahá'u'lláh, The Seven Valleys, The Call of the Divine Beloved.

[194] Bahá'u'lláh, The Seven Valleys, The Call of the Divine Beloved. Emphasis added.

[195] Bahá'u'lláh, The Hidden Words

desired vision practicing the 11 principles of unity taught by The Master?
- A magnet has an attractant force and a repelling one. To contribute to controlling damaging insects, are there any attractant volatiles of beneficial insects and repelling ones of the harmful pest?

Third: The qualitative step, focusing on the specific properties of matter
- How can matter's particular properties increase our arrangement's harmony, equity, and beauty?
- What spiritual qualities or cultural values contribute the most to increasing harmony, equity, and beauty?
- What can I do to polish my mirror to become devoted to Bahá'u'lláh?
- How can I polish my mirror to become devoted to unity and love for humankind?

Fourth: The Root Cause (to be considered collectively or privately by an individual. Negligence, imitation, lack of self-accountability of faults, transgressions, omissions, failing, oversight, inadvertence, or mere ignorance are examples of Root Causes. Not confession or finger-pointing. Please search for the root causes(s) in reaching the set goals of the strategy and its relationship to our collective high destiny, determining the root causes of the problem's facts and symptoms
- Which of the mentioned root causes may affect harmony, equity, unity, and beauty among us?
- Was there any failing in the accompaniment to perceiving the way the strategy contributes to reaching the vision?
- What was the source of any difference in communicating how the strategy contributes to reaching the vision?
- How will I deal with the obstacles that I established, to dedicating time to contemplation, and then reflect, comprehend, and deal with *widespread* problems and difficulties? Is it procrastination? Is it my attachment to the past?
- What would be the root cause and the consequences of not permitting women to address the farmers' unfair share in the added value of what they produced?

Fifth: The Explanatory Step
- What are the substantial risks when modifying the general properties of matter?
- What are the significant risks to the ecosystem's biodiversity when modifying the arrangement or any other general property of matter?
- Should we agree on how we will protect the new arrangement pact during the consultation process?
- Why do we perceive beauty in a law such as profit sharing? In the elimination of prejudice? On equal opportunities for men and women?

Sixth: The Experimental Step
- Which experiment or experience tells us that the new arrangement is sensible addressing augmented justice, unity, harmony, and beauty?
- How can we increase our capacity to perceive more religious and scientific principles related to each matter's general properties?
- How can we increase our capacity to love, practice moderation, and consult looking for justice when science and religion are joined and welded in reality?

- How the diversity of functions of the different organisms (birds, plants, fish, insects, microorganisms, etc.) contribute to the ecosystem's equilibrium?
- How rhythm and coordination can augment harmony, equity, unity, and beauty?

Seventh: The Propositional Step
- What would be the effects of the new arrangement on the different kingdoms' biodiversity?
- Which parameters or indicators of harmony, equity, and beauty do we want to discard, improve, or define and formulate anew?
- How can we test the attraction of beneficial insects and the rejection of harmful ones in the context of the changes made to the arrangement and the design?
How the formulation of the indicator responds to the defined parameter?

4.3 The formative phase of the farmers' situation

Let us then continue with our example of the farmers using the Formative Method. A historical fact about farmers and merchants:

> The aristocratic Athenian State was based upon land-ownership, slavery, and the *entire* freedom of the land-owning class from *all* but family and State duties, from *all* need of engaging in productive industry. So long as the chief wealth of the State consisted in land and its produce, so long as the population was *divided into two classes, the rich and the poor*, and so long as the former had *little* difficulty in keeping *all power* in its own hands. But no sooner did the *growth of commerce* throw wealth into the hands of a class that owned *no land*, and was not above engaging in industry, than this class began to claim a *share* in political power.

> There were *now two wealthy classes, standing opposed to each other*; a proud, conservative one, with "old wealth and worth," and a vain, radical one, with *new wealth and wants*, both bidding for the favor of the class that had *little wealth, little worth, and many wants*, and thus making it feel its importance. Such is the origin of Athenian democracy.

> It is the child of trade and productive industry. It owed its final consecration to the Persian Wars, and especially to the battle of Salamis, in which Athens was saved by her fleet, manned chiefly by marines from *the lower classes, the upper classes*, as we have seen, being trained only for land-service. Thus, the battle of Salamis was not only a victory of Greece over Persia, but *of foreign trade over home agriculture*, of democracy over aristocracy.[196]

Let us continue with our example. To understand how a rural population became so disillusioned with the viability of its farming businesses and the reasons it migrated to the misery belts around big cities, we should analyze:

1. The *soil topography in the region and on the farm,* and the principles of preservation of natural resources.
2. The principles of agriculture and crop layout, especially concerning environmental sanitation and the *disposition of sick plants and farm waste*.
3. The correlation between the various kinds of *discrimination*: dark skin, women, poorly educated, poorly dressed, and of course, being part of the farmers' social class; and its *cumulative* effect on *how much* society values peasants' harvest.

The International Labor Organization (ILO), which is a United Nations agency, says:

---

[196] *Aristotle and Ancient Educational Ideals* by Thomas Davidson. Chapter V: Education as influenced by time, place, and circumstances. Emphasis added.

> *About 40 percent of the world's three-billion strong labor force, some 1.2 billion workers*, are employed in agriculture as self-employed farmers, unpaid family workers, and hired workers. The ILO puts the number of 'waged' or hired workers at *450 million, 38 percent of all persons employed in agriculture equivalent to the entire labor force of the high-income countries*. The ILO includes as waged workers those who receive in-kind payments, and notes that some workers have *several statuses*, such as a person who is a farmer at *some times of the year* and hired worker at others.
>
> ... *Many of the world's hired farm workers* are employed on *plantations, some owned by multinationals*, that provide housing and wages according to the terms of collective bargaining agreements. Plantation workers in 'breakfast commodities' such as banana, coffee, and sugar are *sometimes among their countries'* working elite.
>
> However, *some multinational producers of breakfast commodities* have replaced *year-round workers* who had housing, schools, and other amenities on the plantation with women hired seasonally and offered few or no benefits.[197]
>
> Rural women constitute 50 percent of the agricultural labor force in Africa; they are responsible for 80 percent of the food production and 50 percent of the agricultural output. Women reinvest almost 90 percent of their income in their children and household. Since women are the keys to improving household food security and nutritional wellbeing, increasing women's access to financial resources ultimately leads to increased investments in human capital.[198]

Proponents of the current model of economic production will have to answer whether they are coherent to the *principles of economics when they consider that at least 1.2 billion farmworkers* are being affected by the problem of *discrimination in one or more ways*.

In a document titled Gender and Value Chain Development, we find:

> A gendered impact assessment of organic certified pineapple and coffee producers in Uganda (Bolwig and Odeke, 2007) also touches open changes in women's *workload* as a consequence of certification. The study, based on a household survey (with control group) and focus group interviews, discusses how organic conversion affects men and women differently in respect of changes in the costs and benefits of farming. The study finds that organic conversion has significantly *increased women's labour effort* in coffee production, while the effect on male labour has been weaker. While men enjoy almost exclusive control of income from organic farming, *it is women who carry out most of the additional farming and processing work* needed to meet organic certification and stricter quality and farm management requirements of the organic exporter. According to the authors, it is very likely that *women's increased effort in coffee farming in recent years* has occurred at the expense of their own income-generating activities. Hence, while men over the *last five years* have enjoyed an *increase in the income they control* (from coffee), *women appear to have experienced the opposite*. The skewed gender distribution of the costs and benefits associated with certification was much more pronounced among coffee farmers.

---

[197] ILO. Global Farm Worker Issues. Emphasis added.

[198] Mpule K. Kwelagobe. Commentary - Investing in Rural Women: Closing the Gender Gap in African Agriculture. Emphasis added.

According to the authors, this seemed to be the result of *differences in gender relations*, in land availability and *farm size*, and in market conditions:

- *gender relations generally seemed more equal among pineapple farmers* thus giving *women better access to pineapple incomes* and *men less command* over women's labour for the purpose of pineapple growing;
- pineapple farmers earned *very high incomes (in local comparative terms)*, due to *larger farm size, high yields*, and favourable market conditions, and this allowed them *to hire more labour* thereby relaxing the demand on women's labour.[199]

In the *same* way that much of the early Green Revolution literature (Feder and O'Mara 1981) focused on limited *small farmer* uptake of improved seeds, fertilizer, irrigation, and other components of "modern" production systems, a *large share* of the emerging literature on modern value chains has been concerned with *smallholder participation* in AVCs (Agricultural Value Chains)[200] and with whether these *same value chains* might be leaving *many poorer farmers behind*.[201]

This is perhaps unsurprising given that, historically, market sales of food have been *heavily concentrated* in the hands of a *small number of producers*, even in *regions and countries in which market participation is broad-based*. Although most of the evidence comes from staple grains markets, a *relatively small group* (i.e., *less than 10 percent*) of relatively well-capitalized farmers *located in more favorable agro- ecological zones account for a significant majority of market sales throughout the world* (Barrett 2008). This suggests that gains from agrifood value chain transformation accruing to net sellers in the form of *higher profits will likely concentrate in the hands of a relatively modest share of the farm population* in the developing world, although there is presently *scant* hard evidence on this important point.[202]

Gender and generic value chain interventions:

Generic value chain interventions can have *positive effects for participating women*. But this is more likely to happen when they take into consideration gendered constraints that apply to *upgrading value chain participation and distribution of value* (both along the value chain and within households). For example, value chain interventions emphasizing product and process upgrading as well as forging/strengthening *horizontal and vertical linkages*, can be strategically applied in *parts of the chain* where women play important roles. But in order to secure *positive impacts* and avoid unintended *negative consequences*, a gendered value chain analysis is needed as part of *project design* and implementation. Assuming that women will automatically *gain* from generic value chain interventions can have unintended *negative consequences*. Furthermore, these interventions are not sufficient in themselves to *secure value chain participation* and meaningful welfare outcomes for women when working in

---

[199] Ministry of Foreign Affairs of Denmark. Gender and Value Chain Development. Emphasis added.

[200] AVC stands for: Supermarkets, specialized wholesalers, and processors and agro-exporters' agricultural value chains have begun to transform the marketing channels into which smallholder farmers sell produce in low-income economies.

[201] Christopher B. Barrett, Smallholder Participation in Agricultural Value Chains: Comparative Evidence from Three Continents. p. 32. Emphasis added.

[202] Idem. p. 32. Emphasis added.

'gender conservative' areas. Finally, the gender impact of generic value chain interventions is likely to be *mainly limited to improving the terms of inclusion* of existing value chain participants, rather than promoting the *participation of more women* in the chain.

Generic value chain interventions can (in specific circumstances) have *positive effects* for participating women; but in order to secure *positive impacts* and avoid unintended *negative consequences*, a gendered value chain analysis is required as part of *project design* and implementation.[203]

The reality of women farmers *not being paid what they deserve* is not just common among the *poorly educated masses*. Academia is also responsible for this situation. An example from personal experience comes to mind. When speaking to a Ph.D. in economics, the professor told me that he firmly believed that work done in the household by my wife could not be *counted* as such because there is not an *economic* transaction. I tried several ways to convince him that housework brings well-being, but he said that viewed from the *economic science perspective* is not work. After several attempts, I told him that work is movement and energy consumption from the physics perspective, finally convincing him. Then, for economist self-employed farmers are invisible, if they do not contribute to the social security system. I feel that the *fragmentation* of knowledge into disciplines has been the cause of *many divisions* in society.

We should analyze the principles of the rural economy at the regional and local levels and the organization of the market.

Let us reflect: "The yield of the coffee tree *peaks after 5 to 7 years*. The fruits are left unpicked until they reach the ideal stage of ripeness, *usually after about seven months*."[204]

Coffee has *low supply elasticity* just as the case with cocoa due to the perennial nature of the crops and the *demand is very inelastic*. A situation of *supply shortage* results in *high coffee prices* without a *significant reduction in consumption*; and *when prices are high* it takes *time* for production to adjust. This is exacerbated by the lag between plantation and harvest, *which varies between 18 and 24 months* for coffee.[205]

"Because of the way the international coffee supply chain works, the link between producers and consumers is lost (Oxfam 2002a). Coffee is traded down a complex line of intermediaries, ranging from local traders, exporters, international traders, roasters, and retailers, who each capture a percentage of the retail value of coffee.

Less than 30 percent of the revenues generated by world coffee sales remain in the coffee-producing countries, and smallholders usually capture *less than 10 percent of the retail price*."[206]

---

[203] Ministry of Foreign Affairs of Denmark. Gender and Value Chain Development. Emphasis added.

[204] Traoré Cocoa and Coffee Value Chains in West and Central Africa: Constraints and Options for Revenue-Raising Diversification. p. 50. Emphasis added.

[205] Idem. p. 55. Emphasis added.

[206] Idem. p. 51. Emphasis added.

Bloomberg News on May 24th, 2018, before interviewing Roberto Velez Vallejo, CEO of Colombia's National Federation of Coffee Growers, says that a cup of coffee at Starbucks costs $3.50, and the Colombian coffee grower receives $0.03.[207] Farmers are receiving less than 1%.

In, *Selections from the Writings of 'Abdu' l-Bahá*, we find: "Although our vision must be world-embracing, the initial stage of economic reconstruction is at the local level, beginning with agricultural reform. 'Abdu'l-Bahá said the solution begins with the village, and when the village is reconstructed, then will the cities be also."[208]

The Universal House of Justice, in a letter to the Bahá'ís of Iran on 2 March 2013, said:

> The deepening environmental crisis, driven by a system that condones the pillage of natural resources to *satisfy an insatiable thirst for more*, suggests how entirely inadequate is the present conception of humanity's relationship with nature; the deterioration of the home environment, with the accompanying rise in the systematic exploitation of women and children worldwide, makes clear how pervasive are the misbegotten notions that define *relations within the family unit*; the persistence of despotism, on the one hand, and the increasing disregard for authority, on the other, reveal how unsatisfactory to a maturing humanity is the *current relationship* between the individual and the institutions of society; the *concentration of material wealth in the hands of a minority of the world's population* gives an indication of how fundamentally ill-conceived are *relationships among the many sectors* of what is now an emerging global community.[209]

When someone asked 'Abdu'l-Bahá about the solution to the *economic* problem he said, "The solution of this problem is one of the fundamental principles of His Holiness Bahá'u'lláh. But it must be solved with *justice* and not with force. If this problem is not solved *lovingly* it will result in war."[210]

---

[207] Steel, Alix. Why Colombia Coffee Farmers Are 'Desperate'.

[208] 'Abdu' l-Bahá, Selections from the Writings of 'Abdu' l-Bahá.

[209] The Universal House of Justice, Non-Involvement in Partisan Politics. Emphasis added.

[210] 'Abdu'l-Bahá qtd. in Compilations, Bahá'í Scriptures. p. 340. Emphasis added.

The following is the need to be addressed:
The farmer takes care of the crop for a longer time than the merchant and the agroindustry. In the value chain farmer, agroindustry, merchant, retailer, who handles the most risks? Who works hardest? Who has the risk of considerable fluctuating prices during the harvest season? Who manages a greater complexity?

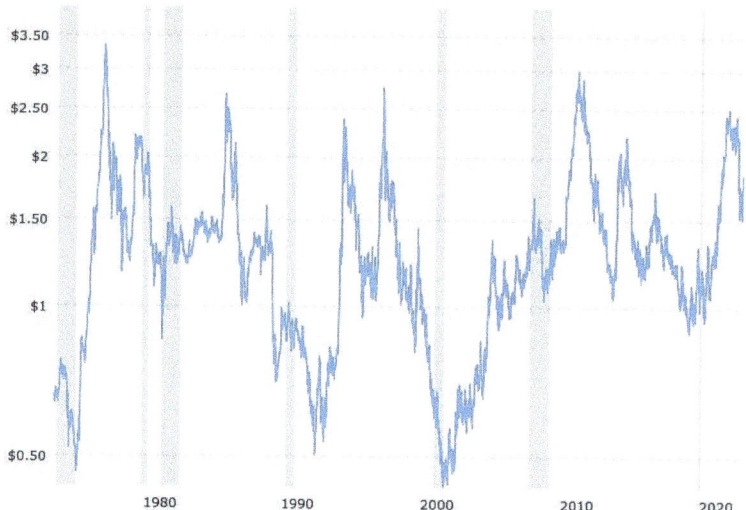

How can a farmer plan his budget to pay loans and mortgages?[211] In Low Coffee Prices – A Dire Call to Action, Anna Canning provides a lucid picture of the situation: in the last 50 years, the average annual coffee price has increased 292% when comparing 12 months starting in 1973 to the most recent 12 months ending on March 3, 2023, and the dollar cumulative price increase of 573.81%.[212] The decline of the dollar's purchasing power has almost doubled.

> Coffee is grown in more than 70 countries, but nearly 70 percent of the world's coffee is produced by just four of them – Brazil, Vietnam, Colombia, and Indonesia. Latin America is the largest regional producer with a 59 percent share, followed by Asia and Oceania (30%) and Africa (11%).
>
> For countries that produce it, coffee exports generate a significant proportion of national income and are a vital source of the foreign exchange earnings that governments rely on to improve health, education, infrastructure, and other social services. For instance, Honduras relies on coffee for nearly a quarter of its export earnings, and in Nicaragua over 15 percent of the labour market is employed in coffee production. In Ethiopia, 15 million smallholders, nearly a fifth of the population, depend on coffee for their livelihood, and it generates around a third of the country's total export earnings. In Uganda, 1.7 million households grow coffee, and it contributes to nearly 20 per cent of the country's export earnings.[213]

---

[211] Anna Canning, Low Coffee Prices – A Dire Call to Action. Coffee Prices - 45 Year Historical Chart.

[212] https://www.in2013dollars.com/us/inflation/1973?amount=1

[213] Fairtrade Foundation, About Coffee.

To propose a solution to the *economic* problem, a macro-economic forecasting software and other means could help anyone interested in solving this issue to examine the implications for farmers and the world as a whole, of a "*uniform* and *universal* system of currency" where:
- "the economic resources of the world will be *organized*,
- its sources of raw materials will be tapped and *fully* utilized,
- its markets will be *coordinated* and developed,
- and the *distribution* of its products will be *equitably* regulated"[214]

There are over one million species of insects. When applied to all eukaryote kingdoms, our approach predicted ~77 million species of animals, ~298,000 species of plants, and approximately 611,000 species of fungus"[215], and there are *many* species of bacteria and viruses.

"In recent years, scientists have discovered that the world of virus diversity — what they sometimes call the virosphere — is unimaginably vast. They have uncovered hundreds of thousands of new species that have yet to be named. And they suspect that there are millions, perhaps even trillions, of species waiting to be found.

"Suffice to say that we have only sampled a minuscule fraction of the virosphere," said Edward Holmes of the University of Sydney in Australia.

Bacteria and other single-celled microbes belong to a group called prokaryotes. In a paper published March 4 in Microbiology and Molecular Biology Reviews, Kuhn and his colleagues argued that there are, at minimum, 100 million species of viruses that infect prokaryotes.

But some researchers suspect there are many more species of prokaryotes in the world — which would mean many more species of viruses. The true figure might be as high as 10 trillion."[216]

Those who labor on a farm are supposed to recognize the different kinds of pests that may affect the crops and report them *immediately*.

Now, compare the complexity of the number of variables that a farmer deals with to the ones of agroindustry when processing and packing the food, and finally to the limited amount of variables that merchants control and tell me which business is more complicated.

Also, they should try to correct agriculture commodities distortions, such as when there is a small number of buyers (oligopoly). In a document of the Food and Agriculture Organization of the United Nations, we find:

West and Central Africa produce about 70 percent of world cocoa. Smallholders produce about 90-95 percent of all cocoa with farm sizes of two to five hectares (Ha).

Big buyers can pick and choose, playing one producer country against the other. In Cote d'Ivoire, just three years after liberalization, there were forty registered exporters, but ten control over 90 percent of the market. Legislation prevents market shares of these exporters from increasing. Concentrated exporters can potentially exercise market power both on farmers and traders in the producing countries and on manufacturers in the consuming countries.

---

[214] Shoghi Effendi in the Introduction of a book by Bahá'u'lláh, The Proclamation of Bahá'u'lláh, pp. xi-xii. Emphasis added.

[215] Camilo, Mora, et al. How Many Species on Earth and in the Ocean?

[216] Carl Zimmer. Welcome to the virosphere, the unimaginably vast world of virus diversity.

Three Transnational Corporations now dominate the processing and supply of the intermediate cocoa product (cocoa butter and powder, and 'industrial' chocolate), accounting for over 35 percent of total worldwide cocoa grinding capacity (Talbot, 2002).

The continuing strong performance of giants on the processed beverage world is in outstanding contrast with the ever-increasing impoverishment of ordinary coffee farmers at a time of low green coffee prices.

Multinationals capture most of the value-added linked with the production of cocoa and coffee. To secure their market share and increase their profit margins, they have made huge investments in branding and advertising, which shelters them from price competition. While coffee prices almost halved between 1999 and 2001, the average retail prices in the US (the largest consumer in volume) decreased by less than 4 percent (Ponte, 2001). This suggests that not only gross margins have increased for roasters, but also profits."[217]

"The relatively greater success for coffee value chain can be attributed to several factors including the fact that consumers buy coffee beans directly, whereas cocoa beans are used as ingredients in recipes and *never purchased* directly by consumers. A second difference is that *there is more Transnational Corporations* involvement in cocoa processing *located in the producing countries than is the case of coffee.*"[218]

With the same forecasting software, examine the consequences of paying women and other groups discriminated against what they justly deserve and the results of fair trade.

---

[217] Traoré Cocoa and Coffee Value Chains in West and Central Africa: Constraints and Options for Revenue-Raising Diversification. p. 52. Emphasis added.

[218] Idem. p. 50. Emphasis added.

5   The Material Cause and the Qualitative Method

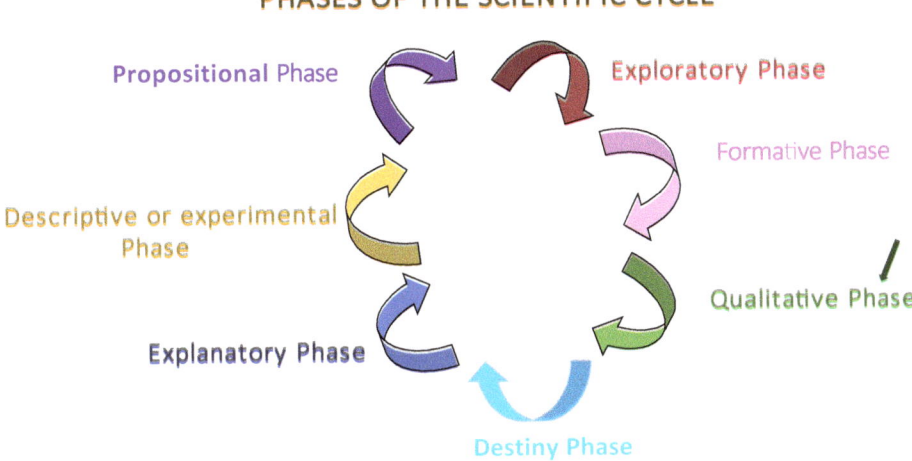

To introduce this chapter, let us listen to what Abdu'l-Baha taught:

> The honour and exaltation of every existing thing are contingent upon certain *causes* and *conditions*.

> The excellence, *adornment*, and *perfection* of the earth consist in this, that through the outpourings of the vernal showers it should become *green* and *verdant*; that plants should spring forth; that flowers and herbs should grow; that blossom-filled trees should produce an *abundant* yield and bring forth *fresh* and *succulent* fruit; that gardens should be *arrayed*; that meadows should be *adorned*; that plains and mountains should don an *emerald* robe; and that fields and bowers, villages and cities should be *decked* forth. This is the *felicity* of the mineral world."[219]

> The *elemental atoms* which constitute all phenomenal existence and being in this illimitable universe are in perpetual motion, undergoing continuous *degrees* of progression. For instance, let us conceive of an *atom* in the *mineral* kingdom progressing upward to the kingdom of the vegetable by entering into the *composition and fibre* of a tree or plant. From thence it is assimilated and transferred into the kingdom of the animal and finally, by the law and process of *composition*, becomes a *part of the body of man*. That is to say, it has traversed the intermediate *degrees* and stations of phenomenal existence, entering into the *composition* of various organisms in its journey. This motion or transference is progressive and perpetual, for after *disintegration* of the human body into which it has entered, it returns to the *mineral* kingdom whence it came and will continue to traverse the kingdoms of phenomena as before. This is an illustration designed to show that the *constituent elemental atoms* of phenomena undergo progressive transference and motion throughout the *material* kingdoms.

---

[219] 'Abdu'l-Bahá, Some Answered Questions, No.15. Emphasis added.

The Material Cause and the Qualitative Method

In its ceaseless progression and journeyings the *atom* becomes imbued with the *virtues* and powers of each *degree* or kingdom it traverses. In the *degree* of the *mineral it possessed mineral affinities*; in the kingdom of the vegetable it manifested the augmentative *virtue* or power of growth; in the animal organism it reflected the intelligence of that degree; and in the kingdom of man it was *qualified* with human *attributes or virtues*.[220]

In the *physical* creation, evolution is from one *degree of perfection* to another. The mineral passes with its *mineral perfections* to the vegetable; the vegetable, with its *perfections*, passes to the animal world, and so on to that of humanity. This world is full of seeming contradictions; in each of these kingdoms (mineral, vegetable and animal) *life* exists in its *degree*; though when compared to the *life* in a man, the earth appears to be dead, yet she, too, lives and has a *life* of her *own*. In this world things live and die, and live again in other forms of life, but in the world of the *spirit* it is quite otherwise.[221]

Let us think about the iron element's prosperity when passes from the mineral kingdom, to the plant, animal, and human kingdoms:

- From being just a solid, that oxidizes quickly. "Iron is by mass the most common element on Earth, forming much of Earth's outer and inner core."[222]
- Reaching the honorable position for iron[223] to be part of fertile soil where "plants grow; flowers and fragrant herbs spring up; fruit-bearing trees[224] become full of blossoms and bring forth fresh and new fruit."[225]
- Iron becoming capable of being part of 194.000 chemical compounds[226] is an example of unity in diversity.

---

[220] 'Abdu'l-Bahá, The Promulgation of Universal Peace.

[221] 'Abdu'l-Bahá, Paris Talks, p. 68. Emphasis added.

[222] Photo 35690593 © Alisbalb | Dreamstime.com

[223] "Iron is a key nutrient in the production of chlorophyll and a critical element in plant respiration." How to Get Iron in Plants. By: Deborah Stephenson

[224] Photo 91989827 © David Izquierdo | Dreamstime.com

[225] 'Abdu'l-Bahá, Some Answered Questions, Page: 78.

[226] Chemistry of Iron, edited by J. Silver.

- The specific role of iron in promoting meristematic cell division during adventitious root formation[227]
- Iron exalted when recognized as a critical component of plants capacity for DNA synthesis, respiration, photosynthesis, and many metabolic pathways[228]
- The prosperity of iron and the inactivation of viruses, and bacteria.[229] [230]
- "The diagram below shows "the binding of oxygen to a heme prosthetic group"[231] Heme is part of hemoglobin, which binds oxygen in the lungs and releases it in cells craving oxygen in animals and humans.

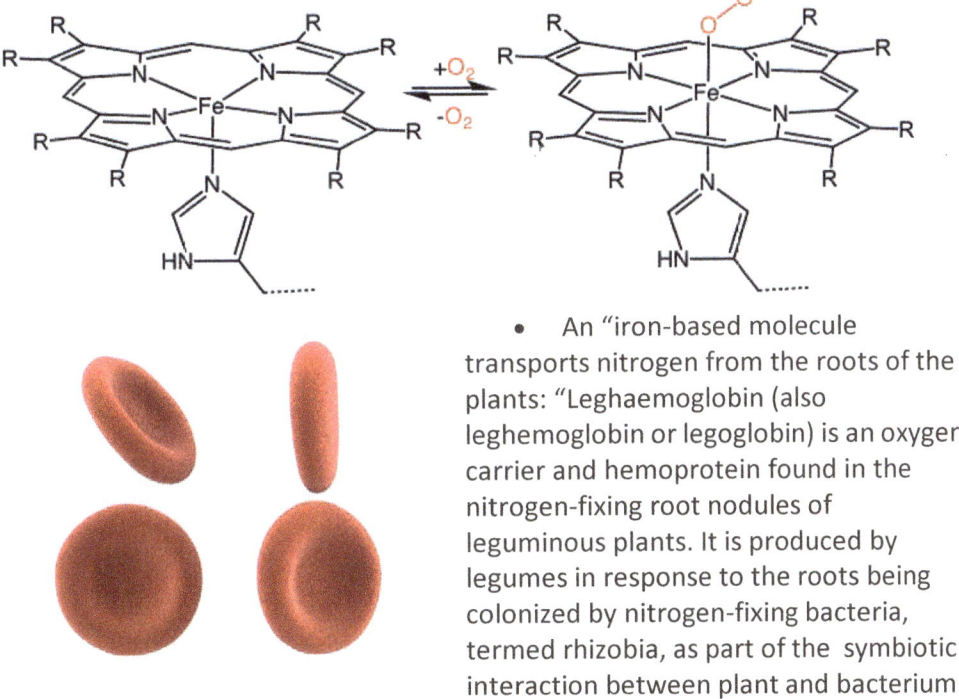

- An "iron-based molecule transports nitrogen from the roots of the plants: "Leghaemoglobin (also leghemoglobin or legoglobin) is an oxygen carrier and hemoprotein found in the nitrogen-fixing root nodules of leguminous plants. It is produced by legumes in response to the roots being colonized by nitrogen-fixing bacteria, termed rhizobia, as part of the symbiotic interaction between plant and bacterium: roots not colonized by Rhizobium do not synthesized leghemoglobin. Leghemoglobin has close chemical and structural similarities to hemoglobin,

---

[227] Alexander Hilo et al. A specific role of iron in promoting meristematic cell division during adventitious root formation.

[228] Antonia María Romero, et al., Global translational repression induced by iron deficiency in . east depeĀds on the Ācn2/eĀ̄2 pathaà y.

[229] Ian M. Bradley, Iron Oxide Amended Biosand Filters for Virus Removal.

[230] José Luis Sánchez-Salas et al., Inactivation of Bacterial Spores and Vegetative Bacterial Cells by Interaction with ZnO-Fe2O3 Nanoparticles and UV Radiation.

[231] By Smokefoot - Own work, CC BY-SA 4.0. Web. November 2022 <https://commons.wikimedia.org/w/index.php?curid=63789703>.

and, like hemoglobin, is red in colour."[232] [233]

The prosperity of iron in its essential role in the transportation of oxygen to all cells of animals. "A human being is typically composed of about 50 trillion cells; the largest Blue Whale[234] 100 quadrillion."[235]

For iron to reach its potential prosperity the plant, animal, and human metabolism requires some cofactors to absorb iron. For example in humans:

> "Only foods derived from animal flesh provide heme iron (though they provide nonheme iron as well). Nonheme iron, on the other hand, is present in vegetables, grains, fortified foods, and supplements. However, despite the fact that heme is better absorbed, most of the iron in our diets is derived from nonheme sources. It is therefore essential to understand some of the factors that enhance and inhibit our absorption of nonheme foods."[236]

The prosperity of iron in guiding animals and humans in their migratory routes[237] [238] [239]

Iron's prosperity, healing attention deficit disorder, and anemia[240]

Iron continues humbly serving as an oxidizing metal, returning to the dust when the organism passes away, waiting for another opportunity to respond to the different

---

[232] Wikipedia, Leghemoglobin.

[233] Photo 80935716 © Kittipong Jirasukhanont | Dreamstime

[234] Photo 148195568 © Mmphotos2017 | Dreamstime.com

[235] Web. January 2019 <http://www.elasmo-research.org/education/topics/r_matter_of_scale.htm>.

[236] Gwen Dewar, Ph.D. Boosting iron absorption: A guide for the science-minded 2009

[237] Long-distance navigation and magnetoreception in migratory animals. By Henrik Mouritsen

[238] Photo 128665815 © gal reichbard | Dreamstime.com

[239] Photo © Jordiferre | Dreamstime.com

[240] Illustration 127143318 © Molokot | Dreamstime.com

kingdoms' needs.

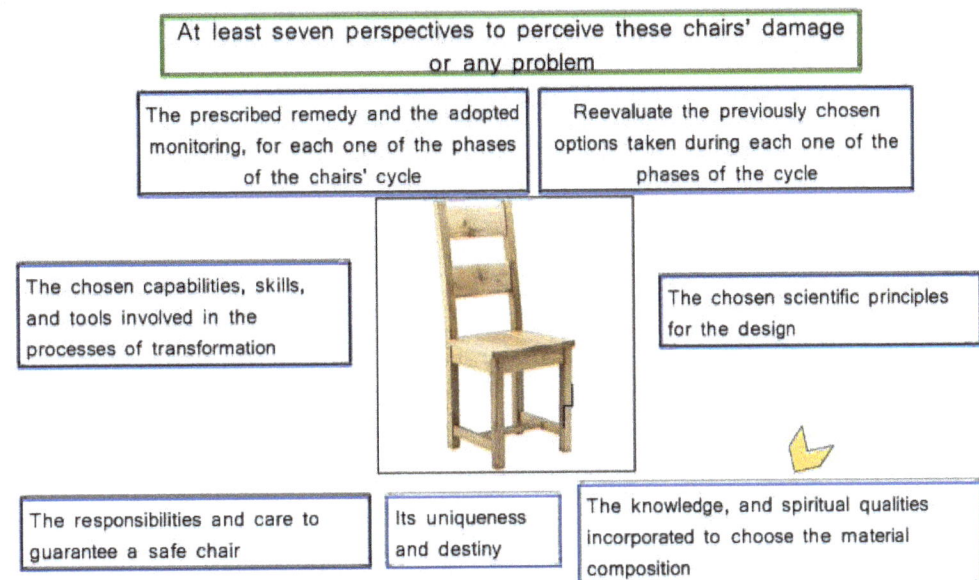

To continue introducing the "Material Cause," let us continue with the chair example. The answer to the question "What is the chair made of?" implies everything that is inherently part of the chair in addition to the raw materials used (wood, fabric, nails, glue, and paint). In other words, we have to include the following:

- The knowledge of the chair's builder,
- the specific properties of the wood utilized in its construction, such as its hardness, resistance, permeability, and susceptibility to attacks by mold,
- and the cultural values that the chair represents, including the harmony it must have with the cultural values of those who will use it.
- To determine the chair's value, the public safety of the place of the chair and the spirituality of the maker complying with the materials pacted are vital factors
- We must perceive that the chair is not just a pile of molecules of different compounds but also an expression of God's generosity that allows us to procure our comfort and welfare. Even before any chair came into existence, the materials that later went into a chair had the potential to be one, and the knowledge of how to transform them into a chair was latent within us.

5.1 Philosophical Argument

The essence of the Spirit, as the material cause originating essence, was the way to conceive the fitting together of the set of notions and the corresponding set of human faculties associated with the Qualitative Method of Science. It will help us to answer the questions: "What am I?"

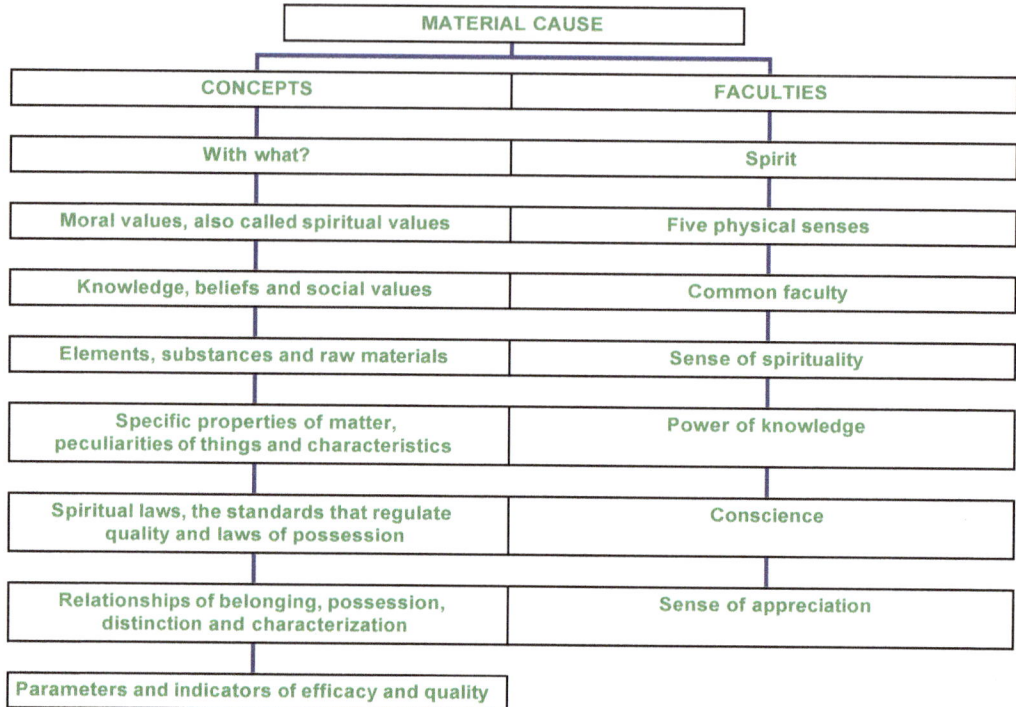

In *A Dictionary of Philosophical Terms and Names,* we find "The Material Cause" defined as: "Basic stuff of which a thing is made" (Kemerling). In an English dictionary, the Material Cause described as something out of which something is made or comes into being (*Merriam Webster Dictionary*, 2012). This Cause is associated with the question: *What is it made of? With what?*

Ian Kluge has a very profound analysis of this topic:

> Because God is 'beyond' the phenomenal realm, both the Baha'i Writings and Aristotle agree that God is *essentially unknowable* and do so for similar reasons. According to Aristotle, God, unlike all phenomena which are composed of *matter* and form, is one because the Divine has no *matter* and is pure form. The Divine is, moreover, pure existence, that is, a non-contingent entity whose nature is to exist; It is also pure thought thinking only on Itself. As time-and-space bound, *composite* beings, we can understand these concepts verbally, but cannot comprehend or understand what it is or means to enjoy this sort of being.[241]

---

[241] Ian Kluge, The Aristotelian Substratum of the Bahá'í Writings.

Abdu'l-Baha says,

> It is evident that whatsoever man understands is a consequence of his existence, and that man is a sign of the All-Merciful: How then can the consequence of the sign encompass the Creator of the sign? That is, how can human understanding, which is a consequence of man's existence, comprehend God? Thus the *reality* of the Divinity *lies hidden* from all understanding and is concealed from the minds of all men, and to ascend to that station is in no wise possible.[242]

By the "Reality of Divinity" Abdu'l-Baha means the *essence* of divinity which is beyond human comprehension. The *attributes* of divinity can, of course, be *known* or comprehended, but not the *essence* of Divinity. As pure form thinking Itself, Aristotle's God also enjoys a form of being whose nature can be deduced by Its *attributes* and actions in the phenomenal realm but cannot be *known* immediately. This is because, according to Aristotle, *true knowledge is knowledge* of causes and not mere description. That, however, is the level at which we must remain with the Unmoved Mover.[243]

From this we can conclude that the Baha'i Writings and Aristotle agree on several key epistemological issues subject to vociferous contemporary debate: first, that *natural reality* is *objectively real* and does not depend on human observers for its existence; second, that *reality* and its laws are given by God, not constructed, and that we must work with what is given; and third, that truth is the correspondence between *reality* and our interpretation of it, or, put otherwise, that *reality* and our interpretation of it are two distinct things and that we must test our interpretations against *reality* to discover whether or not they are in agreement. From this follows that reality is discovered and that there is such a thing as error, that is, an erroneous or inadequate understanding of reality that can be cured by abandoning it in order to change from *ignorant* to more *knowledgeable*. In other words, the Baha'i Writings and Aristotle share a *realist* epistemology. Without these premises, the entire Aristotelian and Baha'i enterprises would collapse, most especially the Baha'i doctrine of progressive revelation which presumes increasingly adequate comprehension of various truths. Finally, the *belief that properties are real* makes the Baha'i Writings and Aristotle incompatible with nominalism, that is, the *belief that properties* are either arbitrary human selections or outright impositions only externally related to their objects and that essences are fictitious. (See Aristotle's refutation of the underlying logic of nominalism in *Metaphysics*, VII, 12.) For its part, realism holds that the relationship between *attributes and substance* is internal, that is, inherent and intrinsic and that *essences are natural and real*.[244]

> *What is it made of? Analyzing the composition of resources and services is the key to setting individuals' and institutions' priorities and objective*s.

ಧಿಧಿ

There are faculties common to all human beings, and with them, we get an idea of the composition of things and appreciate the essential aspects of reality. Besides the power of spirit, human beings have:

---

[242] 'Abdu'l-Bahá, Some Answered Questions, No. 37. Emphasis added.

[243] Kluge, Ian. The Aristotelian Substratum of the Baha'i Writings. Emphasis added.

[244] Kluge, Ian. The Aristotelian Substratum of the Baha'i Writings. Emphasis added.

5.1.1 Senses

With our outer senses, we appreciate the composition and perceive the lack of quality, efficacy, prosperity, and the specific properties of matter.

5.1.2 The Spirit Pervades Creation

Are the spiritual qualities of the carpenter inherent to the chair? The Bahá'í teachings are a source of wisdom to determine the composition of things; in other words, what is inherent to them:

> There are spiritual principles, or what some call human values, by which solutions can be found for every social problem. Any well-intentioned group can in a general sense devise practical solutions to its problems, but good intentions and practical knowledge are usually not enough. The essential merit of spiritual principles is that they not only present a perspective which harmonizes with that which is *inherent in human nature*, it also induces an attitude, a dynamic, a will, an aspiration, which facilitate the discovery and implementation of practical measures. Leaders of governments and all in authority would be well served in their efforts to solve problems if they would first seek to identify the principles involved and then be guided by them.[245]

The author decided to choose "the spirit" as the essential origination of the Material Cause because 'Abdu'l-Bahá is reported to have said:

> *The greatest power in the realm and range of human existence is spirit—the divine breath which animates and pervades all things.* It is manifested throughout creation in different degrees or kingdoms. In the vegetable kingdom it is the augmentative spirit or power of growth, the animus of life and development in plants, trees and organisms of the floral world. In this degree of its manifestation spirit is unconscious of the powers which qualify the kingdom of the animal. The distinctive virtue or plus of the animal is sense perception; it sees, hears, smells, tastes and feels but is incapable, in turn, of conscious ideation or reflection which characterizes and differentiates the human kingdom. The animal neither exercises nor apprehends this distinctive human power and gift. From the visible it cannot draw conclusions regarding the invisible, whereas the human mind from visible and known premises attains knowledge of the unknown and invisible. For instance, Christopher Columbus from information based upon known and provable facts drew conclusions which led him unerringly across the vast ocean to the unknown continent of America. Such power of accomplishment is beyond the range of animal intelligence. Therefore, this power is a distinctive attribute of the human spirit and kingdom. The animal spirit cannot penetrate and discover the mysteries of things. It is a captive of the senses.[246]

It has also been said that the spirit, according to scientific philosophy, is made up of a simple, and therefore indivisible substance:

> Therefore, it is evident that life is the expression of composition, and mortality, or death, is equivalent to decomposition. As the *spirit of man* is not composed of material elements, it is not subject to decomposition and, therefore, has no death. It is self-evident that the human

---

[245] The Universal House of Justice, The Promise of World Peace, pp. 15-16. Emphasis added.

[246] 'Abdu'l-Bahá, The Promulgation of Universal Peace. Emphasis added.

spirit is simple, single and not composed in order that it may come to immortality, and it is a philosophical axiom that the individual or indivisible atom is indestructible.²⁴⁷

Elsewhere we find: "… knowledge is a human attribute but so is ignorance; truthfulness is a human attribute but so is falsehood; and the same holds true of trustworthiness and treachery, justice and tyranny, and so forth. In brief, every perfection and virtue, as well as every vice, is an attribute of man."²⁴⁸

The chair's value depends on the hardwood that the carpenter chooses to make it. Then his knowledge and honesty are reflected in the selected wood. It is verifiable by a person knowledgeable of different types of wood. *Culture* is to human collectivity what *personality* is to an individual. *Personality* has been defined by Guilford (1959) as 'the interactive aggregate of personal *characteristics* that influence the individual's response to its environment'. *Culture* could be defined as the interactive aggregated of *common characteristics* that influences a human group's response to its environment. *Culture* determines the *identity* of a human group in the same way as *personality* determines the *identity* of an individual.²⁴⁹

Are our convictions and cultural values inherent to the chair we bought? Hofstede expressed his perception of culture in the following terms: "I treat culture as 'the collective programming of the mind which distinguishes the members of one group from another.' This is not a complete definition …, but it covers what I have been able to measure.

Culture, in this sense, includes systems of values: and *values are among the building blocks of culture*".²⁵⁰

For instance, if the Guambiano indigenous community in Silvia, Colombia, were to stop using its traditional skirts and vests and start buying thread, fabrics, and woolen cloth from the Colombian textile industry, it would cause in just one generation: the loss of the technological culture related to raising wool-producing animals. They would also lose the knowledge of how to spin, dye, and weave the wool. They would also lose the aggregate economic value of producing their wool and skirts and vests, ultimately leading to the loss of their cultural identity.

---

²⁴⁷ 'Abdu'l-Bahá, The Promulgation of Universal Peace, p. 306. Emphasis added.

²⁴⁸ 'Abdu'l-Bahá, Some Answered Questions, No. 64.

²⁴⁹ Hofstede, Culture's Consequences, p. 21. Emphasis added.

²⁵⁰ Photo 123204124 © Fedecandoniphoto | Dreamstime.com

Was this not precisely the argument made valiantly and peacefully by Mahatma Gandhi when the Indian nation decided not to purchase English cloth and other products made in England because it would be conducive to poverty? In truth, the interdependence of the world's regions is necessary, but this should not prevent us from finding ways to strengthen the local and regional economies by looking for more efficient technologies.[251]

Fundamentals, convictions, cultural values, spiritual values, social and economic values, and knowledge are notions linked to the Material Cause.

Conscience

With the sensibilities of our conscience, we appreciate cultural values and our knowledge or ignorance about the level of trustworthiness, confidence, generosity, courtesy, honesty, and many other spiritual principles in ourselves; and we grasp the degree of perfection of the specific properties of our own health and wellbeing.

Five quotes suffice in this respect:

> According to the great ancient philosophers the words soul, mind and spirit implied the underlying principles of life; the essence was expressed under different names and these three terms designated the various functions of the absolute reality, or the operations of the one single essence; for instance, when they dealt with the sensations of emotion they called it the soul; when they desired to express that power which discovers the reality of phenomena they gave it the appellation of mind and *when they discussed the consciousness which pervades the world of creation they gave it the title of spirit.*[252]

> ... the *conscience* of man is sacred and to be respected; and that liberty thereof produces widening of ideas, amendment of morals, improvement of conduct, disclosure of the secrets of creation, and manifestation of the hidden verities of the contingent world. Moreover, if interrogation of conscience, which is one of the private possessions of the heart and the soul, take place in this world, what further recompense remains for man in the court of divine justice at the day of general resurrection? Convictions and ideas are within the scope of the comprehension of the King of kings, not of kings; and soul and *conscience* are between the fingers of control of the Lord of hearts, not of [His] servants.[253]

> "The principle of obedience to government does not place any Bahá'í under the obligation of *identifying* the teaching of His Faith with the *political* program enforced by the government. For such an *identification*, besides being erroneous and contrary to both the *spirit* as well as the form of the Bahá'í Message, would necessarily create a conflict within the *conscience* of every loyal believer."[254]

> "Man is the discoverer of the mysteries of nature; nature is not *conscious* of those mysteries herself. It is evident, therefore, that man is dual in aspect: as an animal he is subject to nature, but in his spiritual or *conscious being* he transcends the world of material existence."[255]

---

[251] Photo 92111087 / Guambiano © Xing Wang | Dreamstime.com

[252] 'Abdu'l-Bahá, Divine Philosophy, p. 120. Emphasis added.

[253] 'Abdu'l-Bahá, A Traveller's Narrative, p. 91. Emphasis added.

[254] Department of the Secretariat, The Universal House of Justice in a letter to an individual on 27 April 2017 Compilations, Lights of Guidance, p. 446. Emphasis added.

[255] The Promulgation of Universal Peace, p. 81. Emphasis added.

"Although he possesses all the virtues of the lower kingdoms, he is further endowed with the spiritual faculty, the heavenly gift of consciousness."[256]

5.1.3 Sense of spirituality

It is so fascinating to acknowledge that all human beings have a sense of spirituality as an endowment that serves as the foundation to improve our own quality of life and that of others.

If we want to attain the best of such endowment, we have to determine the best approach to strengthen our sense of spirituality. The guidance to reach this goal is in explaining mysticism, Shoghi Effendi says:

> For the core of religious faith is that mystic feeling which unites Man with God. This state of spiritual communion can be brought about and maintained by means of meditation and prayer. And this is the reason why Bahá'u'lláh has so much stressed the importance of worship. It is not sufficient for a believer merely to accept and observe the teachings. He should, in addition, cultivate *the sense of spirituality* which he can acquire chiefly by means of prayer.[257]

5.1.4 Power of knowledge

We determined the composition of an object by knowing two things: first, the specific properties of matter, such as density, hardness, heath, color, and many others; and second, the cultural values and the spiritual qualities reflected from it.

Each leaf has its own particular identity—so to speak, its own individuality as a leaf. Therefore, each atom of the innumerable elemental atoms, during its ceaseless motion through the kingdoms of existence as a constituent of organic composition, not only becomes imbued with the powers and virtues of the kingdoms it traverses but also reflects the attributes and qualities of the forms and organisms of those kingdoms.[258]

Let me include other quotes from the Master: "O my dear Mr. . . .! I beseech God to grant thee the power of knowledge and understanding so that thou mayest fathom the mysteries of the teachings descended from the presence of the Glorious Lord."[259]

"The virtues of humanity are many, but science is the most noble of them all. The distinction which man enjoys above and beyond the station of the animal is due to this paramount virtue. It is a bestowal of God; it is not material; it is divine."[260]

---

[256] The Promulgation of Universal Peace.

[257] Directives from the Guardian, p. 86. Emphasis added.

[258] The Promulgation of Universal Peace

[259] 'Abdu'l-Bahá, Tablets of 'Abdu'l-Bahá, v2 p. 454.

[260] The Promulgation of Universal Peace

"The most noble and praiseworthy accomplishment of man, therefore, is scientific knowledge and attainment."[261]

"The Lord of the Kingdom hath invited, chosen and guided you through His pure favor, feeding you from the heavenly table of divine knowledge! Know ye the value of this favor and bounty and loosen your tongues in praise; showing forth the power of knowledge and assurance and breathing the spirit of guidance into the hearts of the seekers."[262]

Additionally, it is knowledge that distinguishes an individual from others. In other words, an individual's scientific, artistic, moral, and technological culture can set that individual apart from others.

To clarify concepts such as culture, true wealth, true necessity, poverty, health, and wellbeing, the conceptual framework is used to inquire about which cultural and spiritual values—and what knowledge—are required to solve the issues to which the framework is to be applied. It is here where we have to apply the forces of the transformational process of society.

As a scientist, I agree with Richard Feynman: "Science alone of all the subjects contains within itself the lesson of the danger of belief in the infallibility of the greatest teachers in the preceeding generation. As a matter of fact, I can also define science another way: Science is the belief in the ignorance of experts."[263] The author assumes that Feynman is not referring to the Manifestations of God.

I also agree with Ridley:

> The fuel on which science runs is ignorance. Science is like a hungry furnace that must be fed logs from the forests of ignorance that surround us. In the process, the clearing that we call knowledge expands, but the more it expands, the longer its perimeter and the more ignorance comes into view   A true scientist is bored by knowledge; it is the assault on ignorance that motivates him - the mysteries that previous discoveries have revealed. The forest is more interesting than the clearing.[264]

Sense of appreciation

I believe that this is the sense to be polished and strengthened to perceive just the good in others and not their defects.

With regard to your question concerning the meaning of the name 'Hidden Words'. It is, indeed, one of the most suggestive titles of the Writings of Bahá'u'lláh. These words are called hidden due to the fact that men have had *neither the knowledge nor a true sense of appreciation* of them before they were revealed by Bahá'u'lláh. It is through Him, Who is the sole Mouthpiece of God in this age, that spiritual realities and truths have been once more reinterpreted and revealed afresh to mankind.

Bahá'u'lláh's Message is thus the only key to a true revealed afresh to mankind. Bahá'u'lláh's Message is thus the only key to a true understanding of the mysteries that envelop man's spiritual life.[265]

---

[261] Idem.

[262] ibid. v2, p. 478.

[263] The Pleasure of Finding Things Out.

[264] Genome: the autobiography of a species in 23 chapters.

[265] From a letter written on behalf of Shoghi Effendi to an individual believer, September 1, 1935.

Shoghi Effendi, the beloved Guardian, said:

> "I shall indeed grieve if the situation in Palestine should prevent our meeting and prevent your pilgrimage to the Holy Shrines. I pray that this may not be the case. I am so eager to meet you, and express in person my deep and abiding *sense of appreciation* of the splendid and historic services you have rendered. I will continue to pray for you from the depths of my heart."[266]

I accept Richard Feynman's statement, "*Religion is a culture of faith; science is a culture of doubt*" and even Popper's "*falsifiability*"[267] for laws, principles, compositions, methods, and cycles, when proposed by human beings. But, when referring to the Holy Scriptures, Richard Feynman's statement "*Religion is a culture of faith; science is a culture of doubt*" and Popper's "*falsifiability*" should be exchange for "*Religion is a culture of faith; science is a culture of confirmation.*"

The author believes that Science's role is to find the facts that explain the teachings of God's Messengers in the context of Progressive Revelation and Their influence in the evolving organizational structure of tribes into cities, then city-states, then nations, and the promise of a world federated government. For example, "In Moses time tribes where nomadic and there where not tribunals, so the law 'Eye for eye, and tooth for tooth' was the only way to be protected until they settled and changed this practice."[268]

This set of human faculties can help us answer the questions: "What am I?" and play a critical role when making ethical choices.

Another one of the phases of the cycle of scientific research is the qualitative phase, and the set of human faculties associated with the Material Cause is appropriate to answer the question: "With what?"

ෆരി

In his search for a single science, Aristotle wrote:

> The first problem concerns the subject which we discussed in our remarks. It is this- whether the investigation of the causes belongs to one or to more sciences, and whether such a science should survey only the first principles of substance, or also the principles on which all men base their proofs, e.g. whether it is possible at the same time to assert and deny one and the same thing or not, and all other such questions; and if the science in question deals with substance, whether one science deals with all substances, or more than one, and if more, whether all are akin, or some of them must be called forms of Wisdom and the others something else. And this itself is also one of the things that must be discussed-whether sensible substances alone should be said to exist or others also besides them, and whether these others are of one kind or there are several classes of substances, as is supposed by those who believe both in Forms and in mathematical objects intermediate between these

---

Compilations, Lights of Guidance, p. 488. Emphasis added.

[266] Shoghi Effendi, Arohanui - Letters to New Zealand, p. 48. Emphasis added.

[267] The belief that for any hypothesis to have credence, it must be inherently disprovable before it can become accepted as a scientific hypothesis or theory. Web. March 2015 <https://explorable.com/falsifiability>.

[268] Morgenstern, Julian. A Jewish Interpretation of the Book of Genesis, p. 74.

and sensible things. Into these questions, then, as we say, we must inquire, and also whether our investigation is concerned only with substances or also with the essential attributes of substances.[269]

Further down, he wrote:

In general, do all substances fall under one science or under more than one? If the latter, to what sort of substance is the present science to be assigned? - On the other hand, it is not reasonable that one science should deal with all. For then there would be one demonstrative science dealing with all attributes. For ever demonstrative science investigates with regard to some subject its essential attributes, starting from the common beliefs. Therefore to investigate the essential attributes of one class of things, starting from one set of beliefs, is the business of one science. For the subject belongs to one science, and the premises belong to one, whether to the same or to another; so that the attributes do so too, whether they are investigated by these sciences or by one compounded out of them.[270]

Let us appreciate what the wise Diotima taught Socrates about her opinion of love in the context of reading a quote about the quality of love, focusing on connecting to the words with added emphasis. Diotima said,

'He who has been instructed thus far in the things of love, and who has learned to see the beautiful in due order and succession, when he comes toward the end will suddenly perceive a *nature* of wondrous beauty (and this, Socrates, is the final cause of all our former toils)-a *nature* which in the first place is everlasting, not growing and decaying, or waxing and waning; secondly, not fair in one point of view and foul in another, or at one time or in one relation or at one place fair, at another time or in another relation or at another place foul, as if fair to some and-foul to others, or in the likeness of a face or hands or any other part of the bodily frame, or in any form of speech or *knowledge*, or existing in any other being, as for example, in an animal, or in heaven or in earth, or in any other place; but beauty absolute, separate, *simple*, and everlasting, which without diminution and without increase, or any change, is imparted to the ever-growing and perishing beauties of all other things. He who from these ascending under the influence of *true love*, begins to perceive that beauty, is not far from the end. And the true order of going, or being led by another, to the things of love, is to begin from the beauties of earth and mount upwards for the sake of that other beauty, using these as steps only, and from one going on to two, and from two to all fair forms, and from fair forms to fair practices, and from fair practices to fair notions, until from fair notions he arrives at the notion of absolute beauty, and at last *knows* what the *essence* of beauty is.

This, my dear Socrates,' said the stranger of Mantineia, 'is that *life* above all others which man should *live*, in the contemplation of beauty absolute; a beauty which if you once beheld, you would see not to be after the measure of *gold*, and *garments*, and fair boys and youths, whose presence now entrances you; and you and many a one would be content to *live* seeing them only and conversing with them without *meat* or *drink*, if that were possible-you only want to look at them and to be with them. But what if man had eyes to see the true beauty-the divine beauty, I mean, *pure and dear and unalloyed, not clogged with the pollutions of mortality and all the colours and vanities of human life*-thither looking, and holding converse

---

[269] Aristotle, The Metaphysics.

[270] ibid.

with the *true beauty simple and divine*? Remember how in that communion only, beholding beauty with the eye of the mind, he will be enabled to bring forth, not images of beauty, but *realities* (for he has hold not of an image *but of a reality*), and bringing forth and *nourishing true virtue* to become the friend of God and be *immortal*, if *mortal* man may. Would that be an *ignoble life*?'[271]

5.1.5 The specific properties of matter

Above, we considered human spiritual qualities when answering the question, "With what are we going to achieve change?" Now, we will take into account the specific properties (color, hue, density, hardness, caloric capacity, the temperature of fusion and ebullition, characteristic odor and taste, etc.) of material things such as minerals, plants, animals, and human beings. The specific properties of matter are those that help us determine, for example, what materials we would choose to make a hammer that serves its purpose. For instance, the steel quality required to make a hammer putting nails in pinewood is different from that used for a hardened substrate such as concrete.

5.1.6 Common faculty

The author initially associated the Common Faculty with the Material Cause by exclusion. Reflecting later on it, I realized that the atoms of all chemical elements have electrons.

In the following quote, 'Abdu'l-Bahá clearly explains the function of the Common Faculty in relation with the inner and outer senses:

> There are five outward material powers in man which are the means of perception that is, five powers whereby man perceives material things. They are sight, which perceives sensible forms; hearing, which perceives audible sounds; smell, which perceives odours; taste, which perceives edible things; and touch, which is distributed throughout the body and which perceives tactile realities. These five powers perceive external objects.

> Man has likewise a number of spiritual powers: the power of imagination, which forms a mental image of things; thought, which reflects upon the realities of things; comprehension, which understands these realities; and memory, which retains whatever man has imagined, thought, and understood. *The intermediary between these five outward powers and the inward powers is a common faculty, a sense which mediates between them and which conveys to the inward powers whatever the outward powers have perceived. It is termed the common faculty as it is shared in common between the outward and inward powers*.

> For instance, sight, which is one of the outward powers, sees and perceives this flower and conveys this perception to the inward power of the common faculty; *the common faculty* transmits it to the power of imagination, which in turn conceives and forms this image and transmits it to the power of thought; the power of thought reflects upon it and, having apprehended its reality, conveys it to the power of comprehension; the comprehension, once it has understood it, delivers the image of the sensible object to the memory, and the memory preserves it in its repository.[272]

---

[271] Plato, Symposium. Emphasis added.

[272] Some Answered Questions, p. 56. Emphasis added

5.1.7 Supplies

This category includes elements, substances, raw materials, and manufacturing supplies incorporated materially in the product.

5.1.8 Relationships

The following rationality helps us determine the distinction that makes a thing or organism different when comparing it to others. The Material Cause is associated with meaningful relationships such as ownership, possession, characterization, identification, distinction, and many others describing the most detailed features. By using these relationships, we find out how the existence of each quality is specifically appreciated.

It is also essential when deciding, With what are we going to make something?

Consider the spiritual laws and the norms that rule quality, standard, possession, or distinctiveness.

5.1.9 Parameters and indicators

Parameters and indicators of possession, identity, quality, and efficacy are useful to monitor and evaluate the situation. Prosperity is also a parameter of the Material Cause:

> When virtues are established and characters refined, progress in all matters is assured. No nation can attain prosperity and success save through the refinement of morals. When characters are improved, sciences and arts flourish; minds become vastly enlightened; hearts are filled with truthfulness and trustworthiness, with zeal, devotion, and determination; statesmanship attaineth its highest degree; new industries become widespread; commerce is expanded; and courage and audacity raise aloft their banner. The nation passeth from one condition to another.[273]

5.2 Research Steps for the Qualitative Method

If we follow the art of constructing the Single Science, we conclude that each method of science intertwines with the other methods.

The author made an effort to relate this method to the Valley of Knowledge, as revealed in The Seven Valleys. These Valleys "mark the wayfarers' journey from their mortal abode to the heavenly homeland"[274]; in the Valley of Knowledge, we find:

> Gazing with the eye of absolute insight, the wayfarer in this valley seeth in God's creation neither contradiction nor incongruity, and at every moment exclaimeth, "No defect canst thou see in the creation of the God of mercy. Repeat the gaze: Seest thou a single flaw?" He beholdeth justice in injustice, and in justice, grace. In ignorance he findeth many a knowledge hidden, and in knowledge a myriad wisdoms manifest. He breaketh the cage of the body and the hold of the passions, and communeth with the denizens of the immortal realm. He scaleth the ladders of inner truth and hasteneth to the heaven of inner meanings. He rideth in the ark of "We will surely show them Our signs in the world and within themselves", and

---

[273] 'Abdu'l-Bahá, Additional Tablets, Extracts and Talks.

[274] Bahá'u'lláh, The Seven Valleys, The Call of the Divine Beloved.

saileth upon the sea of "until it become plain to them that it is the truth". And if he meeteth with injustice he shall have patience, and if he cometh upon wrath he shall manifest love.[275]

The following are just general ideas that may serve as guiding steps for the qualitative method of science. The suggested steps should link the problem's facts and symptoms to quality, efficacy, or prosperity parameters. Also, let us always remember the connection of the problem with the culture, values, virtues, the specific properties of matter, spiritual laws, laws of possession and identification, and quality standards.

First: The exploratory step. Search for the qualitative modifications that caused the problem's facts and symptoms, affecting the different constituents' quality, prosperity, or efficacy.

- What were the monitoring and evaluation results of the qualitative processes carried out during the last cycle?
- Which feature, symptom, or characteristic shows that the quality, the prosperity, or the efficacy is deficient?
- Which decision might cause the problem's facts and symptoms' to impact the different constituents' quality, efficacy, and prosperity?
- Which features of his(her) character need improvement? Which values of his(her) culture are worthy of being kept or rejected?
- To contribute to the control of damaging insects, are there any parasitoids that may weaken their physiology? In addition, are there any natural remedies to stop their damaging effects?
- Which is the curricular strategy to develop the capacity for reaching what the following quote indicates?

The virtues of humanity are many, but science is the most noble of them all. The distinction which man enjoys above and beyond the station of the animal is due to this paramount virtue. It is a bestowal of God; it is not material; it is divine. Science is an effulgence of the Sun of Reality, the power of investigating and discovering the verities of the universe, the means by which man finds a pathway to God. All the powers and attributes of man are human and hereditary in origin—outcomes of nature's processes—except the intellect, which is supernatural. Through intellectual and intelligent inquiry science is the discoverer of all things. It unites present and past, reveals the history of bygone nations and events, and confers upon man today the essence of all human knowledge and attainment throughout the ages. By intellectual processes and logical deductions of reason this superpower in man can penetrate the mysteries of the future and anticipate its happenings.

Science is the first emanation from God toward man. All created beings embody the potentiality of material perfection, but the power of intellectual investigation and scientific acquisition is a higher virtue specialized to man alone.[276]

Second: The formative step, focusing on the general properties of matter and the faculties closely related to the Formative Cause.

- How do the general properties of matter affect the desired quality, prosperity,

---

[275] Bahá'u'lláh, The Seven Valleys, The Call of the Divine Beloved.

[276] 'Abdu'l-Bahá, The Promulgation of Universal Peace

or efficacy?

Third: The qualitative step, focusing on the specific properties of matter

- What is deficient or in excess within the composition of the service or product offered?
- How a compound or a mix of ingredients may improve the quality?
- What level of perfection is within our reach to set as a short-term goal?
- What seemingly conduct will reach a higher level of perfection in the individual's character?
- What composition, knowledge, virtues, and values may contribute to reaching the expected quality, prosperity, or efficacy of the service or product to offer?

Fourth: The Root Cause

- What could affect reaching your high destiny when presenting the features of the product or service offered to your clients?
- What stumbling blocks did I put as an impediment to strengthen my belief in the value of meditation? Is it a conscious omission or mere ignorance? Is it one of my weaknesses?
- What features of your endowment have you to look for and excel in developing to reach a higher destiny?

Fifth: The Explanatory Step

"Each for all and all for each" is the only principle on which a community can prosper."[277]

- What can we do to comply with the standards required to reach the expected quality, prosperity, or efficacy?
- Are you sure of not causing any damage to the ecosystems with the qualitative changes proposed?

Sixth: The Experimental Step

Please search for the role of biodiversity in providing access to resources and nutrients not usually available to monoculture crops: *Rhizobium leguminosarum* fixates nitrogen in the soil, badly needed by other plants and tree roots bring to the surface nutrients located deep in the ground.

Bacteria is the crucial workforce of soils. They are the final stage of breaking down nutrients and releasing them to the root zone for the plant. In fact, the Food and Agriculture Organization once said "Bacteria may well be the most valuable of life forms in the soil."

Actinomycetes were once classified as fungi, and act similarly in the soil. However, some actinomycetes are predators and will harm the plant, while others living in the soil can act as antibiotics for the plant.

Like bacteria, **fungi** also live in the root zone and helps make nutrients available to plants. For example, Mycorrhizae is a fungus that facilitate water and nutrient uptake by the roots and plants to provide sugars, amino acids, and other nutrients.

Protozoa are larger microbes that love to consume and be surrounded by bacteria. In fact, nutrients that are eaten by bacteria are released when protozoa in turn eat the bacteria.

---

[277] John E. Esslemont, Bahá'u'lláh and the New Era.

Nematodes are microscopic worms that live around or inside the plant. Some nematodes are predators, while others are beneficial, eating pathogenic nematodes and secreting nutrients to the plant.[278]

Which educational experiences further develop the following faculties: the five senses to perceive special qualitative features, sense of appreciation, sense of spirituality, conscience, common faculty, and the power of knowledge?

Which experiment or experience tells us that the individual's character has improved?

Seventh: The Propositional Step
- What test is best to verify the validity of the changes made to the product's composition?
- What test is best to verify the validity of the conscious incorporation of values to the service offered?
- What personal test verifies the credibility of reaching a higher level of perfection in my character, polishing those virtues that I need the most?
- What would be the effects of the modified quality on the different kingdoms?
- Which parameters or indicators of quality, prosperity, or efficacy do we want to discard, improve, or define and formulate anew?

5.3 The qualitative phase of the farmers' situation

To help the reader understand the farmers' example, the author emphasizes the discourse's fundamental notions reasonably linked to each one of the Causes. Let us then, continue with our example of the farmers using the Qualitative Method.

What questions should we consider if we believe that it is possible to find the profound reasons why a rural population is *adamant* in assuming it is not viable to keep trying to continue laboring the land?

Will this question help us: What did we have? What do we have? What do we want to have? And what do we need to achieve it?

To answer these questions, we have to examine:

1. The cultural and spiritual values about the worth of the farmers' work
2. Their knowledge of the specific properties of the cultivated species and those in the environment
3. Their knowledge about the specific properties of soil, natural resources, elements, substances, raw materials, equipment, nutrients, products, services, and wastes concerning agriculture and cattle-raising
4. The cultural and spiritual values, convictions, and knowledge about agriculture transmitted by the educational establishment
5. The policies and budgetary dispensations of the Ministry of Agriculture[279]

---

[278] Kaitlyn Ersek, 5 Types Of Soil Microbes And What They Do For Plants.

[279] It is obvious that there will be government policies related to the five causes, for example: policies of plant sanitation and environmental protection; pricing policies; commercialization macro-economic policies; transference of technology and educational policies, and regulatory policies to allow fish reproduction cycle.

The Material Cause and the Qualitative Method

6. The availability of affordable financing and credit for farmers

I suggest to the reader to appreciate the way politicians value farmers, primarily reflecting on the needs of small ones in a country and the world at large:

Latha Reddy Musukula was making tea on a recent morning when she spotted the *money lenders* walking down the dirt path toward *her house*. They came in a phalanx of 15 men, by her estimate. She *knew their faces*, because they had walked down the path before.

After each visit, her husband, a farmer named Veera Reddy, sank deeper into silence, frozen by some terror he would not explain. Three times he cut his wrists. He tied a noose to a tree, relenting when the family surrounded him, weeping. In the end he waited until Ms. Musukula stepped out, and then he hanged himself from a pipe supporting their roof, leaving a *careful list of each debt he owed* to each money lender. She learned the full sum then: 400,000 rupees, or $6,430.

…. India's small farmers, once the country's economic backbone and most reliable vote bank, are increasingly being left behind. With global competition and rising costs cutting into their lean profits, their ranks are dwindling, as is their contribution to the gross domestic product. If rural voters once made their plight into front-page news around election time, this year the large parties are jockeying for the votes of the urban middle class, and the farmers' voices are all but silent.

Even death is a stopgap solution, when farmers like Mr. Reddy take their *own lives*, *their debts* pass from husband to widow, from father to children. Ms. Musukula is now trying to *scrape a living* from the four acres that defeated her husband. Around her she sees a country transformed by economic growth, full of opportunities to break out of *poverty*, if only her son or daughter could grasp one.

But the trap that closed on her husband is tightening around her. Like nearly every one of her neighbors, she is locked into a bond with village money lenders — an intimate bond, and sometimes a menacing one. No sooner did they cut *her husband's body* down than one of them was in her house, threatening to block the cremation unless she paid.

Her appeals to officials for help have been met with indifference. Lately, her fear has been getting the better of her.

'Sure, they will pay, otherwise it would be as if someone has broken into our house and stolen *our money*,' said Sudhakar Ravula, a slight man who lives in a village about two miles away. He introduces himself as a fisherman, but, under questioning, fishes out a pair of gold-rimmed reading glasses and unfolds a *promissory note* signed by Veera Reddy.

Four years ago, he said, he used *borrowed money to lend Mr. Reddy $800*, at an annual interest rate of 24 percent. Reminded of Mr. Reddy's suicide, Mr. Ravula looked impatient. 'I always feel sad for the man,' he said, 'but committing suicide is not the right way to go about it.'[280]

Seventy years ago, the majority of the world's population lived in rural areas. Once the proportion of the rural regions forcefully became reversed by different factors, such as political conflicts, violence, drug trafficking, natural disasters, high-interest rates, structural racism, and development models, politicians diminished the funding to schools, institutions, and

---

[280] Barry, Ellen., After Farmers Commit Suicide, Debts Fall on Families in India. Emphasis added.

infrastructure. They shifted the priorities to urban areas, and the farmers' votes became irrelevant, and their sufferings invisible.

I hypothesize that the majority of politicians and economists seem to share the same appreciation of economic value in issues such as:

- In USA "In 2007, 1.73 billion tons of *topsoil was lost to erosion*, equal to about 200,000 tons each hour."[281]
- "Despite a tenfold increase in insecticide use between 1945 and 1989, *crop losses due to insect damage nearly doubled*. In 2007, the U.S. agriculture sector used 877 million pounds of pesticides."[282]
- "*Nutrient runoff* in the agricultural upper regions of the Mississippi River creates a *hypoxic "dead zone"* in the Gulf of Mexico. The average size of the region was more than 5,684 square miles from 2007 to 2012."[283]
- "Many parts of the U.S., including agricultural regions, are experiencing *groundwater depletion* (withdrawal exceeds recharge rate) at increasing rates."[284]
- "In 2011, farmers were reliant on income sources outside the farm to make up 83% of their household income, on average."[285]
- "Just 16¢ of every dollar *spent on food* in 2011 went back to the farm; in 1975, it was 40¢."[286]

In, *The political economy of labor relations in agriculture and food*, by Alessandro Bonanno discusses the recent evolution of labor relations. He contends that the 20th Century Fordist era was characterized by stable and pacified labor relations regulated by the tacit, yet effective, "management-labor accord." In this context, agricultural labor represented a *fundamental factor* in the increase of production that generated *abundant* and *inexpensive* food for urban dwellers and a reservoir of labor for the growth of urban manufacturing. While agricultural employment remained seasonal, more unstable and *less remunerated* that its urban counterpart, Fordist *wealth* redistribution mechanisms allowed a progressive *betterment* of labor's living and working *conditions*. Neoliberal globalization changed this situation. The author discusses this change by stressing seven relevant aspects. **First**, there is a flexibilization of labor relations that consists of *reduced* wages, *precarious* and unstable employment, *enhanced* exploitation and the *political weakness* of agri-food workers. **Second**, the mobilization of a large reserve army of labor allows the *enhanced* control of labor's wages and claims. Through decentralization of production and *global sourcing*, distant labor pools are placed in direct competition with each other resulting in the creation of a *docile and inexpensive* labor force. **Third**, similar results are accomplished through the use of immigrant labor. As neoliberalization promotes the free circulation of goods and services,

---

[281] University of Michigan, Center for Sustainable Systems. Emphasis added.

[282] ibid. Emphasis added.

[283] ibid. Emphasis added.

[284] ibid. Emphasis added.

[285] ibid. Emphasis added.

[286] ibid. Emphasis added.

labor mobility remains highly controlled mostly through *political* mechanisms (i.e., citizenship, residency, work permits and work programs). Moreover, the criminalization of immigrant labor further allows the control of this *component* of the global reserve army of labor. **Fourth**, agri-food labor is feminized. The use of female labor is promoted not only in terms of the use of *inexpensive* workers but also in terms of a discourse that legitimizes labor exploitation. The presentation of flexible labor as "convenient" for women conceals the many *disadvantages* associated with this form of employment. **Fifth**, the development of third party production contracts allows firms to circumvent labor laws and regulations and further the exploitation of labor. **Sixth**, the crisis of labor unions *weakens* resistance. This crisis, it is argued, is not only the result of a deliberate corporate attack on unions but also the result of *ineffective* union strategies. Finally, the author contends, resistance centers on spheres that do not consider labor as a central *component.* While successful in some instances, overall, these efforts have not been able to halt the power of corporate agri-food and the *worsening* of the *conditions* of labor.[287]

James O'Connor's opinion in "Political economy of ecology of socialism and capitalism" is the following:

Centrally planned socialist countries may have used up non renewable resources as fast as or faster than capitalist countries in the last three decades. Socialist economies also may have threatened the quality of renewable resources and polluted the air, water, and land as much as if not more than their capitalist counterparts. Environmental problems in the industrial and semi-industrial Eastern European countries, for example, are arguably worse than in comparable economies in the West. Many economists and environmentalists therefore conclude that it is not capitalism and socialism that deserve the onus for causing environmental degradation. Rather they attribute blame to "industrialization," "urbanization," "technology," "bureaucracy," and a "production at all costs" mentality - all of which appear to be common in both the capitalist and socialist worlds.[288]

Most important is what Vandana Shiva, in a chapter Science and Politics in the Green Revolution of her book, *The Violence of the Green Revolution: Third World Agriculture, Ecology, and Politics*, says:

The Green Revolution has been heralded as a political and technological achievement, unprecedented in human history. It was designed as a strategy for peace, through the creation of abundance by breaking out of nature's limits and variabilities. In its very genesis, the science of the Green Revolution was put forward as a political project for creating a social order based on peace and stability.

However, when violence was the outcome of social engineering , the domain of science was artificially insulated from the domain of politics and social processes. The science of the Green Revolution was offered as a 'miracle' recipe for prosperity. But when discontent and new scarcities emerged, science was delinked from economic processes. On the one hand, contemporary society perceives itself as a science-based civilisation, with science providing both the logic as well as propulsion for social transformation. In this aspect science is self-consciously embedded in society. On the other hand, unlike all other forms of social

---

[287] Alessandro Bonanno and Lawrence Busch, The international political economy of agriculture and food: An introduction. Emphasis added.

[288] James O'Connor, Political economy of ecology of socialism and capitalism.

organisation and social production, science is placed above Science and Politics in the Green Revolution society. It cannot be judged, it cannot be questioned, it cannot be evaluated in the public domain. As Harding has observed, 'Neither God nor tradition is privileged with the same credibility as scientific rationality in modern cultures... The project that sciences sacredness makes taboo is the examination of science in just the ways any other institution or set of social practices can be examined.' While science itself is a product of social forces, and has a social agenda determined by those who can mobilise scientific production, in contemporary times scientific activity has been assigned a privileged epistemological position of being socially and politically neutral. Thus science takes on a dual character. It offers technological fixes for social and political problems, but delinks itself from the new social and political problems it creates. Reflecting the priorities and perceptions of particular class, gender, or cultural interests, scientific thought organizes and transforms the natural and social order. However, since both nature and society have their own organisation, the superimposition of a new order does not necessarily take place perfectly and smoothly. There is often resistance from people and nature, a resistance which is externalized as 'unanticipated side effects'. Science stays immune from social assessment, and insulated from its own impacts. Through this split identity is created the 'sacredness of science. Within the structure of modern science itself are characteristics which prevent the perception of linkages. Fragmented into narrow disciplines and reductionist categories, scientific knowledge has a blind spot with respect to relational properties and relational impacts.

The Political Ecology of Technological Change in the dominant paradigm, technology is seen as being above society both in its structure and its evolution, in its offering technological fixes, and in its technological determinism. It is seen as a source of solution to problems that lie in society, and is rarely perceived as a source of new social problems. Its course is viewed as being self-determined. In periods of rapid technological transformation, it is assumed that society and people must adjust to that change, instead of technological change adjusting to the social values of equity, sustainability and participation. There is, however, another perspective which treats technological change as a process that is shaped by and serves the priorities of whoever controls it. In this perspective, a narrow social base of technological choice excludes human concerns and public participation. The interests of that base are protected in the name of sustaining an inherently progressive and socially neutral technology.[289]

In Carmen G. Gonzalez, Trade Liberalization, Food Security and the Environment: The Neoliberal Threat to Sustainable Rural Development, we find a detailed output of the Green Revolution's approach in agriculture:

> The neoliberal trade regime institutionalizes a double standard that permits protectionism in developed countries while requiring developing countries to open their markets to highly subsidized foreign competition. This double standard reinforces pre-existing patterns of trade and production that undermine the livelihoods of rural smallholders, degrade the natural resource base necessary for food production, and impede the economic diversification necessary for food security at the national level. ..., even if the neoliberal model were applied

---

[289] Vandana Shiva, The Violence of the Green Revolution: Third World Agriculture, Ecology, and Politics. 2016.

in an even-handed manner to both developed and developing countries, it would nevertheless have a negative impact on food security and ecological sustainability.

From an environmental and food security perspective, the most significant impact of the Green Revolution was the loss of crop genetic diversity. As a consequence of the Green Revolution, indigenous wheat varieties had virtually disappeared by the 1970s in North Africa, the Himalayas, Turkey, Spain, and Pakistan. Staples such as "barley, rice, millet, sorghum, [and] potatoes" also sustained erosion of genetic diversity.

Genetic erosion occurred even in export crops, such as coffee, bananas, cacao, and cotton, as uniform varieties replaced traditional, diverse varieties.

The loss of crop genetic diversity resulted in outbreaks of pests and disease causing severe damage to food crops. The application of pesticides often exacerbated the problem by destroying the pests' natural enemies and by enabling pests and pathogens to develop pesticide resistance. Finally, genetic erosion resulted in the loss of the very genetic material that might confer resistance in the event of catastrophic pest and disease infestations, thus increasing the vulnerability of the world's food supply.

The Green Revolution contributed to micronutrient malnutrition in the developing world by reducing the absorption of vital minerals into fruit, vegetables, and grains. The intensive monocropping of Green Revolution varieties depleted the soil of important minerals such as "zinc, iron, copper, manganese, magnesium, molybdenum, [and] boron,"' and the application of synthetic pesticides and fertilizers (along with soil compaction) destroyed the microorganisms needed to make these minerals available to food crops.

While organic fertilizers counteract this problem because the organic matter contains and replenishes these micronutrients, synthetic fertilizers generally contain few or none of these minerals. A result of the Green Revolution is that billions of people consume diets deficient in essential micronutrients.

Micronutrient malnutrition can produce serious impacts on human health, learning ability, and productivity.

The Green Revolution also displaced traditional food crops in the developing world, thereby impoverishing the diets of many individuals and communities. As a result of the Green Revolution, monocultures of wheat and corn replaced thousands of nutritious and robust traditional food crops, such as the Senegalese cereal known as fonio and the Indian ragi and jowar grains. The immediate impact of this conversion from polycultural to monocultural production was a decline in the variety of foods consumed, increased reliance on frequently unaffordable and less nutritious purchased foods, and the loss of foods essential to a balanced diet.

The adoption of new seed varieties, and the irrigation systems, pesticides, and fertilizers needed for their cultivation, displaced ecologically sustainable farming practices (such as intercropping, crop rotation, and agroforestry).

Moreover, it often resulted in the loss of local knowledge about traditional agroecological practices. Pesticide use displaced traditional pest control techniques (such as crop rotation and fallowing) that also contributed to soil fertility. Chemical fertilizers replaced the use of animal manure and crop residues. The continuous planting of uniform crops in a given area replaced crop rotation and intercropping. Green Revolution varieties displaced indigenous and traditional crops that required far less irrigation. The ecological consequences were often severe. The heavy use of agrochemicals destroyed beneficial soil organisms and degraded soil quality. The Green Revolution monocultures removed vital micro-nutrients from the soil, resulting in the long-term decline in agricultural yields. Intensive irrigation resulted in water-logging and salinization of soils. In sum, soil quality deteriorated, leading to a loss of agricultural productivity.

The use of pesticides to protect genetically uniform crops harmed the environment and increased pesticide-related deaths and illnesses in developing countries.

Pesticide and fertilizer runoff contaminated drinking water supplies and resulted in the eutrophication of rivers, lakes, and coastal waters. Excessive pesticide use also killed cattle and livestock and eradicated important sources of protein for poor farmers, such as fish, shrimp, and crabs in rice paddies. Nitrous oxide emissions from synthetic fertilizers contributed to global warming and to the depletion of the ozone layer. In addition, fertilizer production, which increased tenfold between 1950 and 1990, required significant inputs of non-renewable petroleum, the extraction and processing of which posed serious environmental risks.

Indeed, industrial agriculture is so dependent on fossil fuels that 9.8 kilocalories (kcal) of fossil fuel energy are required to produce one kcal of food energy.[290]

Carmen G. Gonzalez also said:

Finally, the conversion of forests, grasslands, and wetlands to monocultural farming systems destroyed or fragmented the habitats of various species of flora and fauna, producing a decline in biodiversity. The loss of biodiversity resulted in a loss of ecosystem services (such as water purification, insect management, and climate regulation) as well as the loss of wildlife habitat products useful for medicine, food, and fodder. Because extinction is irreversible, such losses have both global and local implications, and affect both present and future generations.[291]

In, *Political Economy, Capitalism and Sustainable Development*, we learn:

The Capitalist Mode of Production makes a great (*qualitative*) difference in what regards both the society—nature relation under capitalism and the conceptualization of the economy (the economic question), which cannot be conceived as divorced but rather deeply embedded in society. It implies that private property and the dispossession of the majority of direct producers entail a growing alienation and estrangement from nature, while wage labour entails the alienation of the working people.

Consumption may have been the direct purpose of production under pre-capitalist modes of production, but under capitalism consumption is significant only insofar as the commodities produced need to be sold, *with profit maximization being the immediate and dominant purpose of production*. This remark has significant, both social and ecological implications.

"The historically specific value form of labour and the specific character of capitalist valorization, which is exclusively *based on the wage labour exploited and largely ignores the contribution of nature*, while considering the natural forces utilized as "*a free gift of nature to capital*", *imply a potentially unlimited free appropriation of nature at essentially no cost.*"...

---

[290] Carmen G. Gonzalez, Trade Liberalization, Food Security and the Environment: The Neoliberal Threat to Sustainable Rural Development.

[291] Idem.

As a consequence of this underestimation of cost, capital tends to over-expand its activities by exhausting productive resources, polluting the environment and shifting a major part of the cost to society.

"The scale of ecological destruction is much larger under capitalism than under previous modes of production, and this is because of the totalizing tendency of the market mechanism, as well as the value augmentation and accumulation tendency in-built in the Capitalist Mode of Production itself, which implies an also augmenting energy and resource throughput and environmental damage.

Coming now to a more detailed explanation of this increasing ecological rift, we might stress that, under capitalism, an increase in labour productivity is essentially tantamount to a reduction in the amount of abstract socially necessary labour required for the production of any particular commodity (including labour power itself), which is a condition for an increased extraction and appropriation of surplus value. This, as I have noted, is the dominant goal of capitalism, and hence all increases in the productivity of labour should serve this goal. Under this context, an increasing productivity of labour does not imply a process economizing on labour or any other productive resources. On the contrary, insofar *as capital can proceed with a free appropriation of nature "as a gift to capital", there will be a permanent bias towards developing a labour-saving technology*, but this technology is conducive to a maximum throughput of natural resources and energy, which further implies a rapidly increasing depletion of natural resources and an increasing pollution contributing to a systemic environmental degradation. *A labour-saving technology, therefore, and a rising productivity of labour do not necessarily imply an increasing social and ecological efficiency, but rather an increasing potential for material and energy throughput, with an enhanced ecologically damaging impact.* What is more, even a resource-saving technological innovation cannot have, under capitalism, an environmentally protective impact insofar as it will, most likely, imply lower commodity prices and hence an increasing market demand, which will result in an increased (rather than decreased) extraction of the natural resource concerned."[292]

How come technology is neutral? It reminds me of Stenmark's "value-free view of science more precisely in this way: The value-free view of science is the standpoint that science should be autonomous, neutral, impartial, non-responsible, and non- normative."[293] That is one of the reasons why Western civilization is collapsing.

I believe both majorities of politicians and economists have become submissive or agree with neoliberalism corporate interest and power. Others may ignore the power and meaning of the potential of biodiversity; or how narrow it is to examine anything only from a short-term corporate economic perspective.

Unfortunately, formal education of current and former generations of politicians, economists, and professionals, took place under a fragmented science and its pervading culture. In this educational approach, it is suggested that all students be aware of the potential of

---

[292] George Liodakis, Political Economy, Capitalism and Sustainable Development. Emphasis added.

[293] Mikael Stenmark, qtd. in LeRon Shults (ed.), The Evolution of Rationality, p. 51.

biodiversity, the knowledge required to be a good gardener, the risks, roles, and wisdom of farmers, and how we will protect biodiversity and peasants.

To support the ideas mentioned above, let me include a reflection about the "agriculture economy" instead of just "economy":

> With the expansion of topics studied, the beneficiaries of agricultural economists have been broadened from farmers and land managers to public policymakers, financial institutions, and food chain companies. ...
>
> Agricultural economics was among the first and largest applied economics discipline that started to use economic principles to problems with farm management and agricultural policies (Runge, 2006).
>
> A brief survey of major journals in agricultural economics (e.g. European Review of Agricultural Economics, American Review of Agricultural Economics and Journal of Agricultural Economics) reveals an ever-expanding field of study carried out by agricultural economists. Historically, the interests of agricultural economists have developed over two axes: the set of theoretical and methodological approaches and the series of evolving focus areas or phenomena being studied.
>
> While this short overview suggests a continuous development and progress in diverging and specialised topics, it also underlines a standstill in connecting, consolidating and reinvigorating the power of economic thinking across different 'hotspots' that is creating an existential problem for the discipline. The discipline has spawned an archipelago of sub-disciplines. The complex challenges faced by the food systems cannot be addressed by these sub-disciplines individually and, seemingly disconnected, they lack a collective identity.[294]

The Green Revolution science principles are very different from those of the single science; in the non-fragmented single science: values, moral principles, interests in seeking justice and harmony, responsibilities, goals, and strategies are fused in each scientific method.

Are the Political Economy principles of the Green Revolution worthy of being preserved? Recently, in July 2021, I became aware that the United States of America is making some welcome inroads to address the problem, encouraging farmers to keep cover crops. Let us wait and see if the next president does not reverse these efforts. Is this the nature of the old-world order?

ႠႠႠ

To come up with a solution and to comprehend the benefits of this proposal, a macro-economic forecasting software can suffice to simulate different ways of:

- valuing the knowledge and features required to be a good farmer
- examining the working conditions of the farmer
- the best possible approach to addressing the needs of credit, supplies, and education for the rural population,
- Most importantly, considering what 'Abdu'l-Bahá said: "The fundamental basis of the community is agriculture, tillage of the soil."[295] What are we going to do with

---

[294] Louise O Fresco, Sustainable food systems: do agricultural economists have a role?

[295] The Promulgation of Universal Peace.

the most fertile soils? Build housing projects, highways, and malls?

"How long does it take to form an inch of topsoil?" This question has many different answers, but most soil scientists agree that it takes at least 100 years, and it varies depending on climate, vegetation, and other factors.[296]

---

[296] Natural Resources Conservation Service. United States Department of Agriculture, Soil Formation

6 The Root Cause and the Unique Method

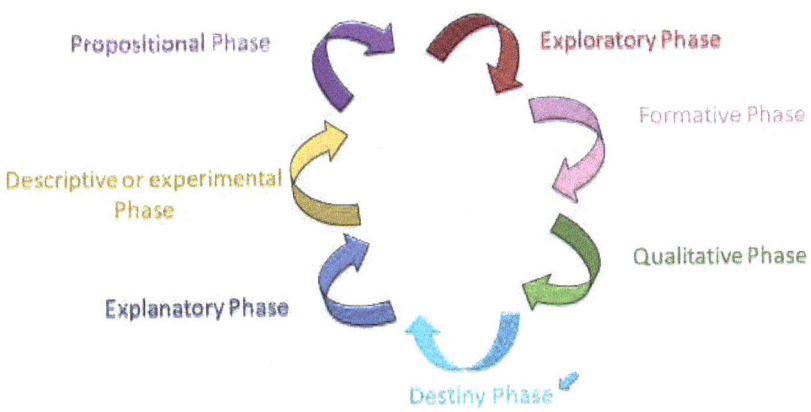

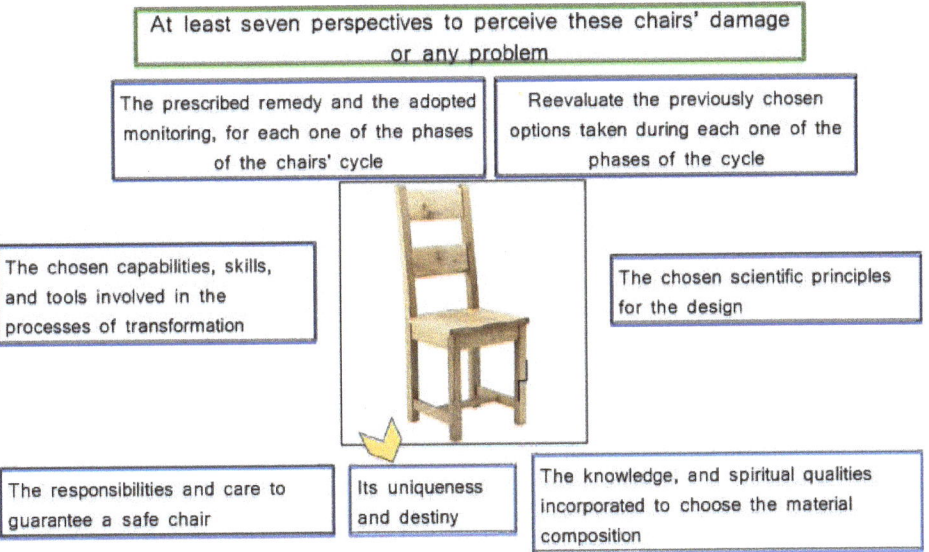

What would happen to the chair if we were negligent in keeping it clean or promptly adjusting the screws if it wobbles

The Root Cause and the Unique Method

The author desires to start this chapter by questioning the validity of the law of entropy.

In *The Science of Snowflakes, and Why No Two Are Alike*, says:

So what's behind the snowflake's unique and elaborate shape?[297] The snowflakes that settle upon our sleeves and scarves during a snowstorm have more variations in shape than you might think. There's the classic snowflake: a flat plate with branchlike, dendritic arms. Some look like hexagonal prisms; others like hollow pencil-shaped columns or tiny needles.

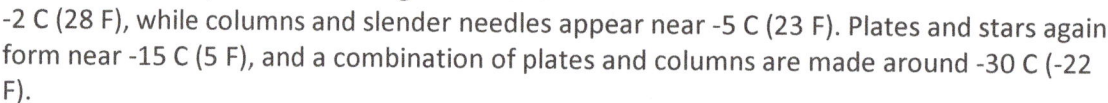

We tracked down two ice experts to help answer the question: Kenneth Libbrecht, a physics professor at the California Institute of Technology and avid snowflake photographer, and John Hallett, director of the Ice Physics Laboratory at the Desert Research Institute in Reno, Nev.

We see that thin plates and stars grow around -2 C (28 F), while columns and slender needles appear near -5 C (23 F). Plates and stars again form near -15 C (5 F), and a combination of plates and columns are made around -30 C (-22 F).

Furthermore, we see from the diagram that snow crystals tend to form simpler shapes when the humidity (supersaturation) is low, while more complex shapes at higher humidities. The most extreme shapes — long needles around -5C and large, thin plates around -15C — form when the humidity is especially high.

But the physics is far from exact, Libbrecht said.

"It's a mystery as to why [snowflake shapes] go from plates to columns to plates to columns as the temperature lowers," he said. "That's one of the things I've been trying to understand. It has been a mystery for about 75 years, and it's still unsolved."

And though it may appear otherwise, the arms of a snowflake are not perfectly symmetrical.

"If you look at the dendritic arms of a snowflake carefully, they usually are a bit different," Hallett said. "The atmosphere is a turbulent place, and crystals tend to oscillate as they are blown around, so even different corners see slightly different environments."

So is it really true that no two snowflakes are alike?

"It's like shuffling a deck and getting the exact same shuffle for 52 cards," Libbrecht said. "You could shuffle every second for the entire life of the universe, and you wouldn't come close to getting two of the same".[298]

In Science Learning Opportunities, we find: The uniqueness of time dimension is that you can travel only forward in it, not backward. This fact has profound implications. It protects causality, that is the law of cause and effect. That is, cause should precede effect and it

---

[297] Photo 105444807 © Chaoticmind | Dreamstime.com

[298] PBS News Hour Science. The Science of Snowflakes, and Why No Two Are Alike.

should not be the other way round. This irreversibility of time is inbuilt through the concept of entropy.

If you study thermodynamics, you will come across the law that entropy or *disorder* in the universe always increases, never can it decrease. You can understand the law of entropy by just observing the irreversible nature of natural phenomena. That is, a cup falling down and breaking, can never be restored to the same condition, with every atom in place, as it was. The irreversibility implied by entropy could be described by the popular line from the Humpty Dumpty nursery rhyme, which says: All the King's horses and all the King's men, couldn't put Humpty together again! For every system, *disorder* always increases. Entropy increase is unidirectional, just as the unidirectionality of time. Thus it is no coincidence that the thermodynamical arrow of time and the arrow of time flow, point in the same direction, as they both preserve causality. As a consequence, traveling back in time impossible, as it would violate causality and the law of entropy. However, special relativity does allow for the possibility of time travel to the future.[299]

In *Evolution and the Snowflake*, by Larry Vardiman, Ph.D., we read:

> Whenever the ordering of a local system results in beauty, symmetry, or function, this requires a pre-designed code, and does not happen by chance. Each physical agent operating at a higher level must function with greater order and power than the effect it produces. The ultimate cause which controls all secondary processes must have infinite power and organizing intelligence. Such a first cause is called God. Thus God either directly or by secondary processes produces order.[300]

Baha'u'llah, quoting the Qur'án, said: "No defect canst thou see in the creation of the God of mercy. Repeat the gaze: Seest thou a single flaw?"[301]

> Likewise, look into this endless universe: a universal power inevitably existeth, which encompasseth all, directing and regulating all the parts of this infinite creation; and were it not for this Director, this Coordinator, the universe would be flawed and deficient. It would be even as a madman; whereas ye can see that this endless creation carrieth out its functions in perfect order, every separate part of it performing its own task with complete reliability, nor is there any flaw to be found in all its workings. Thus it is clear that a Universal Power existeth, directing and regulating this infinite universe. Every rational mind can grasp this fact.[302]

'Abdu'l-Bahá is reported to have said: "The individuality of each created thing is based upon divine wisdom, for in the creation of God there is no defect."[303]

Let us listen to what The Báb revealed:

> No created thing shall ever attain its paradise unless it appeareth in its highest prescribed degree of perfection. For instance, this crystal representeth the paradise of the stone whereof its substance is composed. Likewise there are various stages in the paradise for the

---

[299] ScienceStruck: Science Learning Opportunities. Emphasis added.

[300] Larry Vardiman, Evolution and the Snowflake.

[301] Baha'u'llah, The Call of the Divine Beloved.

[302] 'Abdu'l-Bahá, Selections from the Writings of 'Abdu'l-Bahá.

[303] 'Abdu'l-Bahá, Divine Philosophy, p. 131.

crystal itself…. So long as it was stone it was worthless, but if it attaineth the excellence of ruby—a potentiality which is latent in it—how much a carat will it be worth? Consider likewise every created thing.[304]

The formation of snow crystals contradicts the concept of "entropy as a measure of disorder" mentioned above. The unique conditions prevailing in the path, followed by each water molecule, were the stimuli received to form a nonidentical snowflake; indeed, were the cause of the spontaneous response to creating a unique snowflake.

Some water molecules may evaporate and form new distinctive snowflakes or form non-identical droplets to travel to the ocean. Obviously, it is not reversible to form the exact snowflake because the water molecules are dispersed in space once the snowflake disintegrates. But calling it chaotic or disorderliness is not correct. The diversity of features of snowflakes is not disorderliness; it is the highest degree of perfection, and its diversity is the utmost beauty!

A snowflake exists within the mineral station H^2O. The infinite potential of beings to move towards their imaginable destined perfections within their station, is another dimension of reality. The universe goes from one state of order to one that is increasingly orderly, from one level of perfection to a higher one—contradicting the law of entropy that says that the universe always moves from order to disorder. Of course, there are cataclysmic events, droughts, and extreme temperatures for minerals, plants, animals; and problems, issues, and tests for humans that seem to be retrocession in the process of evolution. But there is also the capacity to respond differentially by the mineral, plant, and animal kingdoms; and intelligent ones by humans to overcome them.

Let us listen to Shogui Effendi in a reflection about disorder and order:

> As we view the world around us, we are compelled to observe the manifold evidences of that universal fermentation which, in every continent of the globe and in every department of human life, be it religious, social, economic or political, is purging and reshaping humanity in anticipation of the Day when the wholeness of the human race will have been recognized and its unity established. A twofold process, however, can be distinguished, each tending, in its own way and with an accelerated momentum, to bring to a climax the forces that are transforming the face of our planet. The first is essentially an integrating process, while the second is fundamentally disruptive. The former, as it steadily evolves, unfolds a System which may well serve as a pattern for that world polity towards which a *strangely-disordered* world is continually advancing; while the latter, as its disintegrating influence deepens, tends to tear down, with increasing violence, the antiquated barriers that seek to block humanity's progress towards its *destined* goal. The constructive process stands associated with the nascent Faith of Bahá'u'lláh, and is the harbinger of the *New World Order* that Faith must erelong establish. The destructive forces that characterize the other should be identified with a civilization that has refused to answer to the expectation of a new age, and is consequently falling into chaos and decline.[305]

There is memory in the DNA where living organisms file and share their good and bad experiences with the next generations:

---

[304] The Báb, Selections from the Writings of the Báb.

[305] The World Order of Bahá'u'lláh. Emphasis added.

Domestication is defined in terms of a coevolutionary mutualism between domesticator and domesticate and is distinguished from related but ultimately different processes of management and agriculture. Domestication results in a range of genotypic, phenotypic, plastic, and contextual impacts that can be used as markers of evolving domesticatory relationships. A consideration of causal scenarios finds greater empirical support for explanatory frameworks grounded in niche-construction theory over those derived from optimal foraging theory".[306]

Plant domestication through artificial selection is one of the best examples of this. When a human tends a plant with favorable attributes, perhaps because it has the largest and sweetest fruits or most resilient husk, and saves the seeds to replant, they are essentially guaranteeing the continuation of that particular organism.

In this way, a farmer can select for the properties they desire by giving special treatment only to the best and most successful plants. Their crop, in turn, starts to take on the desirable properties the farmer selected for, and disadvantageous attributes extinguish over time.[307]

"[T]he innate immune system is an alternate defense strategy and is the dominant immune system response found in plants, fungi, insects, and primitive multicellular organisms,"[308] and "The adaptive immune system (AIS) in mammals, which is centred on lymphocytes bearing antigen receptors that are generated by somatic recombination, arose approximately 500 million years ago in jawed fish,"[309] are still being transfer from one generation to the next. Where is the disorder?

"Peruse thou the Kitáb-i-Íqán and that which the All-Merciful hath sent down unto the King of Paris and to such as are like him, that thou mayest be made aware of the things that have happened in the past, and be persuaded that We have not sought to spread *disorder* in the land after it had been *well-ordered*."[310]

Of course, humans can cause a lot of disorder. Let me quote a message on October 1985 from The Universal House of Justice about *The Promise of World Peace*.

To the Peoples of the World:

The Great Peace towards which people of good will throughout the centuries have inclined their hearts, of which seers and poets for countless generations have expressed their vision, and for which from age to age the sacred scriptures of mankind have constantly held the promise, is now at long last within the reach of the nations. For the first time in history it is possible for everyone to view the entire planet, with all its myriad diversified peoples, in one perspective. World peace is not only possible but inevitable. It is the next stage in the evolution of this planet—in the words of one great thinker, "the planetization of mankind".

---

[306] Melinda A. Zeder, Core questions in domestication research. Emphasis added.

[307] K. Kris Hirst, Plant Domestication. Dates and Locations of Human Farming Advances.

[308] Wikipedia, Innate Immune System.

[309] Martin F. Flajnik and Masanori Kasahara, Origin and evolution of the adaptive immune system: genetic events and selective pressures

[310] Bahá'u'lláh, Tablets of Bahá'u'lláh. Emphasis added.

Whether peace is to be reached only after unimaginable horrors precipitated by humanity's stubborn clinging to old patterns of behaviour, or is to be embraced now by an act of consultative will, is the choice before all who inhabit the earth. At this critical juncture when the intractable problems confronting nations have been fused into one common concern for the whole world, failure to stem the tide of conflict and *disorder* would be unconscionably irresponsible.[311]

✦

The discovery in 2004 of graphene and an alternative method (2017) to produce it from soybean oil is a good example. It shows the moral strength to endure fear and difficulty. The author believes that it was easy to fear failing and bear in such a difficult task:

> The new method involves heating soybean oil in ambient air until it breaks down into "carbon building units that are essential for the synthesis of graphene."
>
> It is then rapidly cooled on nickel foil into a thin rectangle.
>
> Dr. Han said the process is simpler and safer than existing methods, which use explosive compressed gases and vacuum processing.
>
> He said under current technology, a high-quality graphene film with a 10cm (4 inches) diameter costs up to A$1000 (£600, $750).
>
> The new method could make it "significantly" cheaper, he said.
>
> Dr. Han said the discovery, detailed in the journal Nature Communications, was made "kind of by accident."
>
> "In our lab, we have very strict safety rules," he told the BBC.
>
> "We were trying to figure out if there was a safer way of making this material without using dangerous or explosive gases."[312]

In a scientific article defining the unique path followed, we find:

> Here we present a single-step, rapid thermal synthesis of uniform and continuous graphene films in an ambient-air environment, using a cheap and renewable form of biomass, soybean oil, as the precursor. To the best of our knowledge, this is the first time that the synthesis of graphene film has been demonstrated in an ambient-air environment without any compressed gases. *Graphene derived from this unique ambient-air process* exhibits good and tuneable film properties, which are comparable to those of graphene synthesized with conventional methods. This ambient-air process for graphene fabrication is fast, simple, safe, potentially scalable and integration-friendly. Importantly, it offers the scope to potentially address the critical roadblocks towards large-scale, efficient graphene manufacturing.
>
> The growth of graphene in an ambient-air environment may initially seem counter-intuitive, as graphene is expected to be destroyed in air at elevated temperatures (above 500 °C). *However, we hypothesize that the unique processing conditions* promote the controlled synthesis of graphene films in an otherwise destructive environment. Specifically, the thermally dissociated precursor material decomposes in the presence of reactive oxygen species from the ambient-air, leading to the formation of water vapour as a by-product. The

---

[311] The Universal House of Justice, The Promise of World Peace. Emphasis added.

[312] Greg Dunlop, reporting by the BBC's, Australian scientists use "soybean oil" to create graphene. Emphasis added.

water vapour may help suppress the deposition of amorphous carbon, promote the thinning of graphene layers and maintain the catalytic ability of the Ni substrate in breaking down the precursor material into smaller building units necessary for the growth of graphene films."[313]

Do only scientists require courage to walk a path nobody ever walked to find a destiny? Religion reaffirms this belief:

"But you have asked me for proofs and arguments of the impossibility of reincarnation, and we must therefore explain the reasons for its impossibility. The first proof is that the outward is the expression of the inward: The earthly realm is the mirror of the heavenly Kingdom, and the material world is in accordance with the spiritual world. *Now observe that in the sensible world the divine appearances are not repeated, for no created thing can be identical with another in every way.* The sign of Divine Unity is present and visible in all things. If all the granaries of the world were filled with grain, you would be hard-pressed to find two grains that are absolutely identical and indistinguishable in every respect: *Some difference or distinction is bound to remain between them.* Now, as the proof of the Divine Unity exists within all things, and the oneness and singleness of God is visible in the realities of all beings, the recurrence of the same divine appearance is in no wise possible.

Therefore reincarnation, which is the repeated manifestation in this world of the same spirit with its former essence and conditions, would be the selfsame appearance and is thus impossible. And since the recurrence of the same divine appearance is impossible for material beings, the repeated assumption of the same station, whether on the arc of descent or on the arc of ascent, is likewise impossible for spiritual beings, for the material world corresponds to the spiritual world.

With respect to the species, however, return and recurrence are plainly visible in material realities; that is, the trees which in years past bore leaves, blossoms, and fruit will in the years to come bear the same leaves, blossoms, and fruit. This is called recurrence of species. Were anyone to object that the leaf, the blossom, and the fruit have decomposed, have descended from the vegetable to the mineral world, and have returned again to the former, and that there has thus been a recurrence, we would reply that the blossom, the leaf, and the fruit of last year were decomposed, and their component elements were disintegrated and dispersed. It is not that the same particles of last year's leaf and fruit that had decomposed have recombined and returned, but that the essence of the species has returned through the combination of new elements. Likewise, the human body is fully disintegrated after the decomposition and dispersion of its constituent parts. Were this body to return from the mineral or vegetable world, it would not comprise the selfsame constituents as the former person, for its elements were decomposed, disintegrated, and dispersed in space. Afterwards other elemental constituents were combined and another body was formed. And while it may be the case that certain constituents of the former body entered into the composition of the latter, those constituents have not been exactly and completely conserved, without any addition or diminution, so as to be composed again and to give rise through their composition and combination to another individual. One cannot deduce, then, that this body has returned with all its constituent parts, that the former individual has become the latter, and hence that a recurrence has taken place—that the very

---

[313] Dong Han Seo, et al., Single-step ambient-air synthesis of graphene from renewable precursors as electrochemical genosensor.

same spirit, like the body, has returned and that after death its essence has regained this world.[314]

As an initial context for our discussion about destiny, let us look into "*Logos and Civilization*" by Nader Saiedi and *"Nietzsche and the Bahá'í Writings: A First Look"* by Ian Kluge:

> Ibnu'l-'Arabi wrote extensively on the concept of destiny. For him, although destiny comes after the divine will in the order of creation, it is actually destiny that determines the divine will and the divine decree. Destiny, here, becomes the capacities and measures of things in their archetypal essences. God's will is determined by those eternal essences. Divine knowledge is seen as dependent on its objects of knowledge, namely the essences of things or their destiny, and God's creative will is dictated and determined by their logical requirements. On the other hand, this means that every event is preordained and that everything happens in accordance with the arbitrary destiny of things: whatever humans do is part of their predetermined destiny, and there is no reward or punishment based on freedom of will and justice. However, a mystic understands that arbitrary reward and punishment are also part of destiny and must be accepted because all things are foreordained.
>
> This concept of destiny calls for a passive attitude of resignation in the wake of predestination. (The Bezels of Wisdom, 163-71).
>
> Aspects of Ibnu'l-'Arabi's view of destiny are directly criticized in the writings of Shaykh Aḥmad, The Báb, Bahá'u'lláh, and 'Abdu'l-Bahá. First of all, Ibnu'l-'Arabi's theory contradicts the principle of the absolute freedom and transcendence of God. By making divine knowledge dependent on the archetypal essences, or the objects of God's knowledge, Ibnu'l-'Arabi has made the essences of contingent things the determinants and dictators of God's will. As 'Abdu'l-Bahá has affirmed, however, divine knowledge is unlike human knowledge. Human knowledge requires and depends on the existence of the objects of knowledge. But God's knowledge does not, which means that by definition we can have no idea of the nature of divine knowledge. God's will, in other words, is absolutely free.[315]
>
> Paradoxically, however, Ibnu'l-'Arabi not only violates the principle of divine freedom, he also eliminates the possibility of human freedom through his excessively passive concept of destiny. The concept of destiny in the Bahá'í Writings, on the other hand, becomes compatible with the relative freedom of human beings. It is clear that Bahá'u'lláh's concept of the mystery of destiny is not a mere acceptance of whatever "is"; on the contrary, it is an active movement toward realizing spiritual values in one's own life and developing the potentialities and perfections hidden, like "gems," in the "mine" of one's own being.[316] The

---

[314] 'Abdu'l-Bahá, Some Answered Questions, No. 81.

[315] The quote says: "In brief, the Prophets and the philosophers are in agreement on one point, namely, that the cause through which all things are realized is but one. The difference is that the Prophets teach that God's knowledge does not require the existence of created things, whereas the knowledge of the creatures requires the existence of objects of knowledge. If the divine knowledge stood in need of aught else, then it would be like the knowledge of the creatures and not that of God; for the Pre-existent is incommensurate with the originated and the originated is opposite to the Pre-existent." 'Abdu'l-Bahá, Some Answered Questions, No. 82.

[316] Bahá'u'lláh, Gleanings 260.

mystery of destiny, then, among other things, precisely implies transcending the opposition between the divine will and the individual will. It represents the actualization of all one's spiritual powers and the maturation of one's potentialities to the degree that one freely chooses spiritual values and the will of God. This is the stage of perfect freedom and moral autonomy, in which human potentialities are actualized in harmony with divine revelation. That is why Bahá'u'lláh defines this valley as both the station of mystery and the secret of maturation. It implies the integration of the approaches based on self, reason, and love.

This gives us a new mystical problematic. First of all, the Four Valleys becomes ultimately an epistemology of harmony among religion, reason, and mystical enlightenment. The Sufi opposition between religious law and spiritual truth is replaced by an ethic of self-actualization, or development of potentialities, encompassing all the various dimensions of human life. Second, the Four Valleys becomes a dialectic of freedom and self-determination. Spiritual journey is conceived as the process of emancipation from unfreedom and enslavement and as the attainment of freedom, self-consciousness, and autonomy.

This dialectic of freedom, however, is the exact opposite of the modern Western conception of freedom and self-realization, rooted in the philosophy of Enlightenment and liberalism, which regards freedom as the elimination and overcoming of any obstacle to the pursuit of human purposes and desires. That concept places spiritual values and divine law in the category of restrictions on freedom rather than freely chosen purposes. It is a freedom from rather than freedom for. The conception of freedom outlined in the Bahá'í Writings is characterized by a level of spiritual development in which the contradiction between the individual will and the divine will is replaced by an active mode of self-perfection, and a resolve to beautify and spiritualize the world. In this view, destiny paradoxically becomes the road to freedom.[317]

Ian Kluge has an accurate reflection:

Aristotle's ontology of potentials — including teleology — is most dramatically illustrated in Zarathustra's command, "Become what you are!" This command only makes sense on the assumption that we have an essence made up of certain potentials unique to us as members of the human race and as human individuals. It requires the distinction between appearance, i.e. what we are now, and reality, i.e. our unactualized potentials. Furthermore, this command requires free will in order to struggle for a goal within our range of potentials. In addition, this command only makes sense if we have an essence that is stable, i.e. is continuous enough to be given instructions that can be meaningfully followed over a period of time, i.e. has continuity through change. In short, it must have identity or 'being' or be a 'substance.' Thus it appears that Nietzsche did, albeit implicitly — or perhaps inadvertently — accept the concepts of potential, essence, goals, free will, cause-and-effect, the distinction between appearance and reality and substance since without them, a significant portion of his philosophy of self-overcoming would lose its logical and ontological foundations.[318]

6.1 Science and destiny

In The Science of destiny: what is the meaning of life, anyway? By David Zeigler, we learn:

---

[317] Saiedi, Logos and Civilization: spirit. history and order in the Writings of Bahá'u'lláh.

[318] Ian Kluge, Nietzsche and the Bahá'í Writings: A First Look.

"Destiny" is a common synonym for what philosophers call teleology--the idea that events follow a set plan that has a purpose. Teleology is primarily defined as a belief in "final ends" or ultimate goals that constrain reality in such a way that those ends or goals will be reached. While teleological ideas and beliefs tire still fairly common, they are decidedly not scientific or rational.

Science deals only with the physical universe and physical causality. If someone believes that the physical universe is controlled and guided toward certain ends by metaphysical forces such as God, destiny, fate, etc., then that person has stepped beyond the realm of rational scientific explanation.[319]

However, there is hope! Georg Toepfer, in Teleology and its constitutive role for biology as the Science of organized systems in nature, says:

'Nothing in biology makes sense, except in the light of teleology'. This could be the first sentence in a textbook about the methodology of biology. The fundamental concepts in biology, e.g. 'organism' and 'ecosystem', are only intelligible given a teleological framework. Since early modern times, teleology has often been considered methodologically unscientific. With the acceptance of evolutionary theory, one popular strategy for accommodating teleological reasoning was to explain it by reference to selection in the past: functions were reconstructed as 'selected effects'. But the theory of evolution obviously presupposes the existence of organisms as organized and regulated, i.e. functional systems. Therefore, evolutionary theory cannot provide the foundation for teleology. The underlying reason for the central methodological role of teleology in biology is not its potential to offer particular forms of (evolutionary) explanations for the presence of parts, but rather an ontological one: organisms and other basic biological entities do not exist as physical bodies do, as amounts of matter with a definite form. Rather, they are dynamic systems in stable equilibrium; despite changes of their matter and form (in metabolism and metamorphosis) they maintain their identity. What remains constant in these kinds of systems is their 'organization', i.e. the causal pattern of interdependence of parts with certain effects of each part being relevant for the working of the system. Teleological analysis consists in the identification of these system-relevant effects and at the same time of the system as a whole. Therefore, the identity of biological systems cannot be specified without teleological reasoning.[320]

6.2 Fate, Predestination, and Destiny

In the following quotes, it is obvious to conclude that beings in all kingdoms have a destiny[321]. It includes mineral, plants, animals, humans, countries, families, businesses, organizations, and institutions. From 'Abdu'l-Bahá, we learn:

---

[319] David Zeigler, The Science of destiny: what is the meaning of life, anyway? Emphasis added.

[320] Georg Toepfer, Teleology and its constitutive role for biology as the Science of organized systems in nature.

[321] Definition of destiny on Merriam Webster Dictionary: A predetermined course of events often held to be an irresistible power or agency

Thou hadst asked about fate, predestination and will. *Fate[322] and predestinatio[3232] consist in the necessary and indispensable relationships which exist in the realities of things.* These relationships have been placed in the realities of existent beings through the power of creation and every incident is a consequence of the necessary relationship. For example, God hath created a relation between the sun and the terrestrial globe that the rays of the sun should shine and the soil should yield. These relationships constitute predestination, and the manifestation thereof in the plane of existence is fate. Will is that active force which controlleth these relationships and these incidents. Such is the epitome of the explanation of fate and predestination. I have no time for a detailed explanation. Ponder over this; the reality of fate, predestination and will shall be made manifest.[324]

'Abdu'l-Bahá is reported to have said: "God in his wisdom has created all things. Nothing has been created without a special destiny, for every creature has an innate station of attainment".[325]

In *Some Answered Questions*, we find a very interesting reflection about Free Will and Predestination:

Question: When an action which someone will perform becomes the object of God's knowledge and is recorded in the "Guarded Tablet" of destiny, is it possible to resist it?

Answer: The knowledge of a thing is not the cause of its occurrence; for the essential knowledge of God encompasses the realities of all things both before and after they come to exist, but it is not the cause of their existence. This is an expression of the perfection of God.

As to the pronouncements which, through divine revelation, have issued from the Prophets regarding the advent of the Promised One of the Torah, these likewise were not the cause of Christ's appearance. But the hidden mysteries of the days to come were revealed to the Prophets, who thus became acquainted with future events and who proclaimed them in turn. This knowledge and proclamation were not the cause of the occurrence of these events. For instance, tonight everyone knows that in seven hours the sun will rise, but this common knowledge does not cause the appearance and rising of the sun.

Likewise, God's knowledge in the contingent world does not produce the forms of things. Rather, that knowledge is freed from the distinctions of past, present, and future, and is identical with the realization of all things without being the cause of that realization.

In the same way, the record and mention of a thing in the Scriptures is not the cause of its existence. The Prophets of God were informed through divine revelation that certain events would come to pass. For instance, through divine revelation they came to know that Christ

---

[322] Definition of fate on Merriam Webster Dictionary: the will or principle or determining cause by which things in general are believed to come to be as they are or events to happen as they do. In the New Oxford American Dictionary says" The development of events beyond a person's control, regarded as determined by a supernatural power.

[323] Definition of predestination on Merriam Webster Dictionary: the doctrine that God in consequence of his foreknowledge of all events infallibly guides those who are destined for salvation.

[324] 'Abdu'l-Bahá, Selections from the Writings of 'Abdu'l-Bahá, No 167. Emphasis added.

[325] 'Abdu'l-Bahá, Divine Philosophy.

would be martyred, which they in turn proclaimed. Now, did their knowledge and awareness cause the martyrdom of Christ? No: This knowledge is a sign of their perfection and not the cause of His martyrdom.

Through astronomical calculations, the mathematicians determine that at a certain time a solar or lunar eclipse will occur. Surely this prediction is not the cause of the eclipse. This of course is merely an analogy and not an exact image.[326]

6.3 Philosophical Argument

Question: Do only scientists require courage to walk a path nobody ever walked to find a destiny? **Essences:** God's Providence guiding us to reach our destiny in the context of humankind's collective destiny. Fear God's repetitive and increasingly more challenging tests. God's grace is represented in progressive knowledge and bounties to help us reach our destiny	**Faculties:** Capability to recognize God when learning to know ourselves. Capability to identify, acknowledge, and overcome our faults on the path to reach our preordained higher destiny. Capability to discover the root cause of the obstacles impeding to reach our enforced destiny. The capability to discover the guidance and follow that unique path to be tread in our lifetime. Perseverance in following the seven steps ? methodology provided by the Master. Freedom to choose what conditions are required to reach our unique high destiny. The capacity to develop those particular potentialities to reach our uniqueness. We can polish those peculiar perfections necessary to achieve our fixed fate and enforced destiny. The capacity to reach the particular aspiration of our soul
Laws: ""We are all subject to His judgment and reckoning."" We must respect others' privacy while guiding and providing clues on finding the root cause of wrongdoings and the steps to be followed.	Capacity to consciously recognize that each one has a unique higher destiny, "" Capability to discover the uniqueness and destiny of the species in other kingdoms
Relationship of confidence in the will, guidance, and mercy of God. **Situations** of guidance and being fearful to God in the context of the time given to us to accomplish our destiny. Challenging situations to find the root cause of a problem or the potential perfections within other beings in nature	Capability to define parameters and formulate indicators to measure success in courageously removing the root causes of our defective behavior or finding the unique path that each has to walk. Capability to define parameters and formulate indicators to measure success in: cultivation care, crossing, grafting, and domestication, and to improve the relationship between the plant or the animal and its domesticator

---

[326] 'Abdu'l-Bahá, Some Answered Questions, No. 35.

The author considers that linked to predestination, there are three originating causes that exist in the realities of all things: Providence, Grace, and Fear of God.

6.3.1 Providence:

The Báb revealed:

> O God our Lord! Protect us through Thy grace from whatsoever may be repugnant unto Thee and vouchsafe unto us that which well beseemeth Thee. Give us more out of Thy bounty and bless us. Pardon us for the things we have done and wash away our sins and forgive us with Thy gracious forgiveness. Verily Thou art the Most Exalted, the Self-Subsisting.
>
> *Thy loving providence*[327] *hath encompassed all created things in the heavens and on the earth* and Thy forgiveness hath surpassed the whole creation. Thine is sovereignty; in Thy hand are the Kingdoms of Creation and Revelation; in Thy right hand Thou holdest all created things and within Thy grasp are the assigned measures of forgiveness. Thou forgivest whomsoever among Thy servants Thou pleasest. Verily Thou art the Ever-Forgiving, the All-Loving. Nothing whatsoever escapeth Thy knowledge, and naught is there which is hidden from Thee.[328]

Bahá'u'lláh revealed:

> Look not upon the creatures of God except with the eye of kindliness and of mercy, for *Our loving providence hath pervaded all created things*, and *Our grace encompassed the earth and the heavens*. This is the Day whereon the true servants of God partake of the life-giving waters of reunion, the Day whereon those that are nigh unto Him are able to drink of the soft-flowing river of immortality, and they who believe in His unity, the wine of His Presence, through their recognition of Him Who is the Highest and Last End of all, in Whom the Tongue of Majesty and Glory voiceth the call: "The Kingdom is Mine. I, Myself, am, of Mine own right, its Ruler".[329]

In *Tabernacle of Unity*, by Bahá'u'lláh, we find the following quotes:

> O servant of God! The day of deeds hath come: Now is not the time for words. The Messenger of God hath appeared: Now is not the hour for hesitation. Open thou thine inner eye that thou mayest behold the face of the Beloved, and hearken thou with thine inner ear that thou mayest hear the sweet murmur of His celestial voice.
>
> O servant of God! The robe of divine bestowal hath been sewn and readied. Take hold of it and attire thyself therewith. Renounce and forsake the people of the world. O wise one! Shouldst thou heed the counsel of thy Lord, thou wouldst be released from the bondage of His servants and behold thyself exalted above all men.
>
> O servant of God! The Tree which We had planted with the Hand of Providence hath borne its destined fruit, and the glad-tidings We had imparted in the Book have appeared in full effect.

---

[327] Definition of providence: God conceived as the power sustaining and guiding human destiny. Web. June 2020 <https://www.merriam-webster.com/dictionary/providence>.

[328] The Báb. Selections from the Writings of the Báb. Emphasis added

[329] Bahá'u'lláh, Gleanings from the Writings of Bahá'u'lláh. Emphasis added.

> O servant of God! Say: O high priests! The Hand of Omnipotence is stretched forth from behind the clouds; behold ye it with new eyes. The tokens of His majesty and greatness are unveiled; gaze ye on them with pure eyes.
>
> O servant of God! The Daystar of the everlasting realm is shining resplendent above the horizon of His will and the Oceans of divine bounty are surging. Bereft indeed is the one who hath failed to behold them, and lifeless the one who hath not attained thereunto. Close thine eyes to this nether world, open them to the countenance of the incomparable Friend, and commune intimately with His Spirit.[330]

Ian C. Semple in *Obedience* said:

> The exercise of one's mind and the use of one's judgment in obeying a law or instruction are also avenues for divine guidance. I was profoundly impressed by something that the Hand of the Cause Paul Haney once related. He said that sometimes when the Universal House of Justice asked him to undertake a task, he was able at the outset neither to see the wisdom of it, nor how he was to carry it out but, confident in the divine guidance given to the House of Justice, he would set out to do it, and he would find that at every step that he took forward a door would open and the next step would become clear, and he would find at the end that he had been enabled to achieve just what he had been requested to do and he could see the reason for it.
>
> This is a perfect example of obedience, faith, wisdom and judgement. The processes of accepting one's personal responsibility, recognizing one's insufficiency, of seeking and validating an external source of authority and thereby finding the Manifestation of God, of understanding His teachings and of using one's intelligence in implementing them are essential for the development of the individual soul and enable it to fulfil its destiny of coming into harmony with the purpose of God and living in perfect obedience to His designs.[331]

6.3.2 Fear of God

Bahá'u'lláh revealed:

> "Set before thine eyes God's unerring Balance and, as one standing in His Presence, weigh in that Balance thine actions every day, every moment of thy life. Bring thyself to account ere thou art summoned to a reckoning, on the Day when no man shall
>
> have strength to stand for fear of God[332], the Day when the hearts of the heedless ones shall be made to tremble."[333]
>
> O My Servants!
>
> Ye are the trees of My garden; ye must give forth goodly and wondrous fruits, that ye yourselves and others may profit therefrom. Thus it is incumbent on every one to engage in

---

[330] Bahá'u'lláh, The Tabernacle of Unity.

[331] Semple, Ian C. Obedience.

[332] Definition of fear of God: "Fear of God refers to fear or a specific sense of respect, awe, and submission to a deity. People subscribing to popular monotheistic religions might fear divine judgment, hell or God's omnipotence. Wikipedia, Fear of God.

[333] Bahá'u'lláh, Gleanings from the Writings of Bahá'u'lláh.

crafts and professions, for therein lies the secret of wealth, O men of understanding! For results depend upon means, and the grace of God shall be all-sufficient unto you. Trees that yield no fruit have been and will ever be for the fire.[334]

In the fourth Ishráq (splendor) of the Ishráqát (Tablet of Splendors) We have mentioned: "Every cause needeth a helper. In this Revelation the hosts which can render it victorious are the hosts of praiseworthy deeds and upright character. *The leader and commander of these hosts hath ever been the fear of God, a fear that encompasseth all things, and reigneth over all things*".[335]

"The most burning fire is to question the signs of God, to dispute idly that which He hath revealed, to deny Him and carry one's self proudly before Him.

The essence of wisdom is the fear of God, the dread of His scourge and punishment, and the apprehension of His justice and decree".[336]

"By God! The Tree of vicegerency hath been planted, the Point of knowledge hath been made plain, and the sovereignty of God, the Help in Peril, the Self-Subsisting, hath been established. Fear ye the Lord. Follow not the promptings of your evil desires, but keep the law of God all your days. Renew the rules of the ways ye follow, that ye may be led by the light of guidance and may hasten in the path of the True One".[337]

'Abdu'l-Bahá, in *The Promulgation of Universal Peace*, taught:

Religionists have considered the world of humanity as two trees: one divine and merciful, the other satanic; they themselves the branches, leaves and fruit of the divine tree and all others who differ from them in belief the product of the tree which is satanic. Therefore, sedition and warfare, bloodshed and strife have been continuous among them. The greatest cause of human alienation has been religion because each party has considered the belief of the other as anathema and deprived of the mercy of God.

The teachings specialized in Bahá'u'lláh are addressed to humanity. He says, "Ye are all the leaves of one tree." He does not say, "Ye are the leaves of two trees: one divine, the other satanic." He has declared that each individual member of the human family is a leaf or branch upon the Adamic tree; that all are sheltered beneath the protecting mercy and providence of God; that all are the children of God, fruit upon the one tree of His love. God is equally compassionate and kind to all the leaves, branches and fruit of this tree. Therefore, there is no satanic tree whatever—Satan being a product of human minds and of instinctive human tendencies toward error.

God alone is Creator, and all are creatures of His might. Therefore, we must love mankind as His creatures, realizing that all are growing upon the tree of His mercy, servants of His omnipotent will and manifestations of His good pleasure.

Even though we find a defective branch or leaf upon this tree of humanity or an imperfect blossom, it, nevertheless, belongs to this tree and not to another.

---

[334] Bahá'u'lláh, The Hidden Words of Bahá'u'lláh.

[335] Bahá'u'lláh. Epistle to the Son of the Wolf, p. 26.

[336] Bahá'u'lláh, Tablets of Bahá'u'lláh Revealed After the Kitáb-i-Aqdas.

[337] Bahá'u'lláh, The Summons of the Lord of Hosts.

Therefore, it is our duty to protect and cultivate this tree until it reaches perfection. If we examine its fruit and find it imperfect, we must strive to make it perfect. There are souls in the human world who are ignorant; we must make them knowing. Some growing upon the tree are weak and ailing; we must assist them toward health and recovery. If they are as infants in development, we must minister to them until they attain maturity. We should never detest and shun them as objectionable and unworthy. We must treat them with honor, respect and kindness; for God has created them and not Satan. *They are not manifestations of the wrath of God but evidences of His divine favor. God, the Creator, has endowed them with physical, mental and spiritual qualities that they may seek to know and do His will; therefore, they are not objects of His wrath and condemnation.* In brief, all humanity must be looked upon with love, kindness and respect; for what we behold in them are none other than the signs and traces of God Himself. All are evidences of God; therefore, how shall we be justified in debasing and belittling them, uttering anathema and preventing them from drawing near unto His mercy? This is ignorance and injustice, displeasing to God; for in His sight all are His servants.[338]

'Abdu'l-Bahá, in *Paris Talks*, taught:

> The tent of the order of the world is raised and established on the two pillars of "Reward and Retribution."
>
> In despotic Governments carried on by men without Divine faith, where no fear of spiritual retribution exists, the execution of the laws is tyrannical and unjust.
>
> There is no greater prevention of oppression than these two sentiments, hope and fear. They have both political and spiritual consequences.
>
> If administrators of the law would take into consideration the spiritual consequences of their decisions, and follow the guidance of religion, "They would be Divine agents in the world of action, the representatives of God for those who are on earth, and they would defend, for the love of God, the interests of His servants as they would defend their own." If a governor realizes his responsibility, and fears to defy the Divine Law, his judgments will be just. Above all, if he believes that the consequences of his actions will follow him beyond his earthly life, and that "as he sows so must he reap," such a man will surely avoid injustice and tyranny.
>
> Should an official, on the contrary, think that all responsibility for his actions must end with his earthly life, knowing and believing nothing of Divine favors and a spiritual kingdom of joy, he will lack the incentive to just dealing, and the inspiration to destroy oppression and unrighteousness.
>
> When a ruler knows that his judgments will be weighed in a balance by the Divine Judge, and that if he be not found wanting he will come into the Celestial Kingdom and that the light of the Heavenly Bounty will shine upon him, then will he surely act with justice and equity. Behold how important it is that Ministers of State should be enlightened by religion!.

The Decline in the Fortunes of Royalty

And as we survey in other fields the decline in the fortunes of royalty, whether in the years immediately preceding the Great War or after, and contemplate the fate that has overtaken the Chinese Empire, the Portuguese and Spanish Monarchies, and more recently the vicissitudes that have afflicted, and are still afflicting, the sovereigns of Norway, of Denmark

---

[338] 'Abdu'l-Bahá, The Promulgation of Universal Peace. Emphasis added.

and of Holland, and observe the impotence of their fellow-sovereigns, and note the fear and trembling that has seized their thrones, may we not associate their plight with the opening passages of the Súriy-i-Mulúk, which, in view of their momentous significance, I feel impelled to quote a second time: *"Fear God, O concourse of kings, and suffer not yourselves to be deprived of this most sublime grace.... Set your hearts towards the face of God, and abandon that which your desires have bidden you to follow, and be not of those who perish.... Ye examined not His [the Báb's] Cause, when so to do had been better for you than all that the sun shineth upon, could ye but perceive it.... Beware that ye be not careless henceforth, as ye have been careless aforetime.... My face hath come forth from the veils, and shed its radiance upon all that is in heaven and on earth, and yet ye turned not towards Him.... Arise then ... and make ye amends for that which hath escaped you.... If ye pay no heed unto the counsels which, in peerless and unequivocal language, We have revealed in this Tablet, Divine chastisement shall assail you from every direction, and the sentence of His justice shall be pronounced against you.... Twenty years have passed, O kings, during which We have, each day, tasted the agony of a fresh tribulation.... Though aware of most of Our afflictions, ye, nevertheless, have failed to stay the hand of the aggressor. For is it not your clear duty to restrain the tyranny of the oppressor, and to deal equitably with your subjects, that your high sense of justice may be fully demonstrated to all mankind?"*.[339]

God, however, as has been pointed out in the very beginning of these pages, does not only punish the wrongdoings of His children. He chastises because He is just, and He chastens because He loves. Having chastened them, He cannot, in His great mercy, leave them to their fate. Indeed, by the very act of chastening them He prepares them for the mission for which He has created them. "My calamity is My providence," He, by the mouth of Bahá'u'lláh, has assured them, "outwardly it is fire and vengeance, but inwardly it is light and mercy".[340]

6.3.3 Grace

Bahá'u'lláh revealed:

"True reliance is for the servant to pursue his profession and calling in this world, to hold fast unto the Lord, to seek naught but His grace[341], inasmuch as in His Hands is the destiny of all His servants."[342]

Can one of sane mind ever seriously imagine that, in view of certain words the meaning of which he cannot comprehend, the portal of God's infinite guidance can ever be closed in the face of men? Can he ever conceive for these Divine Luminaries, these resplendent Lights either a beginning or an end? *What outpouring flood can compare with the stream of His all-embracing grace, and what blessing can excel the evidences of so great and pervasive a mercy? There can be no doubt whatever that if for one moment the tide of His mercy and grace were to be withheld from the world, it would completely perish.* For this reason, from

---

[339] Shoghi Effendi. The Promised Day is Come. Emphasis added.

[340] Idem. Emphasis added.

[341] Definition of grace: unmerited divine assistance given to humans for their regeneration or sanctification; a virtue coming from God; a state of sanctification enjoyed through divine assistance. Web. June 2020 <https://www.merriam-webster.com/dictionary/grace>.

[342] Bahá'u'lláh, Tablets of Bahá'u'lláh.

the beginning that hath no beginning the portals of Divine mercy have been flung open to the face of all created things, and the clouds of Truth will continue to the end that hath no end to rain on the soil of human capacity, reality and personality their favors and bounties. Such hath been God's method continued from everlasting to everlasting.[343]

I swear by the beauty of the Well-Beloved! This is the Mercy that hath encompassed the entire creation, *the Day whereon the grace of God hath permeated and pervaded all things*. The living waters of My mercy, O Ali, are fast pouring down, and Mine heart is melting with the heat of My tenderness and love. At no time have I been able to reconcile Myself to the afflictions befalling My loved ones, or to any trouble that could becloud the joy of their hearts.

Verily the Cause is great and the Lord is Merciful and Clement. *Trust in the Grace of Thy Lord, and be firm in love for Him who has created thee and made thee. The veils shall be removed, the shining lamp shall beam, the clouds shall be dispelled, the lights of the Sun of love shall appear on the horizons and God shall grant thy wishes and give thee the power of deeds*.[344]

Deal Thou, therefore, O my God, my Beloved, my supreme Desire, with Thy servants and with all that were created by Thee as would beseem Thy beauty and Thy greatness, and would be worthy of Thy generosity and gifts. Thou art, in truth, He Whose mercy hath encompassed all the worlds, *and Whose grace hath embraced all that dwell on earth and in heaven*. Who is there that hath cried after Thee, and whose prayer hath remained unanswered? Where is he to be found who hath reached forth towards Thee, and whom Thou hast failed to approach? Who is he that can claim to have fixed his gaze upon Thee, and toward whom the eye of Thy loving-kindness hath not been directed? I bear witness that Thou hadst turned toward Thy servants ere they had turned toward Thee, and hadst remembered them ere they had remembered Thee. *All grace is Thine, O Thou in Whose hand is the kingdom of Divine gifts and the source of every irrevocable decree*.

Send down, therefore, O my God, upon all that seek Thee that which will entirely strip them of all that pertaineth not unto Thee, and will draw them nigh unto Thy Self.

Assist them, by *Thy grace, to love Thee and to conform unto that which shall please Thee*. Grant, then, that they may go straight on in the path of Thy Cause, the path wherein have slipped the footsteps of the doubters among Thy people and the froward among Thy servants. Thou art, verily, the All-Powerful, the Almighty, the Most Great.[345]

O people! Fear God, and disbelieve not in Him Whose grace hath surrounded all things, Whose mercy hath pervaded the contingent world, and the sovereign potency of Whose Cause hath encompassed both your inner and your outer beings, both your beginning and your end. Stand ye in awe of the Lord, and be of them that act uprightly. Beware lest ye be accounted among those who allow the verses of their Lord to pass them by unheard and unrecognized; these, truly, are of the wayward.[346]

---

[343] Bahá'u'lláh, Gleanings from the Writings of Bahá'u'lláh, p. 68. Emphasis added.

[344] Bahá'u'lláh, Gleanings from the Writings of Bahá'u'lláh. Emphasis added.

[345] Bahá'u'lláh, Prayers and Meditations by Bahá'u'lláh. Emphasis added.

[346] Bahá'u'lláh, The Summons of the Lord of Hosts, Emphasis added.

'Abdu'l-Bahá is quoted in Bahai Education: "O Company of God! To each created thing, the Ancient Sovereignty hath portioned out its own perfection, its particular virtue and special excellence, so that each in its degree may become a symbol denoting the sublimity of the true Educator of humankind, and that each, even as a crystalline mirror, may tell of the grace and splendour of the Sun of Truth."[347]

6.3.4 Our Destiny is Fixed, and the Whole Planet's too

In *The Summons of the Lord of Hosts*, Bahá'u'lláh revealed:

> I swear by Him Who hath fashioned Me from the light of His own Beauty! None have I ever seen that surpasseth you in heedlessness or exceedeth you in blindness. Ye seek to prove your faith in God through such holy Tablets as ye possess, yet when the verses of God were revealed and His Lamp was lighted, ye disbelieved in Him Whose very Pen *hath fixed the destinies of all things* in the Preserved Tablet.[348]

In "The Call of the Divine Beloved" by Bahá'u'lláh, we find:

> It is clear and evident that, in this Dispensation wherein the banner of utterance hath been raised aloft and the candle of discernment hath been lit, there is no Lord but the Exalted One. He it is Who is one in His essence and one in His attributes, single in the kingdom of names and peerless in the realm of actions. It is by virtue of His blessed name that the seas of Divine Unity have been made to surge; it is through the power of His resistless command that *the immutable decrees of destiny have been enforced*; it is through the potency of His sovereign might that *the dictates of fate have been fixed*. Who hath the power to soar in that exalted atmosphere or to cherish another beloved than Him? We all abide beneath His shadow and seek our portion from the ocean of His grace. However far the gnat may fly, it can never traverse the length and breadth of heaven, and however high the sparrow may soar, it can never attain the tree of immortality.[349]

Shoghi Effendi, in *The Promised Day is Come*, dated March 1941, was written for the Bahá'ís of the West, amid the Second World War, said:

> To the general character, the implications and features of this world commonwealth, *destined* to emerge, sooner or later, out of the carnage, agony, and havoc of this great world convulsion, I have already referred in my previous communications.

> Suffice it to say that this consummation will, by its very nature, be a gradual process, and must, as Bahá'u'lláh has Himself anticipated, lead at first to the establishment of that Lesser Peace which the nations of the earth, as yet unconscious of His Revelation and yet unwittingly enforcing the general principles which He has enunciated, will themselves establish. This momentous and historic step, involving the reconstruction of mankind, as the result of the universal recognition of its oneness and wholeness, will bring in its wake the spiritualization of the masses, consequent to the recognition of the character, and the acknowledgment of the claims, of the Faith of Bahá'u'lláh— the essential condition to that ultimate fusion of all

---

[347] 'Abdu'l-Bahá, Bahá'u'lláh, Shoghi Effendi. Bahai Education.

[348] Bahá'u'lláh, The Summons of the Lord of Hosts. Emphasis added.

[349] Bahá'u'lláh, The Call of the Divine Beloved. Emphasis added.

races, creeds, classes, and nations which must signalize the emergence of His New World Order.

Then will the coming of age of the entire human race be proclaimed and celebrated by all the peoples and nations of the earth. Then will the banner of the Most Great Peace be hoisted. Then will the worldwide sovereignty of Bahá'u'lláh—the Establisher of the Kingdom of the Father foretold by the Son, and anticipated by the Prophets of God before Him and after Him—be recognized, acclaimed, and firmly established.

Then will a world civilization be born, flourish, and perpetuate itself, a civilization with a fullness of life such as the world has never seen nor can as yet conceive. Then will the Everlasting Covenant be fulfilled in its completeness. Then will the promise enshrined in all the Books of God be redeemed, and all the prophecies uttered by the Prophets of old come to pass, and the vision of seers and poets be realized. Then will the planet, galvanized through the universal belief of its dwellers in one God, and their allegiance to one common Revelation, mirror, within the limitations imposed upon it, the effulgent glories of the sovereignty of Bahá'u'lláh, shining in the plenitude of its splendor in the Abhá Paradise, and be made the footstool of His Throne on high, and acclaimed as the earthly heaven, *capable of fulfilling that ineffable destiny fixed for it, from time immemorial, by the love and wisdom of its Creator.*[350]

6.3.5 Are we supposed to reach a high destiny?

Stanwood Cobb, in *The Life Beyond*, shares the following condition to encourage change:

It is impossible to consider this life apart from the future life. It is all one great whole. The thought of what is to come after death is not only a great comfort in times of earthly stress and suffering, but is also a powerful influence toward right conduct in this life.

'Abdu'l-Bahá has said that without this vision of the next life there cannot be enough incentive to ethical action here. The rewards and punishments which are assigned here for our actions are as nothing to the more important results of our earthly deeds which come to us in the hereafter.[351]

Someone said that 'Abdu'l-Bahá expressed:

"Divine perfection is infinite; therefore, the progress of the soul is also infinite. From the very birth of a human being the soul progresses, the intellect grows and knowledge increases. When the body dies, the soul lives on. All the differing degrees of created physical beings are limited, but the soul is limitless!

In all religions the belief exists that the soul survives the death of the body. Intercessions are sent up for the beloved dead, prayers are said for their progress and for the forgiveness of their sins. If the soul perished with the body all this would have no meaning. Further, if it were not possible for the soul to advance towards perfection after it had been released from the body, of what avail are all these loving prayers, of devotion?

We read in the sacred writings that "all good works are found again." Now, if the soul did not survive, this also would mean nothing!

---

[350] Shoghi Effendi, The Promised Day is Come. Emphasis added.

[351] Stanwood Cobb, The life beyond. February 1924. Star of the West - 8, p. 330.

The very fact that our spiritual instinct, surely never given in vain, prompts us to pray for the welfare of those, our loved ones, who have passed out of the material world: does it not bear witness to the continuance of their existence?

"In the world of the spirit there is no retrogression. The world of mortality is a world of contradictions, of opposites; motion being compulsory, everything must either go forward or retreat. *In the realm of spirit there is no retreat possible; all movement is bound to be toward a perfect state. Progress is the expression of spirit in the world of matter. The intelligence of man, his reasoning powers, his knowledge, his scientific achievements, all these being manifestations of the spirit, partake of the inevitable law of spiritual progress and are, therefore, of necessity, immortal.*"

"My hope for you is that you will progress in the world of spirit, as well as in the world of matter, that your intelligence will develop, your knowledge will augment, and your understanding be widened".[352]

In the following quotes, let us discern and try to link our predestination, our irrevocable and impending fate, and the forging of our destiny. In Gleanings, He revealed:

> O thou who art the fruit of My Tree and the leaf thereof! On thee be My glory and My mercy. Let not thine heart grieve over what hath befallen thee. Wert thou to scan the pages of the Book of Life, thou wouldst, most certainly, discover that which would dissipate thy sorrows and dissolve thine anguish.
>
> Know thou, O fruit of My Tree, that the decrees of the Sovereign Ordainer, as related to fate and predestination, are of two kinds. Both are to be obeyed and accepted. The one is irrevocable[353], the other is, as termed by men, impending.[354] *To the former all must unreservedly submit, inasmuch as it is fixed and settled. God, however, is able to alter or repeal it. As the harm that must result from such a change will be greater than if the decree had remained unaltered, all, therefore, should willingly acquiesce in what God hath willed and confidently abide by the same.*
>
> The decree that is impending, however, is such that prayer and entreaty can succeed in averting it.
>
> God grant that thou who art the fruit of My Tree, and they that are associated with thee, may be shielded from its evil consequences.[355]

'Abdu'l-Bahá is reported to have said:

> God in his wisdom has created all things. Nothing has been created without a special destiny, for every creature has an innate station of attainment. This flower has been created to mirror forth a harmonious ensemble of color and perfume. Each kingdom of nature holds potentialities and each must be cultivated in order to reach its fulfillment. The divine teachers desire man to be educated that he may attain to the high rank of his own reality, the deprivation of which is the rank of perdition. The flower needs light that it may achieve its

---

[352] Paris Talks, p. 89. Emphasis added.

[353] Definition on Merriam Webster Dictionary: not possible to revoke. Inalterable: not capable of being altered or changed.

[354] Definition on Merriam Webster Dictionary: occurring or likely to occur soon.

[355] Bahá'u'lláh, Gleanings from the Writings of Bahá'u'lláh. Emphasis added.

fruitage; man needs the light of the Holy Spirit, and the measure of illumination throughout creation is proportionate to the different kingdoms.

When we come to the estate of man, we find his kingdom is vested with a divine superiority. Compared to the animal, his perfection or his imperfection is superior. In comparison with man the perfection of a flower is insignificant. Yet if man remain content in an undeveloped state viewed from the point of capacity he is the lowest of creatures. If he attains unto his heritage through divine wisdom, then he becomes a clear mirror in which the beauty of God is reflected; he has eternal life and becomes a participator of the sun of truth. This is to show you how considerable are the degrees of human achievement.

The aim of the prophet of God is to raise man to the degree of knowledge of his potentiality and to illumine him through the light of the kingdom, to transform ignorance into wisdom, injustice into justice, error into knowledge, cruelty into affection and incapability into progress. In short, to make all the attainments of existence resplendent in him.[356]

Bahá'u'lláh revealed:

The purpose underlying the revelation of every heavenly Book, nay, of every divinely-revealed verse, is to endue all men with righteousness and understanding, so that peace and tranquillity may be firmly established amongst them. Whatsoever instilleth assurance into the hearts of men, whatsoever exalteth their station or promoteth their contentment, is acceptable in the sight of God. *How lofty is the station which man, if he but choose to fulfill his high destiny, can attain!* To what depths of degradation he can sink, depths which the meanest of creatures have never reached! Seize, O friends, the chance which this Day offereth you, and deprive not yourselves of the liberal effusions of His grace. I beseech God that He may graciously enable every one of you to adorn himself, in this blessed Day, with the ornament of pure and holy deeds. He, verily, doeth whatsoever He willeth.[357]

In conclusion, we are all predestined to a higher destiny, but it is a matter of choices in the exercising of our free will.

6.3.6 The inner being:

Let us learn about the "fire within" from the wisdom of an indigenous Elder, Lillian Pitawanakwat:

Pitawanakwat recalls that as a child, her father would ask at the end of the day, "My daughter, how is your fire burning?" In recalling the events of the day, she would reflect on whether she had been offensive to anyone, or whether or not she had been offended. This was an important part of nurturing the fire within as children were taught to let go of any distractions of the day and make peace within ourselves in order to nurture and maintain that inner fire.

The story of the Rose, as told by Pitawanakwat (2006), serves as a reminder of the value of nurturance and the essence of life. According to this story, the Creator asked the flower people, "Who among you will bring a reminder to the two-legged about the essence of life?" The buttercup offered but the Creator refused on the basis that the buttercup was 'too bright.' All of the flowers offered their help but were refused. The rose finally offered, stating

---

[356] 'Abdu'l-Bahá, Divine Philosophy. Emphasis added.

[357] Bahá'u'lláh, Gleanings from the Writings of Bahá'u'lláh. Emphasis added.

"Let me remind them with my essence, so that in times of sadness, and in times of joy, they will remember how to be kind to themselves." So the Creator, planted the seed of the rose and as it grew a little, it sprouted very, very sharp little thorns and eventually it bloomed into a full rose. This teaching reminds us that life is like a rose with the thorns representing life's journey; the experiences that make us who we are and the rose representing the many times in life when we decay and die only to bounce back again through reflection, meditation, awareness, acceptance and surrender.[358]

There are many quotes about the inner being, the author selected and added emphasis to a few to reflect on them:

O servants! Eyes are needed if one is to see, and ears, if one is to hear. Whoso in this blessed Day hath not heard the divine call hath indeed no ear. By this is not meant that bodily ear that is perceived by the eye. Open your *inner eye*, that ye may behold the celestial *Fire*, and listen with the *ear of inner understanding*, that ye may hear the delightsome words of the Beloved.[359]

As this physical frame is the throne of the *inner temple*, whatever occurs to the former is felt by the latter. In reality that which takes delight in joy or is saddened by pain is the *inner temple* of the body, not the body itself. Since this physical body is the throne whereon the inner temple is established, God hath ordained that the body be preserved to the extent possible, so that nothing that causeth repugnance may be experienced. The *inner temple* beholdeth its physical frame, which is its throne. Thus, if the latter is accorded respect, it is as if the former is the recipient. The converse is likewise true.

Therefore, it hath been ordained that the *dead body* should be treated with the utmost honor and respect.[360]

"O friends of God! Incline your *inner ears* to the voice of the peerless and self-subsisting Lord, that He may deliver you from the bonds of entanglement and the depths of darkness and enable you to attain the *eternal light*."[361]

O people! *Fear God*, and disbelieve not in Him Whose *grace* hath surrounded all things, Whose mercy hath pervaded the contingent world, and the sovereign potency of Whose Cause hath encompassed both *your inner and your outer beings*, both your beginning and your end. Stand ye in awe of the Lord, and be of them that act uprightly. Beware lest ye be accounted among those who allow the verses of their Lord to pass them by unheard and unrecognized; these, truly, are of the wayward.[362]

Some of the great questions unfolding from the rays of the Sun of Reality upon the mind of man are: the problem of the reality of the spirit of man; of the birth of the spirit; of its birth from this world into the world of God; the question of the *inner life of the spirit and of its fate after its ascension from the body*.

They also *meditate* upon the scientific questions of the day, and these are likewise solved.

---

[358] Lillian Pitawanakwat, The Medicine Wheel Teachings.

[359] Bahá'u'lláh, The Tabernacle of Unity, Two Other Tablets. Emphasis added.

[360] The Báb, Selections from the Writings of the Báb. Emphasis added.

[361] Idem. Emphasis added.

[362] Bahá'u'lláh, The Summons of the Lord of Hosts. Emphasis added.

These people, who are called "Followers of the *inner light*," attain to a superlative degree of power, and are entirely freed from *blind dogmas and imitations*. Men rely on the statements of these people: by themselves—within themselves—they solve all mysteries.

If they find a solution with the assistance of the *inner light*, they accept it, and afterwards they declare it: otherwise they would consider it a matter of *blind imitation*. They go so far as to reflect upon the essential nature of the Divinity, of the Divine revelation, of the manifestation of the Deity in this world. All the divine and scientific questions are solved by them through the power of the spirit.

Bahá'u'lláh says there is a sign (from God) in every phenomenon: *the sign of the intellect is contemplation and the sign of contemplation is silence*, because it is impossible for a man to do two things at one time—he cannot both speak and meditate.

It is an axiomatic fact that while you *meditate* you are speaking with your own spirit. In that state of mind you put certain questions to your spirit and the spirit answers: the light breaks forth and the reality is revealed.

You cannot apply the name "man" to any being void of this faculty of *meditation*; without it he would be a mere animal, lower than the beasts.

Through the faculty of *meditation* man attains to *eternal life*; through it he receives the breath of the Holy Spirit—the bestowal of the Spirit is given in *reflection* and *meditation*.

The spirit of man is itself informed and strengthened during *meditation*; through it affairs of which man knew nothing are unfolded before his view. Through it he receives Divine inspiration, through it he receives heavenly food.

Meditation is the key for opening the doors of mysteries. In that state man abstracts himself: in that state man withdraws himself from all outside objects; in that subjective mood he is immersed in the ocean of spiritual life and can unfold the secrets of things-in-themselves. To illustrate this, think of man as endowed with two kinds of sight; when the power of insight is being used the outward power of vision does not see.

This faculty of *meditation* frees man from the animal nature, discerns the reality of things, puts man in touch with God.[363]

Tread, therefore, the path of acquiescence and resignation. Let no hardship sadden thy heart, nor set thy hope upon any worldly gifts. Be happy and content with whatsoever God hath willed, that thy heart and soul may find tranquillity and t*hine inner being and conscience* may experience true joy. Erelong shall this hardship and tribulation pass away and inner peace and joy be attained.[364]

Humanity, through suffering and turmoil, is swiftly moving on towards its *destiny*; if we be loiterers, if we fail to play our part surely others will be called upon to take up our task as ministers to the crying needs of this afflicted world. Not by the force of numbers, not by the mere exposition of a set of new and noble principles, not by an organized campaign of teaching—no matter how worldwide and elaborate in its character—not even by the staunchness of our faith or the exaltation of our enthusiasm, can we ultimately hope to vindicate in the eyes of a critical and sceptical age the supreme claim of the Abhá Revelation.

---

[363] 'Abdu'l-Bahá, Paris Talks. Emphasis added.

[364] Compilation for the 2018 Counsellors' Conference, From a Tablet of 'Abdu'l-Bahá—translated from the Persian. Emphasis added.

> One thing and only one thing will unfailingly and alone secure the undoubted triumph of this sacred Cause, namely, the extent to which our *own inner life* and private character mirror forth in their manifold aspects the splendor of those eternal principles proclaimed by Bahá'u'lláh.[365]

> Against the conspicuous signs of moral decadence which daily is corroding the foundations of civilized life, these graphic words of Bahá'u'lláh assume an acute urgency: "The vitality of men's belief in God is dying out in every land; nothing short of His wholesome medicine can ever restore it. The corrosion of ungodliness is eating into the vitals of human society; what else but the Elixir of His potent Revelation can cleanse and revive it?" Such words have particular implications for the actions of anyone who has recognized the Lord of the Age. A crucial consequence of this recognition is a belief that impels acceptance of His commandments. Depth of belief is assured by the *inner transformation*, that salutary acquisition of spiritual and moral character, which is the outcome of obedience to the divine laws and principles.

> Towards this end the release of the annotated Kitáb-i-Aqdas in English, and its anticipated early publication in other major languages, provide a mighty infusion of divine guidance for realizing the vitality of faith which is essential to the spiritual well- being and happiness of individuals and the strengthening of the fabric of the community. No less essential to nourishing this vitality is the cultivation of a sense of spirituality, that mystic feeling which unites the individual with God and is achieved through *meditation* and *prayer*.[366]

6.3.7 Transformation

To start this discussion, let me clarify this cause relationship of individuals, organizations, and institutions with self-accountability of faults, transgressions, omissions, wrongdoings, carelessness, failing, oversight, inadvertence, neglect, and offenses. It becomes a powerful tool to reach our mission, vision, goals, and actions. It also becomes a fortress of improvement supporting all choices made during the propositional phase of the cycle.

Our call to reach a higher destiny implies a process of transmutation[367] to get there. Let us study some quotes. In *Gleanings from the Writings of Bahá'u'lláh*, we find:

> The vitality of men's belief in God is dying out in every land; nothing short of His wholesome medicine can ever restore it. The corrosion of ungodliness is eating into the vitals of human society; what else but the Elixir of His potent Revelation can cleanse and revive it? *Is it within human power, O Ḥakím, to effect in the constituent elements of any of the minute and indivisible particles of matter so complete a transformation as to transmute it into purest gold?* Perplexing and difficult as this may appear, the still greater task of converting satanic strength into heavenly power is one that We have been empowered to accomplish. The Force capable of such a transformation transcendeth the potency of the Elixir itself. The Word

---

[365] Shoghi Effendi, Bahá'í Administration. Emphasis added.

[366] The Universal House of Justice. Riḍván 150 – To the Bahá'ís of the World.

[367] Merriam-Webster dictionary definition of transmutation: a) the conversion of base metals into gold or silver b) the conversion of one element or nuclide into another either naturally or artificially. Transmutation: a) refers to the change of one substance into another. b) the alteration of one species into another. Wikipedia, Transmutation.

of God, alone, can claim the distinction of being endowed with the capacity required for so great and far-reaching a change.[368]

Bahá'u'lláh, in *The Kitáb-i-Íqán The Book of Certitude*, revealed:

Behold, how, notwithstanding these and similar traditions, they idly contend that the laws formerly revealed, must in no wise be altered. *And yet, is not the object of every Revelation to effect a transformation[369] in the whole character of mankind, a transformation that shall manifest itself both outwardly and inwardly, that shall affect both its inner life and external conditions?* For if the character of mankind be not changed, the futility of God's universal Manifestations would be apparent.[370]

The Universal House of Justice said:

"What the friends need to remember in this respect is that, in their efforts to achieve personal growth and to uphold Bahá'í ideals, they are not isolated individuals, withstanding alone the onslaught of the forces of moral decay operating in society.

They are members of a purposeful community, global in scope, pursuing a bold spiritual mission — working to establish a pattern of activity and administrative structures suited to a humanity entering its age of maturity. Giving shape to the community's efforts is a framework for action defined by the global Plans of the Faith.

This framework promotes the transformation of the individual in conjunction with social transformation, as two inseparable processes. Specifically, the courses of the institute are intended to set the individual on a path in which qualities and attitudes, skills and abilities, are gradually acquired through service — service intended to quell the insistent self, helping to lift the individual out of its confines and placing him or her in a dynamic process of community building.

In this context, then, every individual finds himself or herself immersed in a community that serves increasingly as an environment conducive to the cultivation of those attributes that are to distinguish a Bahá'í life — an environment in which a spirit of unity animates one and all; in which the ties of fellowship bind them; in which mistakes are treated with tolerance and fear of failure is diminished; in which criticism of others is avoided and backbiting and gossip give way to mutual support and encouragement; in which young and old work shoulder to shoulder, studying the Creative Word together and accompanying one another in their efforts to serve; in which children are reared through an educational process that strives to sharpen their spiritual faculties and imbue them with the spirit of the Faith; in which young people are helped to detect the false messages spread by society, recognize its fruitless preoccupations, and resist its pressures, directing their energies instead towards its betterment. The institutions of the Faith, for their part, strive to ensure that such an environment is fostered. They do not pry into the personal lives of individuals. Nor are they vindictive and judgemental, eager to punish those who fall short of the Bahá'í standard.[371]

---

[368] Gleanings from the Writings of Bahá'u'lláh. Emphasis added.

[369] In the Merriam Webster dictionary, we find the following definition of transformation: a) to change in composition or structure b) to change the outward form or appearance of c) to change in character or condition.

[370] Bahá'u'lláh, The Kitáb-i-Íqán The Book of Certitude. Emphasis added.

[371] The Universal House of Justice. Challenges for Bahá'í Youth in a Western Way of Life.

A fundamental teaching of the Bahá'í Faith is the equality of the sexes. The question of sex and gender is addressed on two fronts. First, the need to abolish culturally produced differences between men and women as one path toward equality because sex stereotyping is spiritually unhealthy. Secondly, in order to do this we must challenge the theoretical inconsistencies inherent in our Bahá'í discussions. Failure to do so has reinforced male supremacist ideology. *Individual inner transformation must precede the outward expression of sexual justice.*[372]

Gandhimohan, in Mahatma Gandhi and the Bahá'ís: Striving towards a Nonviolent Civilization, says:

"Man is the supreme Talisman. Lack of a proper education hath, however, deprived him of that which he doth inherently possess... The Great Being saith: Regard man as a mine rich in gems of inestimable value. Education can, alone, cause it to reveal its treasures, and enable mankind to benefit therefrom..."[373]

Education as referred to here is not identical to book learning. Esslemont explains:

At present a really well educated man is the rarest of phenomena, for nearly everyone has false prejudices, wrong ideals, erroneous conceptions and bad habits drilled into him from babyhood. How few are taught from their earliest childhood to love God with all their hearts and dedicate their lives to Him; to regard service to humanity as the highest aim in life; to develop their powers to the best advantage for the general good of all! Yet surely these are the essential elements of a good education. Mere cramming of the memory with facts about arithmetic, grammar, geography, languages, etc., has comparatively little effect in producing noble and useful lives.

Similarly, Gandhi held the view that education is not the mere training of the intellect, but rather that it is a process with the potential to transform our very essence, by helping to bring out in the individual the finest attributes of humanity:

- *Real education consists in drawing the best out of yourself.*
- *By education I mean an all-round drawing out of the best in child and man — body, mind, and spirit.*[374]

Let us now reflect on what is the Destiny of America in the words of 'Abdu'l-Bahá, and the degree of transformation required to reach its destiny:

One more word in conclusion. Among some of the most momentous and thought-provoking pronouncements ever made by 'Abdu'l-Bahá, in the course of His epoch-making travels in the North American continent, are the following: "May this American Democracy be the first nation to establish the foundation of international agreement. May it be the first nation to proclaim the unity of mankind. May it be the first to unfurl the Standard of the Most Great Peace." And again: "The American people are indeed worthy of being the first to build the Tabernacle of the Great Peace, and proclaim the oneness of mankind…. For America hath

---

Emphasis added.

[372] Ta'eed, Lata. Sex, Gender, and New Age Stereotyping. Emphasis added.

[373] Bahá'u'lláh, Gleanings from the Writings of Bahá'u'lláh.

[374] Gandhimohan, M. V., Mahatma Gandhi and the Bahá'ís: Striving towards a Nonviolent Civilization. Emphasis added.

developed powers and capacities greater and more wonderful than other nations.... The American nation is equipped and empowered to accomplish that which will adorn the pages of history, to become the envy of the world, and be blest in both the East and the West for the triumph of its people. The American continent gives signs and evidences of very great advancement. Its future is even more promising, for its influence and illumination are far-reaching. It will lead all nations spiritually."[375]

The American Bahá'í Community, the leaven *destined* to leaven the whole, cannot hope, at this critical juncture in the fortunes of a struggling, perilously situated, spiritually moribund nation, to either escape the trials with which this nation is confronted, nor claim to be wholly immune from the evils that stain its character. More pointed critiques were sometimes made. The weeding out of negative moral and cultural traits was necessary if the two North American nations were to fulfil their *high destiny*.[376]

6.3.8 How to find the root cause of the problem?

"O SON OF THE SUPREME!

To the eternal I call thee, yet thou dost seek that which perisheth. What hath made thee turn away from Our desire and seek thine own?"[377]

Following the advice of the Hidden Word mentioned above is the first step to finding the root cause of our wrong behavior.

Destiny and the development of the capacity to figure out the root cause of our imperfections:

Definition of root cause:

"A root cause is an initiating cause of either a condition or a causal chain that leads to an outcome or effect of interest. The term denotes the earliest, most basic, 'deepest', cause for a given behavior; most often a fault. The idea is that you can only see an error by its manifest signs. Those signs can be widespread, multitudinous, and convoluted, whereas the root cause leading to them often is a lot simpler."[378]

Another reflection about the definition of the root cause is the following:

Despite the dictionary definition, there is no standard definition for the term root cause. Why? Because if asked, most people have a definition. But if you ask ten different people, you get ten different definitions!

Therefore, a basic problem faced by people performing root cause analysis is that...

People don't even agree on the definition of "root cause.

In 1985 the **Standard Root cause Definition was:** The most basic cause (or causes) that can reasonably be identified that management has control to fix.

---

[375] Shogui Effendi, The Advent of Divine Justice. Emphasis added.

[376] Jack McLean, The Art of Rhetoric in the Writings of Shoghi Effendi. Emphasis added.

[377] Bahá'u'lláh, The Hidden Words of Bahá'u'lláh. Emphasis Added.

[378] Wikipedia, Root Cause.

Allowing for multiple root causes stops arguments over which cause is the "rootiest" of the root causes. Any cause for a problem that fits the definition is one of the problem's root causes.

In the early 1990s the TapRooT® System suggested: "The most basic cause (or causes) that can reasonably be identified that management has control to fix and, when fixed, will prevent (or significantly reduce the likelihood of) the problem's recurrence.

In 2006 they suggested a modern definition: The absence of a best practice or the failure to apply knowledge that would have prevented the problem.[379]

When studying this Cause, we will understand how critical it is for individuals and institutions to search for the root cause of a problem and provide or recommend the remedy. The author believes that this method can meaningfully contribute to "the correction of criminal and disorderly elements", because of the incentive of reaching a higher destiny searching for the root cause of the behavior. In crime, what the justice system looks for is the motive, such as greed, jealousy, revenge, hate, a mental illness, or an addiction. A mental illness such as being anxious or suffering from Post Traumatic Stress Disorder may be the motives of crime, but the root causes are something else.

Beth E. Maultsby, in *High Conflict Family Law Matters and Personality Disorders*, says:

> The majority of the cases in family law courts that require court intervention involve an intense, high level of conflict. In almost every high conflict case, you will find at least one party who has a high conflict personality that is driving the litigation train.
>
> Unfortunately, the high conflict personalities that most often appear in family law cases are often not recognized by the professionals involved in the case.
>
> And, even if the high conflict personality is recognized, many professionals fail to understand how to manage the high conflict case and implement the actions needed to stop or at least minimize any further destruction by the high conflict driven client. Instead, attorneys, judges, mental health professionals and mediators have a tendency to handle the case from the viewpoint of how clients ought to behave rather than understanding how it really is. As professionals, we often want the parties to just find a way to get along. However, because the high level of conflict observed in a family law case is often an extension of the conflict that existed in the marriage that was driven by a party with a high conflict personality, it is unrealistic for any of the professionals to think the conflict is going to stop because the parties have separated and a divorce action has been filed. In fact, many, if not most, the high conflict people will stay in the high conflict cycle their entire life.
>
> Frequently, professionals involved in high conflict cases often make the situation worse by not understanding that there are personality traits and characteristics that make individuals refuse to disengage and get along. It is, therefore, nearly impossible to effectively stop or slow down the high conflict train if all of the professionals involved in the case do not take the time to educate themselves and develop a coordinated plan of action to deal with these families. By failing to be able to identify, manage, represent, and treat the real problem in a high conflict case, attorneys, judges, mental health professionals, and mediators contribute

---

[379] Paradies, Mark, Definition of a Root Cause.

to perpetuating high conflict cases and litigation. The end result is often irreparable damage to families.[380]

The Mayo Clinic describes a good number of personality disorders:

> **Cluster A** personality disorders are characterized by odd, eccentric thinking or behavior. They include paranoid personality disorder, schizoid personality disorder and schizotypal personality disorder.
>
> **Cluster B** personality disorders are characterized by dramatic, overly emotional or unpredictable thinking or behavior. They include antisocial personality disorder, borderline personality disorder, histrionic personality disorder and narcissistic personality disorder.
>
> **Cluster C** personality disorders are characterized by anxious, fearful thinking or behavior. They include avoidant personality disorder, dependent personality disorder and obsessive-compulsive personality disorder.[381]

But, when discussing the causes, it says:

> Although the precise cause of personality disorders is not known, certain factors seem to increase the risk of developing or triggering personality disorders, including:
>
> - Family history of personality disorders or other mental illness
> - Abusive, unstable or chaotic family life during childhood
> - Being diagnosed with childhood conduct disorder
> - Variations in brain chemistry and structure[382]

How can we prevent getting in a business relationship, a marriage, or even voting for someone with a personality disorder to the highest government position? Let us deepen a little bit more on this situation by focusing our discussion on the "Narcissist personality disorder":

- Belief that you're special and more important than others
- Fantasies about power, success, and attractiveness
- Failure to recognize others' needs and feelings
- Exaggeration of achievements or talents
- The expectation of constant praise and admiration
- Arrogance
- Unreasonable expectations of favors and advantages, often taking advantage of others
- Envy of others or belief that others envy you"[383]

"Signs and symptoms of narcissistic personality disorder and the severity of symptoms vary. People with the disorder can:

---

[380] Beth E. Maultsby, High Conflict Family Law Matters and Personality Disorders.

[381] Mayo Clinic, Personality Disorders.

[382] Idem.

[383] Idem.

- Have an exaggerated sense of self-importance
- Have a sense of entitlement and require constant, excessive admiration
- Expect to be recognized as superior even without achievements that warrant it
- Exaggerate achievements and talents
- Be preoccupied with fantasies about success, power, brilliance, beauty or the perfect mate
- Believe they are superior and can only associate with equally special people
- Monopolize conversations and belittle or look down on people they perceive as inferior
- Expect special favors and unquestioning compliance with their expectations
- Take advantage of others to get what they want
- Have an inability or unwillingness to recognize the needs and feelings of others
- Be envious of others and believe others envy them
- Behave in an arrogant or haughty manner, coming across as conceited, boastful and pretentious
- Insist on having the best of everything — for instance, the best car or office

At the same time, people with narcissistic personality disorder have trouble handling anything they perceive as criticism, and they can:

- Become impatient or angry when they don't receive special treatment
- Have significant interpersonal problems and easily feel slighted
- React with rage or contempt and try to belittle the other person to make themselves appear superior
- Have difficulty regulating emotions and behavior
- Experience major problems dealing with stress and adapting to change
- Feel depressed and moody because they fall short of perfection
- Have secret feelings of insecurity, shame, vulnerability and humiliation".[384]

All of these are just manifestations of the disorder. But the root cause is not established yet.

"It's not known what causes narcissistic personality disorder. As with personality development and other mental health disorders, the cause of narcissistic personality disorder is likely complex. Narcissistic personality disorder may be linked to:

- Environment — mismatches in parent-child relationships with either excessive adoration or excessive criticism that is poorly attuned to the child's experience
- Genetics — inherited characteristics
- Neurobiology — the connection between the brain and behavior and thinking."[385]

---

[384] Mayo Clinic. Narcissistic personality disorder.

[385] Idem.

What would be the destiny of a narcissist individual or with any other mental health disorder? What level of courage is required by the individual to seriously confront the root cause of his (her) conduct?

Cluster A, B and C personality disorders are each associated with different types of offenses. Although rates of personality disorder are high in all serious offenders, the role played by personality disorder may be greater in some offences than others, for example, in rapists compared with child molesters, men who kill their fathers rather than their mothers, men who kill their children compared with mothers who kill their children; and in less severe stalking behaviour compared with those who get convictions.[386]

In another study, *The Relationship between Personality Disorders and the Type of Crime Committed and Substance Used among Prisoners*, we learn:

> The results showed that 87.3% of women and 83.3% of men had a personality disorder at the time of committing the crime. Moreover, 46.5% of the target population had developed substance dependence at the time of committing the crime. The highest percentage of substance abuse in both women and men was related to opium, especially in the age group of 18-28 years. The highest rates of mental disorders were related to major depressive disorder (MDD), dependent personality disorder (DPD), borderline personality disorder (BPD), and antisocial personality disorder (ASPD), respectively.[387]

If we start a relationship to get into a business or marriage, which could be the steps to identify possible personality disorders to prevent entanglements and frustrations that may last for decades? One piece of advice: in a quote from a letter written on behalf of the Guardian to an individual believer, May 12, 1925, in "Living the Life", we find: "On no subject are the Bahá'í teachings more emphatic than on the necessity to abstain from faultfinding and backbiting while being ever eager to discover and root out our own faults and overcome our own failings".[388]

How can we handle the situation, obeying the Guardian but also obeying 'Abdu'l-Bahá:

> "As for the question regarding marriage under the Law of God: first thou must choose one who is pleasing to thee, and then the matter is subject to the consent of father and mother. Before thou makest thy choice, they have no right to interfere.
>
> Bahá'í marriage is the commitment of the two parties one to the other, and their mutual attachment of mind and heart. *Each must, however, exercise the utmost care to become thoroughly acquainted with the character of the other*, that the binding covenant between them may be a tie that will endure forever".[389]

The following quote is very pertinent when dealing with this issue:

---

[386] Sophie Davison and Aleksandar Janca, Personality disorder and criminal behaviour: what is the nature of the relationship?

[387] Shahin Fakhrzadegan, et al., The Relationship between Personality Disorders and the Type of Crime Committed and Substance Used among Prisoners.

[388] 'Abdu'l-Bahá, Bahá'u'lláh, Shoghi Effendi, Living the Life. From a letter 12 May 1925 written on behalf of Shoghi Effendi to an individual believer, p. 3.

[389] 'Abdu'l Bahá, Selections from the Writings of 'Abdu'l Bahá. Emphasis added.

The distinction between the purpose of consultation and therapeutic endeavours is made explicit in the following extracts from letters written by or on behalf of the Universal House of Justice: "It should be borne in mind that all consultation is aimed at arriving at a solution to a problem and is quite different from the sort of group baring of the soul that is popular in some circles these days and which borders on the kind of confession that is forbidden in the Faith."[390]

The following quote is from 'Abdu'l-Bahá:

> ... the conscience of man is sacred and to be respected; and that liberty thereof produces widening of ideas, amendment of morals, improvement of conduct, disclosure of the secrets of creation, and manifestation of the hidden verities of the contingent world. Moreover, if interrogation of conscience, which is one of the private possessions of the heart and the soul, take place in this world, what further recompense remains for man in the court of divine justice at the day of general resurrection? Convictions and ideas are within the scope of the comprehension of the King of kings, not of kings; and soul and conscience are between the fingers of control of the Lord of hearts, not of [His] servants.[391]

Our commitment is to help the individual to become interested in reaching his (her) high destiny. Identifying the problem and looking for the root cause is exclusively his (hers). However, if the individual invites you to consult about a solution to his(her) personality disorder, there is no problem if you accept it after careful consideration.

6.3.9 Examples of root causes:

Bahá'u'lláh revealed:

> "The source of all evil is for man to turn away from his Lord and set his heart on things ungodly.
>
> The essence of abasement is to pass out from under the shadow of the Merciful and seek the shelter of the Evil One".[392]

Abdu'l-Bahá, in The Promulgation of Universal Peace, taught: "The prophets of God have founded the laws of divine civilization. They have been the root and fundamental source of all knowledge".[393]

The Master, in Some Answered Questions, guides us:

> Experience has shown that crime is less prevalent among civilized peoples—that is, among those who have acquired true civilization. And true civilization is divine civilization, the civilization of those who combine material and spiritual perfections. *As ignorance is the root*

---

[390] From a letter dated 19 March 1973 from the Universal House of Justice to a National Spiritual Assembly, published in "Consultation: A Compilation" (Wilmette: Bahá'í Publishing Trust), p. 22 [Ed. - sel. 45]. Shoghi Effendi, Universal House of Justice, Research Department of the Universal House of Justice, "Community Functioning, Issues Concerning: Fostering the Development of Bahá'í Communities". Web. June 2021 <https://bahai library.com/uhj_issues_community_functioning>.

[391] 'Abdu'l-Bahá, A Traveller's Narrative, p. 91.

[392] Bahá'u'lláh, Tablets of Bahá'u'lláh.

[393] 'Abdu'l-Bahá, The Promulgation of Universal Peace, No. 52. Emphasis added.

cause of crime, the more knowledge and learning advance, the less crime will be committed. Consider the lawless tribes of Africa: How often they kill one another and even consume each other's flesh and blood! Why do such savageries not take place in Switzerland? The reason, clearly, is education and virtue.

Therefore, the body politic must seek to prevent crimes from being committed in the first place, rather than devise harsh punishments and penalties.[394]

In *Selections from the Writings of 'Abdu'l-Bahá,* he interpreted:

You have asked about strikes. Great difficulties have arisen and will continue to arise from this issue. The origin of these difficulties is twofold: One is the excessive greed and rapacity of the factory owners, and the other is the gratuitous demands, the greed, and the intransigence of the workers. One must therefore seek to address both.

The root cause of wrongdoing is ignorance, and we must therefore hold fast to the tools of perception and knowledge. Good character must be taught. Light must be spread afar, so that, in the school of humanity, all may acquire the heavenly characteristics of the spirit, and see for themselves beyond any doubt that there is no fiercer hell, no more fiery abyss, than to possess a character that is evil and unsound; no more darksome pit nor loathsome torment than to show forth qualities which deserve to be condemned.[395]

Now, the root cause of these difficulties lies in the law of nature that governs present-day civilization, for it results in a handful of people accumulating vast fortunes that far exceed their needs, while the greater number remain naked, destitute, and helpless.

This is at once contrary to justice, to humanity, and to fairness; it is the very height of inequity and runs counter to the good-pleasure of the All-Merciful.[396]

And among the teachings of Bahá'u'lláh is man's freedom, that through the ideal Power he should be free and emancipated from the captivity of the world of nature; for as long as man is captive to nature he is a ferocious animal, as the struggle for existence is one of the exigencies of the world of nature. This matter of the struggle for existence is the *fountain-head* of all calamities and is the supreme affliction.[397]

And among the teachings of Bahá'u'lláh is that religious, racial, political, economic and patriotic prejudices destroy the edifice of humanity. As long as these prejudices prevail, the world of humanity will not have rest. For a period of 6,000 years history informs us about the world of humanity. During these 6,000 years the world of humanity has not been free from war, strife, murder and bloodthirstiness. In every period war has been waged in one country or another and that war was due to either religious prejudice, racial prejudice, political prejudice or patriotic prejudice. It has therefore been ascertained and proved that all prejudices are destructive of the human edifice. As long as these prejudices persist, the struggle for existence must remain dominant, and bloodthirstiness and rapacity continue. Therefore, even as was the case in the past, the world of humanity cannot be saved from the

---

[394] 'Abdu'l-Bahá, Some Answered Questions, No 77. Emphasis added.

[395] 'Abdu'l-Bahá, Selections from the Writings of 'Abdu'l-Bahá, No. 111. Emphasis added.

[396] Idem. No. 78.2. Emphasis added.

[397] 'Abdu'l-Bahá, Tablets to The Hague. Emphasis added.

darkness of nature and cannot attain illumination except through the abandonment of prejudices and the acquisition of the morals of the Kingdom.[398]

Shoghi Effendi, wrote in 1942, in the middle of World War II (1939-1945) in which between 70-85 million people died, the following:

> If we could perceive the true reality of things we would see that the greatest of all battles raging in the world today is the spiritual battle. If the believers like yourself, young and eager and full of life, desire to win laurels for true and undying heroism, then let them join in the spiritual battle — whatever their physical occupation may be— which involves the very soul of man. The hardest and the noblest task in the world today is to be a true Bahá'í; this requires that we defeat not only the current evils prevailing all over the world, but the *weaknesses, attachments to the past, prejudices, and selfishnesses* that may be inherited and acquired within our own characters; that we give forth a shining and incorruptible example to our fellow-men.[399]

Shoghi Effendi in *The World Order of Bahá'u'lláh*, interpreted:

The Impotence of Statesmanship

Dearly-beloved friends! Humanity, whether viewed in the light of man's individual conduct or in the existing relationships between organized communities and nations, has, alas, strayed too far and suffered too great a decline to be redeemed through the unaided efforts of the best among its recognized rulers and statesmen— however disinterested their motives, however concerted their action, however unsparing in their zeal and devotion to its cause. No scheme which the calculations of the highest statesmanship may yet devise; no doctrine which the most distinguished exponents of economic theory may hope to advance; no principle which the most ardent of moralists may strive to inculcate, can provide, in the last resort, adequate foundations upon which the future of a distracted world can be built. No appeal for mutual tolerance which the worldly-wise might raise, however compelling and insistent, can calm its passions or help restore its vigor. *Nor would any general scheme of mere organized international coöperation, in whatever sphere of human activity, however ingenious in conception, or extensive in scope, succeed in removing the root cause of the evil that has so rudely upset the equilibrium of present-day society.* Not even, I venture to assert, would the very act of devising the machinery required for the political and economic unification of the world— a principle that has been increasingly advocated in recent times— provide in itself the antidote against the poison that is steadily undermining the vigor of organized peoples and nations.

What else, might we not confidently affirm, but the unreserved acceptance of the Divine Program enunciated, with such simplicity and force as far back as sixty years ago, by Bahá'u'lláh, embodying in its essentials God's divinely appointed scheme for the unification of mankind in this age, coupled with an indomitable conviction in the unfailing efficacy of each and all of its provisions, is eventually capable of withstanding the forces of internal disintegration which, if unchecked, must needs continue to eat into the vitals of a despairing society. It is towards this goal— the goal of a new World Order, Divine in origin, all-embracing

---

[398] 'Abdu'l-Bahá, Selections from the Writings of 'Abdu'l-Bahá, , No. 227.

[399] 'Abdu'l-Bahá, Bahá'u'lláh, Shoghi Effendi, Excellence in All Things. On behalf of Shoghi Effendi, 5 April 1942 to an individual believer. Emphasis added.

in scope, equitable in principle, challenging in its features— that a harassed humanity must strive.[400]

The Universal House of Justice, in its *One Common Faith* guidance, in 2005, guides us by saying:

> The power through which these goals will be progressively realized is that of unity. Although to Bahá'ís the most obvious of truths, its implications for the current crisis of civilization appear to escape most contemporary discourse. Few will disagree that the universal disease sapping the health of the body of humankind is that of disunity. Its manifestations everywhere cripple political will, debilitate the collective urge to change, and poison national and religious relationships. How strange, then, that unity is regarded as a goal to be attained, if at all, in a distant future, after a host of disorders in social, political, economic and moral life have been addressed and somehow or other resolved. Yet the latter are essentially symptoms and side effects of the problem, *not its root cause*. Why has so fundamental an inversion of reality come to be widely accepted? The answer is presumably because the achievement of genuine unity of mind and heart among peoples whose experiences are deeply at variance is thought to be entirely beyond the capacity of society's existing institutions. While this tacit admission is a welcome advance over the understanding of processes of social evolution that prevailed a few decades ago, it is of limited practical assistance in responding to the challenge.[401]

In, *Examination of the Environmental Crisis*, by Paul Fieldhouse and Chris Jones Kavelin, we find:

> To summarize again, deep ecology postulates anthropocentrism as the root cause; social ecology postulates a dysfunctional social hierarchical structure; while ecofeminism sees a parallel relationship between the variety of principles governing the dualistic domination and objectification of male over female with the same principles governing the domination and objectification of humanity over nature.[402]

Paul Fieldhouse, in Food, Justice, and the Baha'i Faith, writes:

> Saiedi, a contemporary Bahá'í writer and thinker, points to an intellectual and moral chaos resulting from what he sees as the collapse of modernity.[403] An all-encompassing vision for restoring a sense of spiritual and social order is needed, he says, to counter despair and relativism on the one hand or narcissism and worship of greed and violence on the other. And, he continues, "While modernist certainties collapse, post-modernism, with its emphasis on difference, is unable to provide an adequate vision of self, society and ethical ideas to meet new global realities". *Wendy Heller adds that whereas secularism was once heralded as the saviour of civilization it is now increasingly identified as the root cause of its disintegration, a conclusion anticipated in the Bahá'í writings which affirm that social and moral deterioration is directly related to the decline of religion as a social force.* Shoghi Effendi particularly criticised the "prevailing spirit of modernism" which by emphasising

---

[400] Shoghi Effendi. The World Order of Bahá'u'lláh. Emphasis added.

[401] The Universal House of Justice, One Common Faith. Emphasis added.

[402] Fieldhouse, Paul, et al. Food, Justice, and the Bahá'í Faith, in Examination of the Environmental Crisis. Emphasis added.

[403] Nader Saiedi, Logos and Civilization: spirit. history and order in the Writings of Bahá'u'lláh.

materialism eroded conceptions of duty and solidarity, reciprocity and loyalty, replacing them with individual selfishness and intolerance leading to the breakdown of family, economic and political structures (Shoghi Effendi 1974 183).[404]

William S. Hatcher in, *Love, Power, and Justice*, has an interesting reflection:

> As Bahá'u'lláh explains, *the root cause of all the major injustices of history has been the pursuit of power and dominance over others.* Here is one passage where He clearly articulates this thesis: "And amongst the realms of unity is the unity of rank and station. It redoundeth to the exaltation of the Cause, glorifying it among all peoples. Ever since the seeking of preference and distinction came into play, the world hath been laid waste. It hath become desolate. Those who have quaffed from the ocean of divine utterance and fixed their gaze upon the Realm of Glory should regard themselves as being on the same level as the others and in the same station. Were this matter to be definitely established and conclusively demonstrated through the power and might of God, the world would become as the Abhá Paradise. Indeed, man is noble, inasmuch as each one is a repository of the sign of God. Nevertheless, to regard oneself as superior in knowledge, learning or virtue, or to exalt oneself or seek preference, is a grievous transgression".[405]

Hooshmand Badee says in his Ph.D. dissertation:

> "*Materialism, therefore, is the root cause of many social illnesses, consumerism being one.* Hanley has explored this subject and noted numerous worrying examples of the present condition of the world associated with materialism. Smart argues that global brands, developed mainly in the West, have successfully penetrated local cultures and attracted people around the world to consume this or that commodity by generating persuasive and appealing promotional lines and marketing messages".[406]
>
> "K_hán believed that a root cause of public indifference was nothing but indolence".[407]
>
> Plato's quotes: "Ignorance is the root cause of all difficulties. Better to be unborn than untaught, for ignorance is the root of all misfortune.
>
> For Plato, the division of labor is the root of all good, while for Marx, it is the root of all evil."[408]

Uniqueness:

Aline Hanle's scientific approach to oneness:

> Love has its own kingdom governed by the heart. As a result, its Science has its own sets of laws, rules and principles. There is a great deal of Wisdom in understanding that what the mind perceives is one world and what the heart knows is another. While humanity has actively worked at exploring the Science of nature from a mindful

---

[404] Fieldhouse, Paul, et al. Food, Justice, and the Bahá'í Faith, in Examination of the Environmental Crisis. Emphasis added.

[405] William S. Hatcher, Love, Power, and Justice. Emphasis added.

[406] Bahá'í Teachings on Economics and Their Implications for the Bahá'í Community and the Wider Society.

[407] Gail, Marzieh. Summon Up Remembrance.

[408] Dematteis, Philip. Was Plato a Libertarian?

approach, here is a new scientific outlook on life from a Soulful perception.

Both are equally valid and as their uniqueness[409] brings different understanding about our existence, their combination gives a new dimension to Life. A dimension that lets us perceive more possibilities with fewer limitations.

Mathematical Constant: 1 is to the heart what Pi is to the mind. In Love, there is one constant number and it is 1. It is, in itself and simultaneously, the number of Infinity and Oneness. It is a transcendental number by excellence. By transcendental, I mean that this is in truth the only number that allows an experience to transcend itself. All must be one before an experience expands.[410]

Bahá'u'lláh revealed:

"The All-Knowing Physician hath His finger on the pulse of mankind. He perceiveth the disease, and prescribeth, in His unerring wisdom, the remedy. *Every age hath its own problem, and every soul its particular aspiration.* The remedy the world needeth in its present-day afflictions can never be the same as that which a subsequent age may require. Be anxiously concerned with the needs of the age ye live in, and center your deliberations on its exigencies and requirements".[411]

'Abdu'l-Bahá taught:

"O Company of God! To each created thing, the Ancient Sovereignty hath portioned out its *own perfection, its particular virtue and special excellence*, so that each in its degree may become a symbol denoting the sublimity of the true Educator of humankind, and that each, even as a crystalline mirror, may tell of the grace and splendour of the Sun of Truth."[412]

And in *The Promulgation of Universal Peace,* 'Abdu'l-Bahá describes uniqueness in the following way:

(T)he forms and organisms of phenomenal being and existence in each of the kingdoms of the universe are myriad and numberless. The vegetable plane or kingdom, for instance, has its infinite variety of types and material structures of plant life— *each distinct and different within itself, no two exactly alike in composition and detail— for there are no repetitions in nature*, and the augmentative virtue cannot be confined to any given image or shape. Each leaf has *its own particular identity—* so to speak, *its own individuality as a leaf*. Therefore, each atom of the innumerable elemental atoms, during its ceaseless motion through the kingdoms of existence as a constituent of organic composition, not only becomes imbued with the powers and virtues of the kingdoms it traverses but also reflects the attributes and qualities of the forms and organisms of those kingdoms. As each of these forms has *its individual and particular virtue*, therefore, each elemental atom of the universe has the opportunity of expressing an infinite variety of those individual virtues. No atom is bereft or

---

[409] Uniqueness: a) existing as the only one or as the sole example; single; solitary in type or characteristics. b) having no like or equal; unparalleled; incomparable. Web. July 2020 <https://www.dictionary.com/browse/unique>.

[410] Hanle, Aline. Love and Science: A New Dimension to Life.

[411] Bahá'u'lláh, Gleanings from the Writings of Bahá'u'lláh. Emphasis added.

[412] 'Abdu'l-Bahá, Bahá'u'lláh, Shoghi Effendi. Bahai Education. Emphasis added.

deprived of this opportunity or right of expression. Nor can it be said of any given atom that it is denied equal opportunities with other atoms; nay, all are privileged to possess the virtues existent in these kingdoms and to reflect the attributes of their organisms. In the various transformations or passages from kingdom to kingdom the virtues expressed by the atoms in each degree are *peculiar* to that degree. For example, in the world of the mineral the atom does not express the vegetable form and organism, and when through the process of transmutation it assumes the virtues of the vegetable degree, it does not reflect the attributes of animal organisms, and so on.[413]

The author chose the Laws of God, the Spirit, the Word of God, Mercy, Love, Providence, Fear of God, and Grace as the essences of all things and the originating causes of scientific methods. Someone may start thinking that we can become one with God. Ian Kluge in, *Reason and the Bahá'í Writings: The Use and Misuse of Logic and Persuasion*, clarifies:

> Meditate on what the poet hath written: "Wonder not, if my Best-Beloved be closer to me than mine own self; wonder at this, that I, despite such nearness, should still be so far from Him."... Considering what God hath revealed, that "We are closer to man than his life-vein," the poet hath, in allusion to this verse, stated that, though the revelation of my Best-Beloved hath so permeated my being that He is closer to me than my life-vein, yet, notwithstanding my certitude of its reality and my recognition of my station, I am still so far removed from Him. By this he meaneth that his heart, which is the seat of the All-Merciful and the throne wherein abideth the splendor of His revelation, is forgetful of its Creator, hath strayed from His path, hath shut out itself from His glory, and is stained with the defilement of earthly desires.[414]

In the previous quotation, we are told God is closer than our "life-vein" while elsewhere the Writings affirm that the distance between God and humankind is infinite and unbridgeable.[415] How can both of these be true?

> In strict, logical terms, distance here can only mean existential distance (since there can be no infinite physical distance), i.e. degrees of dependence. Existentially, (in the order of being) God is 100% independent of humankind - which is existentially 100% dependent on God, just as a drawing depends on the artist. However, while the artist personally is 100% independent of or distant from the drawing, s/he is also ontologically present in the drawing insofar as the drawing is a direct manifestation of, and would not exist without, the artist's power. Every artist is both infinitely distant from, independent, and, through his/her power, intimately present in his or her work. Logically speaking, we can resolve this apparent contradiction by distinguishing between God's (a) existential independence and His (b) ontological presence just as we distinguish the existence of the artist from his/her presence through a manifestation of his/her powers.[416]

Kluge, in another of his profound articles, has a complementary answer:

---

[413] The Promulgation of Universal Peace in Writings and Utterances of 'Abdu'l-Bahá. Emphasis added.

[414] Bahá'u'lláh, Gleanings from the Writings of Bahá'u'lláh, XCIII, p.185.

[415] Idem, XXVII, p. 66.

[416] Ian Kluge, Reason and the Bahá'í Writings: The Use and Misuse of Logic and Persuasion.

This statement makes it clear that there is absolutely no standpoint from which the soul can alter its essentially human condition and become ontologically one with God. To become one with God, also violates Baha'u'llah's injunction not to "transgress the limits of one's own rank and station".[417] This re- emphasizes the dualist position: man is always man and God is always God. In other words, we always remain in one of the three stations of existence: "Know that the degrees of existence are finite—the degrees of servitude, of prophethood, and of Divinity—but that the perfections of God and of creation are infinite."[418] Man is always in the (ontological) condition of servitude and nothing can change that, either in this life or the life to come.

Furthermore, in the "Commentary on the Islamic Tradition, 'I was a Hidden Treasure …' ", 'Abdu'l-Bahá categorically states his own position that "the path to knowing the innermost Essence of the Absolute is closed to all beings … How can the reality of non-existence ever understand the ipseity[419] of being?[420] Since the knowledge of God is utterly impossible, then no one – regardless of spiritual condition – can attain the necessary and sufficient conditions for obtaining such knowledge which in effect denies the possibility of unity with God.[421]

'Abdu'l-Bahá, in *Selections from the Writings of 'Abdu'l-Bahá*, interpreted: "For instance, in the originated we see ignorance; in the Pre-existent we affirm knowledge. In the originated we see weakness; in the Pre-existent we affirm power. In the originated we see poverty; in the Pre-existent we affirm wealth. Hence the originated is the source of all imperfections, and the Pre-existent is the sum of all perfections."[422]

6.3.10 The Divine Teachers are Gardeners

If the mountains, hills and plains of the material world are left wild and uncultivated under the rule of nature, they will remain an unbroken wilderness, no fruitful tree to be found anywhere upon them. A true cultivator changes this forest and jungle into a garden, training its trees to bring forth fruit and causing flowers to grow in place of thorns and thistles. The holy Manifestations are the ideal Gardeners of human souls, the divine Cultivators of human hearts. The world of existence is but a jungle of *disorder* and confusion, a state of nature producing nothing but fruitless, useless trees. The ideal Gardeners train these wild, uncultivated human trees, cause them to become fruitful, water and cultivate them day by day so that they adorn the world of existence and continue to flourish in the utmost beauty."[423]

In Some Answered Questions, we find:

---

[417] Bahá'u'lláh, Gleanings from the Writings of Bahá'u'lláh, XCIII, 188.

[418] 'Abdu'l-Bahá, Some Answered Questions, No. 62.

[419] individual identity: selfhood. Definition in The Merriam Webster Dictionary.

[420] 'Abdu'l-Bahá , "Commentary on the Islamic Tradition: 'I Was a Hidden Treasure' "; emphasis added.

[421] Ian Kluge, Relativism and the Bahá'í Writings.

[422] 'Abdu'l-Bahá, Selections from the Writings of 'Abdu'l-Bahá.

[423] 'Abdu'l-Bahá, The Promulgation of Universal Peace. Emphasis added.

To summarize: Just as man progresses, evolves, and is *transformed* from one form and appearance to another in the womb of the mother, while remaining from the beginning a human embryo, so too has man remained a distinct essence—that is, the human species—from the beginning of his formation in the matrix of the world, and has passed gradually from form to form. It follows that this change of appearance, this evolution of organs, and this growth and development do not preclude the originality of the species. Now, even accepting the reality of evolution and progress, nevertheless, from the moment of his appearance man has possessed perfect composition, and has had the capacity and potential to acquire both material and spiritual perfections and to become the embodiment of the verse, "*Let Us make man in Our image, after Our likeness.*" At most, he has become more pleasing, more refined and graceful, and *by virtue of civilization he has emerged from his wild state, just as the wild fruits become finer and sweeter under the cultivation of the gardener, and acquire ever greater delicacy and vitality.*

The gardeners of the world of humanity are the Prophets of God.[424]

… That is why man is said to be the greatest sign of God—that is, he is the Book of Creation—for all the mysteries of the universe are found in him. Should he come under the shadow of the true Educator and be rightly trained, he becomes the gem of gems, the light of lights, and the spirit of spirits; he becomes the focal centre of divine blessings, the wellspring of spiritual attributes, the dawning-place of heavenly lights, and the recipient of divine inspirations. Should he, however, be deprived of this education, he becomes the embodiment of satanic attributes, the epitome of animal vices, and the source of all that is oppressive and dark.[425]

In *The Promulgation of Universal Peace*, we find:

In nature there is the law of the survival of the fittest. Even if man be not educated, then according to the natural institutes this natural law will demand of man supremacy. The purpose and object of schools, colleges and universities is to educate man and thereby rescue and redeem him from the exigencies and defects of nature and to awaken within him the capability of controlling and appropriating nature's bounties. If we should relegate this plot of ground to its natural state, allow it to return to its original condition, it would become a field of thorns and useless weeds, but by cultivation it will become fertile soil, yielding a harvest. Deprived of cultivation, the mountain slopes would be jungles and forests without fruitful trees. The gardens bring forth fruits and flowers in proportion to the care and tillage bestowed upon them by the gardener. Therefore, it is not intended that the world of humanity should be left to its natural state. It is in need of the education divinely provided for it. The holy, heavenly Manifestations of God have been the Teachers. They are the divine Gardeners Who transform the jungles of human nature into fruitful orchards and make the thorny places blossom as the rose. It is evident, then, that the intended and especial function of man is to rescue and redeem himself from the inherent defects of nature and become qualified with the ideal virtues of Divinity.[426]

---

[424] 'Abdu'l-Bahá, Some Answered Questions, No 49. Emphasis added.

[425] 'Abdu'l-Bahá, Some Answered Questions. No 64.

[426] 'Abdu'l Bahá, The Promulgation of Universal Peace.

6.3.11 Oneness is a law

Bahá'u'lláh, within the Epistle to the Son of the Wolf, revealed:

> Through Him the light of unity hath shone forth above the horizon of the world, and *the law of oneness* hath been revealed amidst the nations, who, with radiant faces, have turned towards the Supreme Horizon, and acknowledged that which the Tongue of Utterance hath spoken in the kingdom of His knowledge: "Earth and heaven, glory and dominion, are God's, the Omnipotent, the Almighty, the Lord of grace abounding!"[427]

Does every atom have a unique path? For deciding if uniqueness is a law that encompasses all kingdoms, please read again. 'Abdu'l-Bahá, in *The Promulgation of Universal Peace,* describes uniqueness in the following way:

> Each leaf has its *own particular identity*— so to, its *own individuality* as a leaf. Therefore, each atom of the innumerable elemental atoms, during its ceaseless motion through the kingdoms of existence as a constituent of organic composition, not only becomes imbued with the powers and virtues of the kingdoms it traverses but also reflects the attributes and qualities of the forms and organisms of those kingdoms. *As each of these forms has its individual and particular virtue, therefore, each elemental atom of the universe has the opportunity of expressing an infinite variety of those individual virtues.* No atom is bereft or deprived of this opportunity or right of expression. Nor can it be said of any given atom that it is denied equal opportunities with other atoms; nay, all are privileged to possess the virtues existent in these kingdoms and to reflect the attributes of their organisms. *In the various transformations or passages from kingdom to kingdom the virtues expressed by the atoms in each degree are peculiar to that degree.* For example, in the world of the mineral the atom does not express the vegetable form and organism, and when through the process of *transmutation* it assumes the virtues of the vegetable degree, it does not reflect the attributes of animal organisms, and so on.[428]

Shoghi Effendi, in *The Promised Day Is Come*, explains:

> That which was applicable to human needs during the early history of the race can neither meet nor satisfy the demands of this day, this period of *newness* and consummation. Humanity has emerged from its former state of limitation and preliminary training. Man must now become imbued with *new* virtues and powers, *new* moral standards, *new* capacities. *New* bounties, perfect *bestowals*, are awaiting and already descending upon him. The gifts and blessings of the period of youth, although timely and sufficient during the adolescence of mankind, are now incapable of meeting the requirements of its maturity." "In every Dispensation," He moreover has written, "the light of *Divine Guidance* has been focused upon one central theme.… In this wondrous Revelation, this glorious century, the foundation of the Faith of God, *and the distinguishing feature of His Law, is the consciousness of the oneness of mankind.*"[429]

> Inasmuch as the fundamental principle of the teaching of Bahá'u'lláh is the oneness of the world of humanity, *I will speak to you upon the intrinsic oneness of all phenomena.* This is one of the abstruse subjects of divine philosophy.

---

[427] Bahá'u'lláh, Epistle to the Son of the Wolf. Emphasis added.

[428] 'Abdu'l-Bahá, The Promulgation of Universal Peace. Emphasis added.

[429] Shoghi Effendi, The Promised Day Is Come. Emphasis added.

Fundamentally all existing things pass through the same degrees and phases of development, and any given phenomenon embodies all others. An ancient statement of the Arabian philosophers declares that all things are involved in all things. It is evident that each material organism is an aggregate expression of single and simple elements, *and a given cellular element or atom has its coursings or journeyings through various and myriad stages of life.* For example, we will say the cellular elements which have entered into the composition of a human organism were at one time a component part of the animal kingdom; at another time they entered into the composition of the vegetable, and prior to that they existed in the kingdom of the mineral. They have been subject to transference from one condition of life to another, passing through various forms and phases, exercising in each existence special functions. Their journeyings through material phenomena are continuous.

Therefore, each phenomenon is the expression in degree of all other phenomena. The difference is one of successive transferences and the period of time involved in evolutionary process.

For example, it has taken a certain length of time for this cellular element in my hand to pass through the various periods of metabolism. At one period it was in the mineral kingdom subject to changes and transferences in the mineral state. Then it was transferred to the vegetable kingdom where it entered into different grades and stations. Afterward it reached the animal plane, appearing in forms of animal organisms until finally in its transferences and coursings it attained to the kingdom of man. Later on it will revert to its primordial elemental state in the mineral kingdom, being subject, as it were, to infinite journeyings from one degree of existence to another, passing through every stage of being and life. Whenever it appears in any distinct form or image, it has its opportunities, virtues and functions. *As each component atom or element in the physical organisms of existence is subject to transference through endless forms and stages, possessing virtues peculiar to those forms and stations, it is evident that all phenomena of material being are fundamentally one.* In the mineral kingdom this component atom or element possesses certain virtues of the mineral; in the kingdom of the vegetable it is imbued with vegetable qualities or virtues; in the plane of animal existence it is empowered with animal virtues—the senses; and in the kingdom of man it manifests qualities peculiar to the human station.[430]

6.3.12 The concept of reality

'Abdu'l-Bahá in *The Promulgation of Universal Peace* describes reality in the following way:

First among the great principles revealed by Him is that of the investigation of reality. The meaning is that every individual member of humankind is exhorted and commanded to set aside *superstitious beliefs, traditions and blind imitation of ancestral forms in religion* and investigate reality for himself. Inasmuch as the *fundamental reality is one*, all religions and nations of the world will become one through investigation of reality. The announcement of this principle is not found in any of the sacred Books of the past.[431]

---

[430] 'Abdu'l-Bahá, The Promulgation of Universal Peace. Emphasis added.

[431] 'Abdu'l Bahá, The Promulgation of Universal Peace. Emphasis added.

The third principle or teaching of Bahá'u'lláh is the *oneness* of religion and science. Any religious belief which is not conformable with scientific proof and investigation is *superstition*, for true science is reason and reality, and religion is essentially reality and pure reason; therefore, the two must correspond. Religious teaching which is at variance with science and reason is *human invention and imagination unworthy of acceptance*, for the antithesis and opposite of knowledge is *superstition born of the ignorance of man*. If we say religion is opposed to science, we lack knowledge of either true science or true religion, for both are founded upon the premises and conclusions of reason, and both must bear its test.[432]

The foundation of the divine religions is reality; were there no reality, there would be no religions. Abraham heralded reality. Moses promulgated reality. Christ established reality. Muḥammad was the Messenger of reality. The Báb was the door of reality.

Bahá'u'lláh was the splendor of reality. Reality is one; it does not admit multiplicity or division. *Reality is as the sun, which shines forth from different dawning points; it is as the light, which has illumined many lanterns.*

Therefore, if the religions investigate reality and seek the essential truth of their own foundations, they will agree and no difference will be found. But inasmuch as religions are submerged in dogmatic imitations, forsaking the original foundations, and as imitations differ widely, therefore, the religions are divergent and antagonistic. These imitations may be likened to clouds which obscure the sunrise; but *reality is the sun*. If the clouds disperse, *the Sun of Reality shines upon all*, and no difference of vision will exist. The religions will then agree, for fundamentally they are the same. The subject is one, but *predicates* are many.[433]

The teachings of Bahá'u'lláh are boundless, innumerable; time will not allow us to mention them in detail. The foundation of progress and real prosperity in the human world is reality, for reality is the divine standard and the bestowal of God. *Reality is reasonableness, and reasonableness is ever conducive to the honorable station of man. Reality is the guidance of God. Reality* is *the cause of illumination of mankind. Reality is love*, ever working for the welfare of humanity. *Reality is the bond which conjoins hearts*. This ever uplifts man toward higher stages of progress and attainment. *Reality is the unity of mankind*, conferring everlasting life. *Reality is perfect equality*, the foundation of agreement between the nations, the first step toward international peace.[434]

In divine questions we must not depend entirely upon the heritage of tradition and former human experience; nay, rather, we must exercise reason, analyze and logically examine the facts presented so that confidence will be inspired and faith attained. Then and then only the reality of things will be revealed to us.[435]

The very fact that the reality of phenomena is limited well indicates that there must needs be an unlimited reality, for were there no unlimited, or infinite, reality in life, the finite being of objects would be inconceivable. To make it plainer for you, if there were no wealth in the world, you would not have poverty. If there were no light in the world, you could not

---

[432] Idem. Emphasis added.

[433] Ibidem. Emphasis added.

[434] Ibidem. Emphasis added.

[435] Ibidem. Emphasis added.

conceive of darkness, for we know things philosophically by their antitheses. We know, for example, that poverty is the lack of wealth. Where there is no knowledge, there is no ignorance. What is ignorance? It is the absence of knowledge. Therefore, our limited existence is a conclusive proof that there is an unlimited reality, and this is a shining proof and evident argument. Many are the proofs concerning this matter, but there is not time to go into the subject further.[436]

Phenomenal, or created, things are known to us only by their attributes. Man discerns only manifestations, or attributes, of objects, while the identity, or reality, of them remains hidden. For example, we call this object a flower. What do we understand by this name and title? We understand that the qualities appertaining to this organism are perceptible to us, but the intrinsic elemental reality, or identity, of it remains unknown. Its external appearance and manifest attributes are knowable; but the inner being, the underlying reality or intrinsic identity, is still beyond the ken and perception of our human powers. Inasmuch as the realities of material phenomena are impenetrable and unknowable and are only apprehended through their properties or qualities, how much more this is true concerning the reality of Divinity, that holy essential reality which transcends the plane and grasp of mind and man? That which comes within human grasp is finite, and in relation to it we are infinite because we can grasp it. Assuredly, the finite is lesser than the infinite; the infinite is ever greater. If the reality of Divinity could be contained within the grasp of human mind, it would after all be possessed of an intellectual existence only—a mere intellectual concept without extraneous existence, an image or likeness which had come within the comprehension of finite intellect. The mind of man would be transcendental thereto. How could it be possible that an image which has only intellectual existence is the reality of Divinity, which is infinite? Therefore, the reality of Divinity in its identity is beyond the range of human intellection because the human mind, the human intellect, the human thought are limited, whereas the reality of Divinity is unlimited. How can the limited grasp the unlimited and transcend it? Impossible. The unlimited always comprehends the limited. The limited can never comprehend, surround nor take in the unlimited. Therefore, every concept of Divinity which has come within the intellection of a human being is finite, or limited, and is a pure product of imagination, whereas the reality of Divinity is holy and sacred above and beyond all such concepts.[437]

6.3.13 Methods

The Universal House of Justice, in a letter to an individual on October 11/1978, said: Dear Bahá'í Friend,

The Universal House of Justice has received your moving appeal for guidance in your letter of 5 September 1978 and has instructed us to convey to you the following advice.

Each individual is unique and has a unique path to tread in his lifetime. In espousing the Bahá'í Faith you have defined the direction of that path, for your recognition of God's Manifestation for this Day and your devotion to His Message provide the spiritual and ethical basis for all aspects of your life of service to mankind, while the continuing guidance that He

---

[436] Idem. Emphasis added.

[437] Idem. Emphasis added.

has provided for the community of His followers enables you to know the directions in which the most effort is required at the present time.

While, during the early years of the development of the Faith, Bahá'u'lláh, 'Abdu'l-Bahá and Shoghi Effendi sometimes gave specific instructions to individual believers on how they should serve the Cause, the Universal House of Justice seldom does this. It is, indeed, the precious privilege of the individual human being to direct the course of his own life. Through exercising this privilege while striving always to conform his conduct to the divine Teachings and devote his talents in the best possible way to the service of the Cause and mankind, a soul deepens his understanding of God and His will.

This does not mean that you are left to make your decisions without guidance. This you will find from several sources. Firstly, in general, you will find it in the Writings. Secondly, and more specifically, in the teaching plans issued by the Universal House of Justice. Thirdly, in the plans and projects of your own National Spiritual Assembly. All these, it would seem from your letter, you have been striving to follow. Fourthly, with regard to your own personal goals and actions, is the guidance you can receive through consultation—with your wife, with friends of your choice whose opinions you value, with your Local Spiritual Assembly, with such committees of your National Assembly as are concerned with the fields of activity towards which your inclinations lie. Fifthly, there is prayer and meditation.

You mention that the answers to your prayers never seem to have come through clearly. Mrs. Ruth Moffett has published her recollection of five steps of prayer for guidance that she was told by the beloved Guardian. When asked about these notes, Shoghi Effendi replied, in letters written by his secretary on his behalf, that the notes should be regarded as "personal suggestions," that he considered them to be "quite sound," but that the friends need not adopt them "strictly and universally." The House of Justice feels that they may be helpful to you and, indeed, you may already be familiar with them. They are as follows:

… use these five steps if we have a problem of any kind for which we desire a solution, or wish help.

1. Pray and meditate about it. Use the prayers of the Manifestations, as they have the greatest power. Learn to remain in the silence of contemplation for a few moments. During this deepest communion take the next step.

2. Arrive at a decision and hold to this. This decision is usually born in a flash at the close or during the contemplation. It may seem almost impossible of accomplishment, but if it seems to be an answer to prayer or a way of solving the problem, then immediately take the next step.

3. Have determination to carry the decision through. Many fail here. The decision, budding into determination, is blighted and instead becomes a wish or a vague longing. When determination is born, immediately take the next step.

4. Have faith and confidence, that the Power of the Holy Spirit will flow through you, the right way will appear, the door will open, the right message, the right principle or the right book will be given to you. Have confidence, and the right thing will come to meet your need. Then as you rise from prayer take immediately the fifth step.

5. Act as though it had all been answered. Then act with tireless, ceaseless energy.

> And, as you act, you yourself will become a magnet which will attract more power to your being, until you become an unobstructed channel for the Divine Power to flow through you.

Also the Guardian's secretary wrote to an individual believer on his behalf: "The Master said guidance was when the doors opened after we tried. We can pray, ask to do God's will only, try hard, and then if we find our plan is not working out, assume it is not the right one, at least for the moment."

The Universal House of Justice deeply appreciates your candor and spirit of devotion, and assures you of its prayers in the Holy Shrines on your behalf.

With loving Bahá'í greetings,

Department of the Secretariat[438]

> "Transformation is the essential purpose of the Cause of Bahá'u'lláh, but it lies in the will and effort of the individual to achieve it in obedience to the Covenant. Necessary to the progress of this life-fulfilling transformation is knowledge of the will and purpose of God through regular reading and study of the Holy Word".[439]

We must try to become aware of what part of our destiny is irrevocable. Because being irrevocable, we should expect tests, over and over again, until we pass the exam. It is a personal obligation when we follow the prescribed responsibility, begging for guidance to discern among the many faults that we own. In *The Hidden Words of Bahá'u'lláh*, we find: "O Son of Being! Bring thyself to account each day ere thou art summoned to a reckoning; for death, unheralded, shall come upon thee and thou shalt be called to give account for thy deeds".[440]

Thornton Chase in, *The Bahai Revelation*, says:

> God asks man to pray to Him. He has given freely everything for the necessities of human life. He has filled the lands and waters with foods and taught man how to cultivate and use them; He has given the forests and all materials for shelter, the cotton and the wool, and enabled man to mould them for his comfort; He has given the reasoning faculties that man may progress in material welfare and exercise the ethical and moral knowledge offered to him; He has given conceptions of beauty that man may seek for more than the physical dimensions of existence; and He has implanted the perception of higher possibilities and a desire for eternal destinies, that man may turn his face toward the Infinite.[441]

'Abdu'l-Bahá's precious gift:

'Abdu'l-Bahá in *The Promulgation of Universal Peace* has given us a precious gift when saying:

> In the world of existence man has traversed successive degrees until he has attained the human kingdom. In each degree of his progression he has developed capacity for advancement to the next station and condition. While in the kingdom of the mineral he was attaining the capacity for promotion into the degree of the vegetable. In the kingdom of the

---

[438] Department of the Secretariat of The Universal House of Justice, in a letter to an individual October 11/1978. Emphasis added.

[439] Riḍván 1989, the Universal House of Justice to the Bahá'ís of the World. Emphasis added.

[440] Bahá'u'lláh, The Hidden Words of Bahá'u'lláh.

[441] Thornton Chase, The Bahai Revelation.

vegetable he underwent preparation for the world of the animal, and from thence he has come onward to the human degree, or kingdom. Throughout this journey of progression he has ever and always been potentially man.

In the beginning of his human life man was embryonic in the world of the matrix. There he received capacity and endowment for the reality of human existence. The forces and powers necessary for this world were bestowed upon him in that limited condition. In this world he needed eyes; he received them potentially in the other. He needed ears; he obtained them there in readiness and preparation for his new existence. The powers requisite in this world were conferred upon him in the world of the matrix so that when he entered this realm of real existence he not only possessed all necessary functions and powers but found provision for his material sustenance awaiting him.

Therefore, in this world he must prepare himself for the life beyond. That which he needs in the world of the Kingdom must be obtained here. Just as he prepared himself in the world of the matrix by acquiring forces necessary in this sphere of existence, so, likewise, the indispensable forces of the divine existence must be potentially attained in this world. "What is he in need of in the Kingdom which transcends the life and limitation of this mortal sphere? That world beyond is a world of sanctity and radiance; therefore, it is necessary that in this world he should acquire these divine attributes. In that world there is need of spirituality, faith, assurance, the knowledge and love of God. These he must attain in this world so that after his ascension from the earthly to the heavenly Kingdom he shall find all that is needful in that eternal life ready for him.

That divine world is manifestly a world of lights; therefore, man has need of illumination here. That is a world of love; the love of God is essential. It is a world of perfections; virtues, or perfections, must be acquired. That world is vivified by the breaths of the Holy Spirit; in this world we must seek them. That is the Kingdom of everlasting life; it must be attained during this vanishing existence.

By what means can man acquire these things? How shall he obtain these merciful gifts and powers? First, through the knowledge of God. Second, through the love of God. Third, through faith. Fourth, through philanthropic deeds. Fifth, through self-sacrifice. Sixth, through severance from this world. Seventh, through sanctity and holiness. Unless he acquires these forces and attains to these requirements, he will surely be deprived of the life that is eternal. But if he possesses the knowledge of God, becomes ignited through the fire of the love of God, witnesses the great and mighty signs of the Kingdom, becomes the cause of love among mankind and lives in the utmost state of sanctity and holiness, he shall surely attain to second birth, be baptized by the Holy Spirit and enjoy everlasting existence.[442]

To identify the possible root cause of our faults and wrongdoings and to address it, we should follow the indicated path given in the seven steps as mentioned above to reach the merciful gifts and powers that we are supposed to receive in the course of accomplishing our high destiny:

---

[442] 'Abdu'l-Bahá, The Promulgation of Universal Peace, in Writings and Utterances of 'Abdu'l-Bahá. -81- Emphasis added.

First, through the knowledge of God

Bahá'u'lláh revealed: "True loss is for him whose days have been spent in utter ignorance of his self".[443] "One must, then, read the book of one's own self, rather than the treatise of some grammarian. Wherefore He hath said, "Read thy Book: There needeth none but thyself to make out an account against thee this day.""[444]

> We will surely show them Our signs in the world and within themselves." Again He saith: "And also in your own selves: will ye not, then, behold the signs of God?" And yet again He revealeth: "And be ye not like those who forget God, and whom He hath therefore caused to forget their own selves." In this connection, He Who is the eternal King -- may the souls of all that dwell within the mystic Tabernacle be a sacrifice unto Him -- hath spoken: "He hath known God who hath known himself.[445]

Bahá'u'lláh in, *"The Valley of Knowledge,"* revealed:

> And if, confirmed by the Creator, the lover escapeth the claws of the eagle of love, he will enter the Realm of Knowledge and come out of doubt into certitude, and turn from the darkness of wayward desire to the guiding light of the fear of God. His inner eye will open and he will privily converse with his Beloved; he will unlock the gates of truth and supplication and shut the doors of idle fancy. He in this realm is content with the divine decree, and seeth war as peace, and in death findeth the meaning of everlasting life. With both inward and outward eyes he witnesseth the mysteries of resurrection in the realms of creation and in the souls of men, and with a spiritual heart apprehendeth the wisdom of God in His endless manifestations. In the sea he findeth a drop, in a drop he beholdeth the secrets of the sea.[446]

> In Gleanings from the Writings of Bahá'u'lláh, we find: "Whatever duty Thou hast prescribed unto Thy servants of extolling to the utmost Thy majesty and glory is but a token of Thy grace unto them, that they may be enabled to ascend unto the station conferred upon their own inmost being, *the station of the knowledge of their own selves*.[447]

William S. Hatcher in, *The Concept of Spirituality*, writes: Here the 'duties' which God has prescribed for man are seen not as ends in themselves but rather as 'tokens,' in other words, as symbols for and means towards another, *ultimate end*. This end is characterized as being a *particular kind of knowledge*, here called self-knowledge.[448]

In the following quote, Bahá'u'lláh speaks similarly of self-knowledge: "O My servants! Could ye apprehend with what wonders of My munificence and bounty *I have willed to entrust your souls, ye would, of a truth, rid yourselves of attachment to all created things, and would*

---

[443] Bahá'u'lláh, Tablets of Bahá'u'lláh Revealed After the Kitáb-i-Aqdas.

[444] Bahá'u'lláh, The Call of the Divine Beloved. The Seven Valleys.

[445] Baha'u'llah, Gleanings from the Writings of Baha'u'llah, p. 178.

[446] Bahá'u'lláh, The Call of the Divine Beloved. The Seven Valleys.

[447] Bahá'u'lláh, Gleanings from the Writings of Bahá'u'lláh. Emphasis added.

[448] William S. Hatcher, The Concept of Spirituality.

gain a true knowledge of your own selves—a knowledge which is the same as the comprehension of Mine own Being".[449]

One significant aspect of the previous passage is knowledge of self is similar to the comprehension of God. That knowledge of God is identical with the fundamental purpose of life for the individual is clearly stated by Bahá'u'lláh in numerous passages. For example:

The purpose of God in creating man hath been, and will ever be, to enable him to know his Creator and to attain His Presence. To this most excellent aim, this supreme objective, all the heavenly Books and the divinely-revealed and weighty Scriptures unequivocally bear witness. Whoso hath recognized the Dayspring of Divine guidance and entered His holy court hath drawn nigh unto God and attained His Presence, a Presence which is the real Paradise, and of which the loftiest mansions of heaven are but a symbol. Such a man hath attained the *knowledge* of the station of Him Who is "at the distance of two bows," Who standeth beyond the Sadratu'l-Muntahá. Whoso hath failed to recognize Him will have condemned himself to the misery of remoteness, a remoteness which is naught but utter nothingness and the essence of the nethermost fire. Such will be his fate, though to outward seeming he may occupy the earth's loftiest seats and be established upon its most exalted throne.[450] [451]

In the treasuries of the knowledge of God there lieth concealed a knowledge which, when applied, will largely, though not wholly, eliminate fear. This knowledge, however, should be taught from childhood, as it will greatly aid in its elimination. Whatever decreaseth fear increaseth courage. Should the Will of God assist Us, there would flow out from the Pen of the Divine Expounder a lengthy exposition of that which hath been mentioned, and there would be revealed, in the field of arts and sciences, what would renew the world and the nations. A word hath, likewise, been written down and recorded by the Pen of the Most High in the Crimson Book which is capable of fully disclosing that force which is hid in men, nay of redoubling its potency. We implore God — exalted and glorified be He — to graciously assist His servants to do that which is pleasing and acceptable unto Him.[452]

'Abdu'l-Bahá's commentary in "By the Fig and the Olive", said:

If we take the Fig literally to refer to the fruit of the tree, it can stand as a symbol of man's destiny in many ways. Under cultivation it can be one of the finest, most delicious, and most wholesome fruits in existence: in its wild state, it is nothing but tiny seeds, and it is insipid, and often full of worms and maggots. So man at his best has a noble destiny: at his worst, he is "the lowest of the low".[453]

In this way, we learn a lot from the Fig plant. This reflection may lead each of us to personal knowledge of our defects and what we must confront. It increases our understanding of the possible perfections that can become actualities. This knowledge of ourselves may increase our courage and lower our fear of God, welcoming the tests that

---

[449] Bahá'u'lláh, Gleanings from the Writings of Bahá'u'lláh. Emphasis added.

[450] Bahá'u'lláh, Gleanings from the Writings of Bahá'u'lláh. Emphasis added.

[451] Hatcher, The Concept of Spirituality. Emphasis added.

[452] Bahá'u'lláh. Epistle to the Son of the Wolf. Emphasis added.

[453] 'Abdu'l-Bahá'. Commentary in Ottoman Turkish on the Qur'ánic Sura 95.

accompanying them. By doing this, we are acknowledging that God's Will is irresistible and increases our trust in Him because He knows better the hidden perfections within us.

According to `Abdu'l-Bahá, our task and destiny is to perfect our human existence by strengthening and developing the spiritual aspects of our nature. This means that human beings share a universal duty and destiny — a struggle to control our unruly animal nature and make it work for the good of the soul and our spiritual development. Both as individuals and collectives we succeed in varying degrees in this process and sometimes slip into complete failure.[454]

Adib Taherzadeh in "*The Covenant of Bahá'u'lláh*", says:

The daily recital of any of the three obligatory prayers can act as a mighty weapon in the spiritual battle against one's own self, a battle that every believer must fight in order to subdue his greatest enemy and drive the 'stranger' away. The recital of the obligatory prayer, which is enjoined upon every believer by Bahá'u'lláh and constitutes one of the most sacred rites of the Faith, is a major factor in enabling a soul to recognise its own importance in relation to its Creator and to acknowledge its own shortcomings.[455]

Nader Saiedi in, *Gate of the Heart*, explains:

The Root Principle and Destiny

The third chapter of the Epistle of Justice[456], the longest section of the book, is devoted to a substantive discussion of the root principles of religion. Since this is the age of the sanctuary of the heart, the root principle of faith in this age is also an affirmation of spiritual knowledge in the sanctuary of unity. Although this principle is explained in terms of its diverse manifestations, its essence is described in the first paragraph of the chapter:

Know thou that the essence of religion is the knowledge of God. The perfection of this knowledge is belief in His unity. The perfection of this belief is the negation of all names and attributes before His sanctified Essence. And the perfection of this negation is to immerse oneself with certain knowledge in the Ocean of oneness and to witness one's attainments to its bounty. And the truth underlying all these stages is the Sign of God alone, whereby the existence of the Lord of might and glory is recognized and known with certitude.[457] [458]

The Epistle of Justice affirms that the seven types of recognition that constitute the root principles of religion need to be crystallized and manifested in the observance of divine ordinances and laws. The text, however, does not deal with the specifics of derivative laws-- with the two notable exceptions of the command to "act in regard to the people of paradise in accordance with affection and mercy," and the duty to treat women "in the utmost manner of love." These two ordinances occupy a prominent position in the writings of the Báb, a fact that is evident in this text. But although the Báb does not address here the details of the

---

[454] Kluge, Some Answered Questions: A Philosophical Perspective.

[455] Adib Taherzade, The Covenant of Bahá'u'lláh. Emphasis added.

[456] Risaliy-i-'Adiyyih (The Epistle of Justice) by The Báb.

[457] Nader Saiedi, Gate of the Heart.

[458] The Báb, Sahifiy-i-'Adliyyih, p. 15. Authorized translation provided by the Research Department of the Universal House of Justice.

derivative laws, He discusses something more fundamental; the moral and spiritual motives for performing actions.

Purification of motive and detachment from all but God become the mediating principle which links the discourse on the root principles of faith to that of the derivative and secondary laws and ordinances.[459]

The later writings of the Báb make it clear that the purification He speaks of in the Epistle of Justice is the same as attaining the sanctuary of the heart and fixing one's gaze on the Supreme Source of all the root principles of the Faith. One who achieves such purity and devotion will be able to recognize the new Manifestation of God together with the new root principles as well as the derivative laws. According to the Epistle of Justice, humanity has arrived at a new stage of spiritual development. The seven types of recognition of God, as the root principles of religion, which are the same as the five stations of the letter Ha', refer to the reality of the Báb, Who is proclaiming the commencement of the sanctuary of unity and the revelation of the inner essence of the divine Mystery.[460]

From Plato, we learn:

> The first and the best victory is to conquer self. The
>
> greatest wealth is to live content with little.
>
> Happiness springs from doing good and helping others.[461]
>
> From Aristotle, we learn:
>
> The hardest victory is the victory over self.
>
> The best tragedies are conflicts between a hero and his destiny.
>
> Excellence is never an accident. It is always the result of high intention, sincere effort, and intelligent execution; it represents the wise choice of many alternatives - choice, not chance, determines your destiny.
>
> Health is a matter of choice, not a mystery of chance
>
> True happiness comes from gaining insight and growing into your best possible self. Otherwise all you're having is immediate gratification pleasure, which is fleeting and doesn't grow you as a person.[462]

Second, through the love of God. Should we beg for tribulations?

"The history of the coming of the Kingdom of God on earth is the story of God's love for mankind. Rejection of it is the greatest of all tragedies, and the root cause of the trials of men and nations".[463]

Bahá'u'lláh revealed:

> "O Son of Spirit! *The best beloved of all things in My sight is Justice;* turn not away therefrom if thou desirest Me, and neglect it not that I may confide in thee. By its aid

---

[459] The Báb, Sahifiy-i-'Adliyyih, pp. 32, 38.

[460] Nader Saiedi, Gate of the Heart.

[461] Plato, Azquotes.

[462] Aristotle, AzQuotes.

[463] Elizabeth Herrick, The Bahá'í Dispensation. 2, May 1923. Star of the West (Vol. 8).

thou shalt see with thine own eyes and not through the eyes of others, and shalt know of thine own knowledge and not through the knowledge of thy neighbor.
Ponder this in thy heart; how it behooveth thee to be. Verily justice is My gift to thee and the sign of My loving-kindness. Set it then before thine eyes."

They that are just and fair-minded in their judgment occupy a sublime station and hold an exalted rank. The light of piety and uprightness shineth resplendent from these souls. We earnestly hope that the peoples and countries of the world may not be deprived of the splendors of these two luminaries.[464]

"The essence of wealth is love for Me; whoso loveth Me is the possessor of all things, and he that loveth Me not is indeed of the poor and needy. This is that which the Finger of Glory and Splendor hath revealed".[465]

The essence of love is for man to turn his heart to the Beloved One, and sever himself from all else but Him, and desire naught save that which is the desire of his Lord.

The essence of all that We have revealed for thee is Justice, is for man to free himself from idle fancy and imitation, discern with the eye of oneness His glorious handiwork, and look into all things with a searching eye.[466]

"Verily, no God is there but Me, the Powerful, the Mighty, the All-Subduing, the Most Exalted, the Omniscient, the All-Wise." In truth, there is no God but Him, the Omnipotent Ruler of the worlds. Were it His Will, He would, through but a single word proceeding from His presence, lay hold on all mankind. Beware lest ye hesitate in your acceptance of this Cause — a Cause before which the Concourse on high and the dwellers of the Cities of Names have bowed down. Fear God, and be not of those who are shut out as by a veil. *Burn ye away the veils with the fire of My love,* and dispel ye the mists of vain imaginings by the power of this Name through which We have subdued the entire creation.[467]

In *The Hidden Words of Bahá'u'lláh*, we read:

O SON OF MAN!

To everything there is a sign; and the sign of Love is patience to endure the trials, *the destiny*, ordained by Me.[468]

O SON OF MAN!

The true lover yearneth for tribulation even as doth the rebel for forgiveness and the sinful for mercy.[469]

O SON OF MAN!

If adversity befall thee not in My path, how canst thou walk in the ways of them that are content with My pleasure? If trials afflict thee not in thy longing to meet Me, how wilt

---

[464] Bahá'u'lláh, Tablets of Bahá'u'lláh Revealed After the Kitáb-i-Aqdas. Emphasis added.

[465] Bahá'u'lláh, Tablets of Bahá'u'lláh Revealed After the Kitáb-i-Aqdas.

[466] Bahá'u'lláh, Tablets of Bahá'u'lláh Revealed After the Kitáb-i-Aqdas. Emphasis added.

[467] Bahá'u'lláh, The Kitáb-i-Aqdas. Emphasis added.

[468] Bahá'u'lláh, The Hidden Words. From the Arabic, No. 48

[469] Bahá'u'lláh, The Hidden Words. From the Arabic, No. 49

thou attain the light in thy love for My beauty? [470]

Third, through faith

Bahá'u'lláh taught: "The essence of faith is fewness of words and abundance of deeds; he whose words exceed his deeds, know verily his death is better than his life".[471]

O friends! Be not careless of the virtues with which ye have been endowed, neither be neglectful of your high destiny. Suffer not your labors to be wasted through the vain imaginations which certain hearts have devised. Ye are the stars of the heaven of understanding, the breeze that stirreth at the break of day, the soft-flowing waters upon which must depend the very life of all men, the letters inscribed upon His sacred scroll. With the utmost unity, and in a spirit of perfect fellowship, exert yourselves, that ye may be enabled to achieve that which beseemeth this Day of God. Verily I say, strife and dissension, and whatsoever the mind of man abhorreth are entirely unworthy of his station. *Center your energies in the propagation of the Faith of God.* Whoso is worthy of so high a calling, let him arise and promote it. Whoso is unable, it is his duty to appoint him who will, in his stead, proclaim this Revelation, whose power hath caused the foundations of the mightiest structures to quake, every mountain to be crushed into dust, and every soul to be dumbfounded. Should the greatness of this Day be revealed in its fullness, every man would forsake a myriad lives in his longing to partake, though it be for one moment, of its great glory—how much more this world and its corruptible treasures![472]

I admire Martin Seligman's work on being optimistic:

> "Pessimistic labels lead to passivity, whereas optimistic ones lead to attempts to change."

> "Authentic happiness derives from raising the bar for yourself, not rating yourself against others."

> "Success requires persistence, the ability to not give up in the face of failure. I believe that optimistic explanatory style is the key to persistence."[473]

> "When it comes to our health, there are essentially four things under our control: the decision not to smoke, a commitment to exercise, the quality of our diet, and our level of optimism. And optimism is at least as beneficial as the others."[474]

Fourth, through philanthropic deeds

Bahá'u'lláh Revealed:

> "The essence of charity is for the servant to recount the blessings of his Lord, and to render thanks unto Him at all times and under all conditions".[475]

---

[470] Bahá'u'lláh, The Hidden Words. From the Arabic, No. 50

[471] Bahá'u'lláh, Tablets of Bahá'u'lláh Revealed After the Kitáb-i-Aqdas.

[472] Bahá'u'lláh, Gleanings from the Writings of Bahá'u'lláh. Emphasis added

[473] Martin Seligman, <https://www.goodreads.com/work/quotes/28610>.

[474] Idem. <https://www.inspiringquotes.us/author/5934-martin-seligman>.

[475] Bahá'u'lláh, Tablets of Bahá'u'lláh Revealed After the Kitáb-i-Aqdas.

"Prove yourselves worthy of his trust and confidence in you, and withhold not from the poor the gifts which the grace of God hath bestowed upon you. He, verily, shall recompense the charitable, and doubly repay them for what they have bestowed."[476]

"The beginning of magnanimity is when man expendeth his wealth on himself, on his family and on the poor among his brethren in his Faith".[477]

'Abdu'l-Bahá taught:

> Man reacheth perfection through good deeds, voluntarily performed, not through good deeds the doing of which was forced upon him. And sharing is a personally chosen righteous act: that is, the rich should extend assistance to the poor, they should expend their substance for the poor, but of their own free will, and not because the poor have gained this end by force. For the harvest of force is turmoil and the ruin of the social order. On the other hand voluntary sharing, the freely chosen expending of one's substance, leadeth to soci-ety's comfort and peace. It lighteth up the world; it bestoweth honor upon humankind.[478]

Fifth, through self-sacrifice

Bahá'u'lláh revealed: "The laborer cuts up the earth with his plough, and from that earth comes the rich and plentiful harvest. The more a man is chastened, the greater is the harvest of spiritual virtues shown forth by him. A soldier is no good General until he has been in the front of the fiercest battle and has received the deepest wounds".[479]

'Abdu'l-Bahá, in *Paris Talks*, said: "Men who suffer not, attain no perfection. The plant most pruned by the gardeners is that one which, when the summer comes, will have the most beautiful blossoms and the most abundant fruit".[480]

'Abdu'l-Bahá is reported to have said:

> First of all, be ready to sacrifice your lives for one another, to prefer the general well-being to your personal well-being. Create relationships that nothing can shake; form an assembly that nothing can break up; have a mind that never ceases acquiring riches that nothing can destroy. If love did not exist, what of reality would remain? It is the fire of the love of God which renders man superior to the animal. Strengthen this superior force through which is attained all the progress in the world.[481]

Sixth, through severance from this world

Bahá'u'lláh revealed:

> "We must therefore labour to destroy the animal condition, till the meaning of humanity cometh to light."[482]

---

[476] Bahá'u'lláh, Gleanings from the Writings of Bahá'u'lláh, p. 109. Emphasis added

[477] Bahá'u'lláh, Tablets of Bahá'u'lláh Revealed After the Kitáb-i-Aqdas.

[478] 'Abdu'l-Bahá, Selections from the Writings of 'Abdu'l-Bahá.

[479] Bahá'u'lláh, The Summons of the Lord of Hosts.

[480] 'Abdu'l-Bahá, Paris Talks.

[481] 'Abdu'l-Bahá, Divine Philosophy.

[482] Bahá'u'lláh, The Call of the Divine Beloved. The Seven Valleys.

> The essence of love is for man to turn his heart to the Beloved One, and sever himself from all else but Him, and desire naught save that which is the desire of his Lord.
>
> The essence of detachment is for man to turn his face towards the courts of the Lord, to enter His Presence, behold His Countenance, and stand as witness before Him.
>
> The essence of understanding is to testify to one's poverty, and submit to the Will of the Lord, the Sovereign, the Gracious, the All-Powerful.
>
> The essence of true safety is to observe silence, to look at the end of things and to renounce the world.
>
> The source of all glory is acceptance of whatsoever the Lord hath bestowed, and contentment with that which God hath ordained.
>
> True remembrance is to make mention of the Lord, the All-Praised, and forget aught else beside Him.[483]

In *The Promulgation of Universal Peace* by 'Abdu'l-Bahá, we read:

> All the Prophets have drawn near to God through severance. We must emulate those Holy Souls and renounce our own wishes and desires. We must purify ourselves from the mire and soil of earthly contact until our hearts become as mirrors in clearness and the light of the most great guidance reveals itself in them.
>
> Bahá'u'lláh proclaims in the Hidden Words that God inspires His servants and is revealed through them. He says, "Thy heart is My home; sanctify it for My descent. Thy spirit is My place of revelation; cleanse it for My manifestation." Therefore, we learn that nearness to God is possible through devotion to Him, through entrance into the Kingdom and service to humanity; it is attained by unity with mankind and through loving-kindness to all; it is dependent upon investigation of truth, acquisition of praiseworthy virtues, service in the cause of universal peace and personal sanctification. In a word, nearness to God necessitates sacrifice of self, severance and the giving up of all to Him. Nearness is likeness.[484]

'Abdu'l-Bahá is reported to have said: "The afflictions which come to humanity sometimes tend to center the consciousness upon the limitations. This is a veritable prison. Release comes by making of the will a door through which the confirmations of the spirit come. They come to a man or woman who accepts his life with Radiant Acquiescence."[485]

Orcella Rexford wrote:

> Acquiescence means to "give in," to drop resistance, to tacitly agree. Divine acquiescence means to be submissive to the divine will. Everything in nature is acquiescent to the plan of the Universe and works in harmony with it except man. "Radiant acquiescence" means not only to give up your will to the Divine Will, but to do so joyfully and with radiance, knowing it is the best way in the end. The ordinary way of meeting the circumstances of life is to have a negative, passive submission to God's will and to blame every circumstance that was unfortunate on the "Will of God" and to be

---

[483] Bahá'u'lláh, Tablets of Bahá'u'lláh Revealed After the Kitáb-i-Aqdas.

[484] 'Abdu'l-Bahá, The Promulgation of Universal Peace in Writings and Utterances of 'Abdu'l-Bahá.

[485] 'Abdu'l-Bahá, Divine Philosophy.

unwillingly resigned to this condition and to do nothing to change it. Many become bitter and at enmity with life because of obstacles and calamities, and their faces register discontent and unhappiness.[486]

Seventh, through sanctity and holiness

Bahá'u'lláh revealed:

> O peoples of the world! Give ear unto the call of Him Who is the Lord of Names, Who proclaimeth unto you from His habitation in the Most Great Prison:
>
> Know thou of a truth that the seeker must, at the beginning of his quest for God, enter the Garden of Search. In this journey it behoveth the wayfarer to detach himself from all save God and to close his eyes to all that is in the heavens and on the earth. There must not linger in his heart either the hate or the love of any soul, to the extent that they would hinder him from attaining the habitation of the celestial Beauty. *He must sanctify his soul from the veils of glory and refrain from boasting of such worldly vanities, outward knowledge, or other gifts as God may have bestowed upon him.* He must search after the truth to the utmost of his ability and exertion, that God may guide him in the paths of His favour and the ways of His mercy. For He, verily, is the best of helpers unto His servants. He saith, and He verily speaketh the truth: *"Whoso maketh efforts for Us, in Our ways shall We assuredly guide him."* And furthermore: *"Fear God and God will give you know-ledge."*[487]

> 'Abdu'l-Bahá said: "Entrance into the Kingdom is through the love of God, through detachment, through sanctity and holiness, through truthfulness and purity, through steadfastness and faithfulness, and through self-sacrifice."[488]

In *The Secret of Divine Civilization*, by 'Abdu'l-Bahá, we read:

> A man should pause and reflect and be just: his Lord, out of measureless grace, has made him a human being and honored him with the words: "Verily, We created man in the goodliest of forms"—and caused His mercy which rises out of the dawn of oneness to shine down upon him, *until he became the wellspring of the words of God and the place where the mysteries of heaven alighted, and on the morning of creation he was covered with the rays of the qualities of perfection and the graces of holiness*. How can he stain this immaculate garment with the filth of selfish desires, or exchange this everlasting honor for infamy? "Dost thou think thyself only a puny form, when the universe is folded up within thee"?[489]

'Abdu'l-Bahá, in *Paris Talks*, said about happiness:

> There is a dialogue about The Progress of the Soul: *"Does the soul progress more through sorrow or through the joy in this world?"*
>
> 'Abdu'l-Bahá: "The mind and spirit of man advance when he is tried by suffering. The more the ground is ploughed the better the seed will grow, the better the harvest will be. Just as the plough furrows the earth deeply, purifying it of weeds and thistles, so suffering and tribulation free man from the petty affairs of this worldly life until he arrives at a state

---

[486] Rexford, Orcella. Radiant Acquiescence.

[487] Bahá'u'lláh, Gems of Divine Mysteries. Emphasis added.

[488] 'Abdu'l-Bahá, Some Answered Questions, No. 67.

[489] 'Abdu'l-Bahá, The Secret of Divine Civilization. Emphasis added.

of complete detachment. His attitude in this world will be that of divine happiness. Man is, so to speak, unripe: the heat of the fire of suffering will mature him. Look back to the times past and you will find that the greatest men have suffered most."

"He who through suffering has attained development, should he fear happiness?"

'Abdu'l-Bahá: "Through suffering he will attain to an eternal happiness which nothing can take from him. The apostles of Christ suffered: they attained eternal happiness."

"Then it is impossible to attain happiness without suffering?"

'Abdu'l-Bahá: "To attain eternal happiness one must suffer. He who has reached the state of self-sacrifice has true joy. Temporal joy will vanish."[490]

Aristotles' concept of happiness:

> Turning to happiness then, the aim of the whole Ethics; according to the original definition of Book I it is the activity or being-at-work chosen for its own sake by a morally serious and virtuous person. This raises the question of why play and bodily pleasures cannot be happiness, because for example tyrants sometimes choose such lifestyles. But Aristotle compares tyrants to children, and argues that play and relaxation are best seen not as ends in themselves, but as activities for the sake of more serious living. Any random person can enjoy bodily pleasures, including a slave, and no one would want to be a slave.
>
> Aristotle says that if perfect happiness is activity in accordance with the highest virtue, then this highest virtue must be the virtue of the highest part, and Aristotle says this must be the intellect (nous) "or whatever else it be that is thought to rule and lead us by nature, and to have cognizance of what is noble and divine". This highest activity, Aristotle says, must be contemplation or speculative thinking (energeia ... theōrētikē). This is also the most sustainable, pleasant, self-sufficient activity; something aimed at for its own sake. (In contrast to politics and warfare it does not involve doing things we'd rather not do, but rather something we do at our leisure.) However, Aristotle says this aim is not strictly human, and that to achieve it means to live in accordance not with our mortal thoughts but with something immortal and divine which is within humans. According to Aristotle, contemplation is the only type of happy activity it would not be ridiculous to imagine the gods having. The intellect is indeed each person's true self, and this type of happiness would be the happiness most suited to humans, with both happiness (eudaimonia) and the intellect (nous) being things other animals do not have. Aristotle also claims that compared to other virtues, contemplation requires the least in terms of possessions and allows the most self-reliance, "though it is true that, being a man and living in the society of others, he chooses to engage in virtuous action, and so will need external goods to carry on his life as a human being.[491]

The supervision of the transformation and cultivation processes

> Here, there are some ideas to start thinking about defining parameters and formulating indicators in the context of The Root Cause and the Unique Method of Science:

---

[490] 'Abdu'l-Bahá, Paris Talks. Emphasis added.

[491] Aristotle, Nicomachean Ethics.

For ourselves, the author believes it must be an individual reflection:
- Of all my faults, which ones seem to be more recidivist?
- Have I been able to find the root cause of my wrongdoing?
- Have I followed the steps suggested by The Master to address my misconduct?
- Have I persevered and overcome my behaviors that seem to be obstacles to reaching my inexorable destiny?
- Have I found my unique path to tread in my life?

    A suggested path to monitor and evaluate our knowledge of reaching the destiny of agriculture:

In *Genetic and molecular regulation of fruit and plant domestication traits in tomato and pepper*, by Ilan Paran and Esther van der Knaap, we find:

Two of the key traits that were selected during domestication of pepper were non-deciduous fruit that remained on the plant until harvest and the change in position from erect to pendant fruit. This latter change may be associated with an increase in fruit size, better protection from sun exposure, and predation by birds. Other changes associated with domestication and variety improvement were fruit appearance and reduced pungency. While wild peppers can be found in several basic shapes including oval, spherical, or elongated, continued selection resulted in a large increase of shape variation and tremendous increases in fruit mass. Selection also resulted in yellow, orange, and brown[492] fruit colours in addition to the wild-type red, which occurs in all cultivated pepper species. Lastly, another important selection was that of non-pungent fruits.[493]

In plants and animals:
- What perfections and potentials of this species seem plausible to become actualities?
- When choosing between the following alternatives: domestication, caring for plants, increasing the number of soil microorganisms, finding mutualistic relationships with animals, grafting, or crossings, what seems to be the best option to reach the destiny of this species?
- Which seems to be the number of planting cycles to detect inheritance of the desired features?

---

[492] Photo 48663888 © Ian Andreiev | Dreamstime.com

[493] Ilan Paran and Esther van der Knaap, Genetic and molecular regulation of fruit and plant domestication traits in tomato and pepper.

6.4 Research Steps for the Unique Method

The author made an effort to relate this method to the First Station of Unity, as revealed in The Seven Valleys. These Valleys "mark the wayfarers' journey from their mortal abode to the heavenly homeland"[494]; in the First Station of Unity, we find: "After passing through the Valley of Knowledge, which is the last station of limitation, the wayfarer cometh to The First Station of Unity and drinketh from the cup of *oneness*, and gazeth upon the manifestations of *singleness*. *In this station he pierceth the veils of plurality*, fleeth the realms of the flesh, and ascendeth unto the heaven of unity."[495]

The Master has delineated the method's seven steps, and that is enough. It is for the individual to find the path to reach his (her) particular aspiration; and for the scientist to find the unique method, as was the case with producing graphene from soybean oil.

6.5 The Root Phase of the Farmers' Situation

To help the reader understand the farmers' example, the author emphasizes the discourse's fundamental notions reasonably linked to each one of the Causes.

From 'Abdu'l-Bahá, who passed away in 1921 not long after the end of World War I, we learn:

> O ye lovers of truth, ye servants of humankind! Out of the flowering of your thoughts and hopes, fragrant emanations have come my way, wherefore an inner sense of obligation compelleth me to pen these words.
>
> Ye observe how the world is divided against itself, how many a land is red with blood and its very dust is caked with human gore. The fires of conflict have blazed so high that never in early times, not in the Middle Ages, not in recent centuries hath there ever been such a hideous war, a war that is even as millstones, taking for grain the skulls of men. Nay, even worse, for flourishing countries have been reduced to rubble, cities have been leveled with the ground, and many a once prosperous village hath been turned into ruin. Fathers have lost their sons, and sons their fathers.
> Mothers have wept away their hearts over dead children. Children have been orphaned, women left to wander, vagrants without a home. From every aspect, humankind hath sunken low. Loud are the piercing cries of fatherless children; loud the mothers' anguished voices, reaching to the skies.
>
> And the breeding ground of all these tragedies is *prejudice: prejudice of race and nation, of religion, of political opinion*; and the *root cause* of *prejudice* is blind *imitation* of the past — *imitation* in religion, in racial attitudes, in national bias, in politics. So long as this aping of the past persisteth, just so long will the foundations of the social order be blown to the four winds, just so long will humanity be continually exposed to direst peril.[496]

Let us examine the situation of farmers and prejudice within the context of the Root Cause and the Unique Method of science: There is a correlation between the various kinds of

---

[494] Bahá'u'lláh, The Seven Valleys, The Call of the Divine Beloved.

[495] Idem. Emphasis added.

[496] Selections from the Writings of 'Abdu'l-Bahá. Emphasis added.

discrimination: dark skin, women, poorly educated, poorly dressed, and of course, being part of the farmers' social class, and its *cumulative* effect on *how much* society values what peasants produce.

What happened in the United States when racial separation prevailed as the leading cause of the civil war (1861-1865)?

For 110 years, the numbers stood as gospel: 618,222 men died in the Civil War, 360,222 from the North, and 258,000 from the South — by far the greatest toll of any war in American history.

But new research shows that the numbers were far too low.

By combing through newly digitized census data from the 19th century, J. David Hacker, a demographic historian from Binghamton University in New York, has recalculated the death toll and increased it by more than 20 percent — to 750,000.[497]

Slavery on plantations was the main reason for the war. Was Rwanda's destiny similar?

The Belgian colonial occupation had a much more lasting effect in Rwanda. The most lasting effect was how the colonial authorities racialized the differences between Hutu, Twa, and Tutsi.

Rwanda gained independence from Belgium in 1962, but the post-colonial period was marred by ethnically motivated violence. This violence culminated into the 1994 Rwandan genocide in which more than 800.000 Tutsi people were killed, including thousands of Hutu people who were either part of the opposition or who had refused to take part in the killings.[498]

Hutus were farmers and Tutsis cattle herders. "Upon first entering the region, German colonizers rationalized subjugation of Rwanda's large Hutu population under the flawed assumption that the Tutsi were more Caucasian and thus more fit to rule."[499]

After the genocide, women of both sides met. They consulted and decided to adopt the orphans of the other tribe — Hutu women raising Tutsy orphans, Tutsy women adopting Hutu orphans (Out of Madness, A Matriarchy by Kimberlee Acquaro and Peter Landesman).

What was the fate of those regimens stubbornly supporting oppressive systems of governance?

ʤʠ

Clues in how to proceed in a new scientific method to approach reality

Because the author believes that this is a new scientific method to approach reality, we can find clues in how to proceed in the following quotations about uniqueness. In, Why is everyone's DNA so unique? The part of DNA which makes us unique, we read:

Recombination is a process where sections of DNA are traded between the chromosomes that make up a pair. After recombination, the chromosomes will look somewhat like a quilt because they are made up of DNA from both parents. The total amount of DNA on each

---

[497] Hacker, J. David. New Estimate Raises Civil War Death Toll.

[498] South African History Online (SAHO)

[499] What Impact Did the Belgian Presence in Rwanda Have to Spark Further Conflict? By Stephen Skok

chromosome should not change in a significant way, because a portion of our mom's chromosome was traded for the same portion on our dad's chromosome.

But how does this create a unique sequence? The human DNA sequence consists of nearly 3 billion DNA base pairs and the order of these base pairs is nearly identical from person to person, but sometimes there are random changes in the sequence and we call these changes variants. The combination of all of our variants make up the 0.1% difference in our DNA and helps give us a unique sequence. This means when chromosome pairs come together, the chromosomes we inherit from our mom are slightly different from the chromosomes we inherit from our dad thanks to the many DNA variants on each of the chromosomes.

When recombination happens, the chromosomes are essentially trading DNA variants amongst themselves. This process helps drive evolution by creating a slightly new version of the DNA: each chromosome we have is a unique quilt of DNA, representing segments of the genome that have been passed down from generation to generation. This shuffling has helped drive evolution through time, and ultimately has helped write our genome and our story.[500]

Now let us think about, "The uniqueness of each living organism is reflected in its DNA pattern. In that sense, it is of core importance for biologists to study the genome of living beings, in order to understand their growth, functioning and reproduction."[501] And we learn from Monica May, "*Surprising science: Not all our cells have the same DNA.*"[502]

In an exciting article about areal uniqueness, *Uniquity: A general metric for biotic uniqueness of sites*, we find:

- Uniquity – a new metric for biotic uniqueness – is presented and validated.
- Uniquity can be applied sites with different areal representativity.
- Uniquity is applicable to classical survey data and DNA data.
- Uniquity has a better correspondence with the number of rare species than competing biodiversity metrics.

Species richness is unrivalled as the most reported biodiversity metric in ecological and conservation research. Unfortunately, species richness ignores the scale- dependency of biodiversity.

We propose the metric uniquity, a quantitative and spatially scalable measure of uniqueness of a site based on a species-by-site matrix and a site-by-habitat type classification with area weights for habitat types correcting for sampling biases.

An example of uniquity is presented using vascular plant data from 130 sites representing a larger region (Denmark). We demonstrate the importance of the scale parameter of uniquity for the prediction of independent uniqueness indices calculated from species distribution data and the number of recorded red-listed species.

---

[500] Riposati, Andrea. Why is everyone's DNA so unique? The part of DNA which makes us unique.

[501] Leandro Gomes, et al. Synchronous searching for DNA patterns.

[502] Monica May, Surprising science: Not all our cells have the same DNA.

We compare the performance of uniquity with the performance of the indices Local Contribution to Beta Diversity (LCBD) and Range Rarity Richness (RRR), and we investigate its sensitivity to small sample size and poorly resolved habitat classification.

We assess the performance of the uniquity metric applied to DNA metabarcoding data for plants, fungi and eukaryotes from the same set of study sites.

Uniquity is a strong predictor of site uniqueness based on national distribution data and also correlates neatly with the observed number of red listed species. Uniquity based on DNA metabarcoding corresponds well with the number of red listed species observed.

Perspective: Uniquity is generally applicable to biotas sampled with comparable effort, including field inventories, trap sampling, and DNA metabarcoding data. To our knowledge uniquity is the first index of uniqueness that explicitly considers spatial scale and sampling biases, while simultaneously accepting non-annotated DNA-data as input. Based on our study we offer general recommendations for further use and testing of uniquity as conservation value metric.[503]

In another article about foliar microbiome uniqueness, The plant is crucial: specific composition and function of the phyllosphere microbiome of indoor ornamentals, we find:

The plant microbiome is a key determinant of plant health. Less is known about the phyllosphere microbiota and its driving factors in built environments. To study the variability of the microbiome in relation to plant genotype and climate under different controlled conditions, we investigated 14 phylogenetically diverse plant species grown in the greenhouses of the Botanical Garden in Graz (Austria). All investigated plants showed *specific bacterial abundances* of up to 106 CFU cm−2 on their leaves. Bacterial diversity (H : 2.4–7.9) and number of putative OTUs (461–2013) were strongly plant species dependent. Statistical analysis showed a significantly higher correlation of community composition to plant genotype in comparison to the ambient climatic variables. In addition to the microbiome structure, we studied the antagonistic potential towards the foliar pathogen Botrytis cinerea as functional indicator. A high proportion of isolates (up to 58%) were able to inhibit pathogen growth by production of volatile organic compounds (VOCs). Data of structure and function were linked: frequently isolated VOCs producers (e.g. Bacillus and Stenotrophomonas) were highly present in phyllosphere communities, which were dominated by members of Firmicutes. This study indicates that indoor ornamentals *feature a distinct, stable microbiota on leaves irrespective of the indoor climate*.[504]

Cultivation

'Abdu'l-Bahá in a Talk at Bahá'í Women's Reception. Hotel La Salle, Chicago, Illinois, mentioned in *The Promulgation of Universal Pease*, said:

When we look upon the kingdoms of creation below man, we find three forms or planes of existence which await education and development. *For instance, the function of a gardener is to till the soil of the mineral kingdom and plant a tree which under his training and cultivation will attain perfection of growth. If it be wild and fruitless, it may be made fruitful and prolific by grafting. If small and unsightly, it will become lofty,*

---

[503] Ejrnæs, Rasmus, Uniquity: A general metric for biotic uniqueness of sites.

[504] Ortega, Rocel Amor, et al., The plant is crucial: specific composition and function of the phyllosphere microbiome of indoor ornamentals.

beautiful and verdant under the gardener's training, whereas a tree bereft of his cultivation retrogresses daily, its fruit grows acrid and bitter as the trees of the jungle, or it may become entirely barren and bereft of its fruitage. Likewise, we observe that animals which have undergone training in their sphere of limitation will progress and advance unmistakably, become more beautiful in appearance and increase in intelligence. For instance, how intelligent and knowing the Arabian horse has become through training, even how polite this horse has become through education. As to the human world: It is more in need of guidance and education than the lower creatures. Reflect upon the vast difference between the inhabitants of Africa and those of America. Here the people have been civilized and uplifted; there they are in the utmost and abject state of savagery. What is the cause of their savagery and the reason of your civilization? It is evident that this difference is due to education and the lack of education. Consider, then, the effectiveness of education in the human kingdom. It makes the ignorant wise, the tyrant merciful, the blind seeing, the deaf attentive, even the imbecile intelligent. How vast this difference. How wide the chasm which separates the educated man from the man who lacks teaching and training. *This is the effect when the teacher is merely an ordinary teacher.*

But—praise be to God!—your Teacher and Instructor is Bahá'u'lláh. He is the Educator of the Orient and Occident. He is the Teacher of the very world of divinity and spirituality, the Sun of Truth, the Word of God. The lights of His education are radiating even as the sun. See what it has accomplished, how it is developing all humanity so that I, a Persian, have come to this meeting of revered souls upon the American continent and am standing here expounding to you in the greatest love.

This is through the training of Bahá'u'lláh, which can unite and has united these hearts. In this way it has enlightened the world. Even so it has breathed the spirit of God into men. Even so it has resuscitated the hearts of men.

Therefore, praise be to God that you have been brought under the education of this One Who is the very Sun of Reality and Who is shining resplendently upon all humankind, endowing all with a life that is everlasting.

Praise be to God a thousand times!"[505]

'Abdu'l-Bahá also said in *The Promulgation of Universal Peace*:

In the spiritual world, the divine bestowals are infinite, for in that realm there is neither separation nor disintegration which characterize the world of material existence. Spiritual existence is absolute immortality, completeness and unchangeable being. Therefore we must thank God that He has created for us both material blessings and spiritual bestowals. He has given us material gifts and spiritual graces, outer sight to view the lights of the sun and inner vision by which we may perceive the glory of God. He has designed the outer ear to enjoy the melodies of sound and the inner hearing wherewith we may hear the voice of our creator. We must strive with energies of heart, soul and mind to develop and manifest the perfections and virtues latent within the realities of the phenomenal world, for the human reality may be compared to a seed. If we sow the seed, a mighty tree appears from it. The virtues of the seed are revealed in the tree; it

---

[505] 'Abdu'l-Bahá, The Promulgation of Universal Peace in Writings and Utterances of 'Abdu'l-Bahá. Emphasis added.

puts forth branches, leaves, blossoms, and produces fruits. *All these virtues were hidden and potential in the seed. Through the blessing and bounty of cultivation these virtues became apparent.*

Similarly the merciful God our creator has deposited within human realities certain virtues latent and potential. Through education and culture, these virtues deposited by the loving God will become apparent in the human reality even as the unfoldment of the tree from within the germinating seed.[506]

In a talk about *The Station of Man and His Progress After Dead*, 'Abdu'l-Bahá said:

Man is in the ultimate degree of materiality and the beginning of spirituality; that is, he is at the end of imperfection and the beginning of perfection. He is at the furthermost degree of darkness and the beginning of the light. That is why the station of man is said to be the end of night and the beginning of day, meaning that he encompasses all the degrees of imperfection and that he potentially possesses all the degrees of perfection. He has both an animal side and an angelic side, and the role of the educator is to so train human souls that the angelic side may overcome the animal. Thus, should the divine powers, which are identical with perfection, overcome in man the satanic powers, which are absolute imperfection, he becomes the noblest of all creatures, but should the converse take place, he becomes the vilest of all beings. That is why he is the end of imperfection and the beginning of perfection.

Further down in the same conversation, we find:

This is the wisdom of the appearance of the Prophets: to educate humanity, that this lump of coal may become a diamond and this barren tree may be grafted and yield fruit of the utmost sweetness and delicacy. And after the noblest stations in the world of humanity have been attained, further progress can be made only in the degrees of perfection, not in station, for the degrees are finite but the divine perfections are infinite.[507]

'Abdu'l-Bahá, in the Need for an Educator:

When we consider existence, we observe that the mineral, the vegetable, the animal, and the human realms, each and all, are in need of an educator.

If the land is deprived of a cultivator, it becomes a thicket of thriving weeds, but if a farmer is found to cultivate it, the resulting harvest provides sustenance for living things. It is therefore evident that the land is in need of the farmer's cultivation. Consider the trees: If they remain uncultivated, they bear no fruit, and without fruit they are of no use. *But when committed to a gardener's care, the barren tree becomes fruitful, and, through cultivation, crossing, and grafting, the tree with bitter fruit yields sweet fruit.* These are rational arguments, which are what the people of the world require in this day.

Consider likewise the animals: If an animal is trained, it becomes *domesticated*, whereas man, if he is left without education, becomes like an animal. Indeed, if man is abandoned to the rule of nature, he sinks even lower than the animal, whereas if he is educated he becomes even as an angel. For most animals do not devour their own kind, but men in the Sudan, in the middle of Africa, rend and eat each other.

Further down in the same talk:

---

[506] Idem. Emphasis added.

[507] 'Abdu'l-Bahá, Some Answered Questions. No. 64. Emphasis added.

Now observe that it is education that brings East and West under man's dominion, produces all these marvellous crafts, promotes these mighty arts and sciences, and gives rise to these new discoveries and undertakings. Were it not for an educator, the means of comfort, civilization, and human virtues could in no wise have been acquired. If a man is left alone in a wilderness where he sees none of his own kind, he will undoubtedly become a mere animal. It is therefore clear that an educator is needed.

But education is of three kinds: material, human, and spiritual. Material education aims at the growth and development of the body, and consists in securing its sustenance and obtaining the means of its ease and comfort. This education is common to both man and animal.

Human education, however, consists in civilization and progress, that is, sound governance, social order, human welfare, commerce and industry, arts and sciences, momentous discoveries, and great undertakings, which are the central features distinguishing man from the animal.

As to divine education, it is the education of the Kingdom and consists in acquiring divine perfections. *This is indeed true education, for by its virtue man becomes the focal centre of divine blessings and the embodiment of the verse "Let Us make man in Our image, after Our likeness." This is the ultimate goal of the world of humanity*.[508]

The Master also said:

To enter the Kingdom is easy, but to remain firm and constant is difficult. *The planting of trees is easy, but their cultivation and training to strengthen their roots and to make them firm is difficult.* Now, as thou art a firm tree, thou shalt certainly grow and send out branches, leaves and blossoms and bear fruits.

These branches, leaves, blossoms and fruits are the souls who may be guided, through the providence of God, by thee. Therefore, thou art confirmed and strengthened.[509]

Domestication

On Human Origins: A Bahá'í Perspective, by Craig Loehle, we learn:

The period around 10,000 years ago represents a *unique crisis and turning point in human history*. During the period 12,000 to 10,000 B.P., rapid global warming caused the retreat of the global ice sheets of the last glaciation. This rapid warming was accompanied by massive shifts in local climates and vegetation such as expansion of the grasslands in the American West. Animals previously adapted to cold climates, particularly larger mammals, were unable to adapt, and many became extinct. By 10,000 B.P. the large mammal herds in most areas outside Africa, upon which early humans had depended, were either reduced in number or extinct. We can think of this time as the historical expulsion from the Garden, in a sense. *This crisis forced people into new food sources including fishing, more sophisticated hunting techniques, and agriculture* (Geist, Life), thus leading directly to the establishment of more advanced culture and technology. In particular, *the earliest dates known for domestication of both plants and animals are in the period 12,000– 10,000 B. P. from the Middle East*

---

[508] 'Abdu'l-Bahá, Some Answered Questions. No 3. Emphasis added.

[509] 'Abdu'l-Bahá, Tablets of 'Abdu'l-Bahá 'Abbás. Emphasis added.

(Simmons et al., "Neolithic"). The period around 10,000 B. P. is when the earliest villages (permanent settlements) were established, also in the Middle East.[510]

The level of the transformation asked from us is not a minor thing. In *"Plant Domestication"* by Nino Brown, we find the following:

There is no better example of how these traits can come together to produce something truly nutritive, than corn and its progenitor teosinte.[511] [512] Teosinte is a weedy looking grass with a small seed head made up of only two rows of several small, hard seeds. It looks as though it came out of a gardener's nightmare. But somehow this wild and bushy bunch grass became the robust, single stalked behemoth we now know as field corn, whose constituents are used as ingredients in a plethora of food products consumed heartily by Americans daily."[513]

In, Plant domestication, a unique opportunity to identify the genetic basis of adaptation, we read:

Plant domestication fundamentally altered the course of human history. The adaptation of plants to cultivation was vital to the shift from hunter–gatherer to agricultural societies, and it stimulated the rise of cities and modern civilization. Humans still rely on crops that were domesticated >10,000 years ago in such diverse places as Central America, New Guinea, and the Fertile Crescent. Nonetheless, modern humans are reliant on a surprisingly small number of crops: Nearly 70% of the calories consumed by humans are supplied by only 15 crops. The cereals are particularly important, with five crops (rice, wheat, maize, sugarcane, and barley) contributing more than half of the calories consumed.

Despite the critical importance of these crops, in most cases little is known about their domestication. Some obvious questions pertain to the domesticators: Who

were they? How did they identify the incipient crop? What were their cultivation methods? Other questions concern crop history:

---

[510] Loehle, Craig. On Human Origins: A Bahá'í Perspective. Emphasis added.

[511] Photo 25535061 © Melica | Dreamstime.com

[512] Nicolle R. Fuller, Sayo Studio. Web January 2023 <www.SayoStudio.com>. Email info@SayoStudio.com

[513] Nino Brown, Plant Domestication.

What was the wild progenitor of the modern crop? Did domestication occur more than once? If so, where? The application of phylogeographic methods is beginning to inform the answers to this latter set of questions, but the picture for any one crop remains far from complete".[514]

Melinda A. Zeder in, *Core questions in domestication research*, says:

"Domestication of plants and animals marks a major transition in human history that represents a vibrant area of interdisciplinary scientific inquiry. Consideration of three central questions about domestication—what it is, what it does, and why it happened—provide a unifying framework for diverse research on the topic.

Domestication is defined in terms of a coevolutionary mutualism between domesticator and domesticate and is distinguished from related but ultimately different processes of management and agriculture. Domestication results in a range of genotypic, phenotypic, plastic, and contextual impacts that can be used as markers of evolving domesticatory relationships. A consideration of causal scenarios finds greater empirical support for explanatory frameworks grounded in niche-construction theory over those derived from optimal foraging theory.[515]

K. Kris Hirst agrees with Melinda A. Zeder, in Plant Domestication. Dates and Locations of Human Farming Advances, says: "Plant domestication is a slow and tiresome process that is only successful when both parties—humans and plants—benefit from each other through a mutualistic relationship. The result of thousands of years of this symbiosis came to be known as coevolution."

Coevolution

Coevolution describes the process of two species evolving to suit each other's needs. Plant domestication through artificial selection is one of the best examples of this.

When a human tends a plant with favorable attributes, perhaps because it has the largest and sweetest fruits or most resilient husk, and saves the seeds to replant, they are essentially guaranteeing the continuation of that particular organism.

In this way, a farmer can select for the properties they desire by giving special treatment only to the best and most successful plants. Their crop, in turn, starts to take on the desirable properties the farmer selected for and disadvantageous attributes are extinguished over time.

Though plant domestication via artificial selection is not foolproof—complications include long-distance trading and uncontrolled seed dispersal, accidental cross- breeding of wild and domesticated plants, and unexpected disease wiping out genetically similar plants—it demonstrates that human and plant behavior can become intertwined. When plants do what is expected of them by humans, humans work to preserve them.[516]

Can these quotations guide us to new approaches of research of plants and animals' potential higher destiny?

---

[514] K. Kris Hirst

[515] Melinda A. Zeder. Core questions in domestication research. Emphasis added.

[516] K. Kris Hirst. Plant Domestication. Dates and Locations of Human Farming Advances

7   The Final Cause and the Explanatory Method

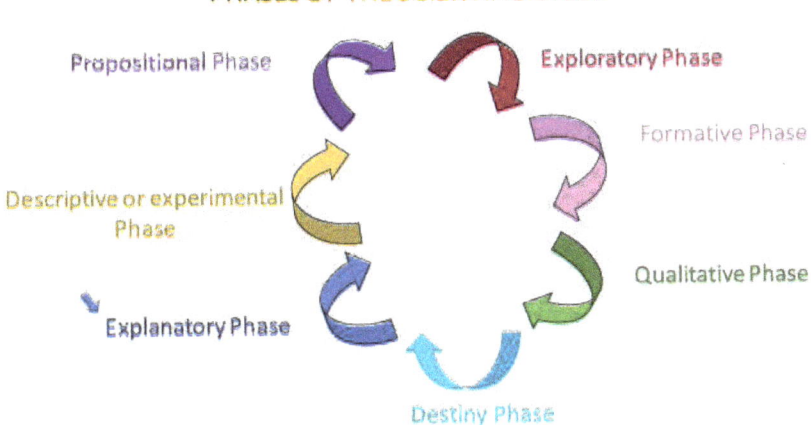

> "The classical concept of final causality is regarded as useless for physicists. And even for biologists, the notion of purpose in the natural order is retained largely for instructional reasons: e.g., what is the function of any evolved adaptation? Certainly not for anything more immediate than survival."[517]

For many biologists there is no purpose in creation, just chance:

> Conway Morris, a paleontologist at Cambridge University, has long argued that life in the universe is probably very rare—but at the same time, where it does take root, must almost certainly lead to the evolution of consciousness.
>
> A committed Darwinian, Conway Morris nevertheless disagreed with his fellow paleontologist, the late Stephen Jay Gould, when the latter famously argued that if you could 'rewind the tape' and start over, the history of life on earth would have been vastly different—and human evolution would never repeat itself. Chance, for Gould, could not be counted on to replay itself for our benefit.
>
> But Conway Morris argues that in the grand scheme, evolution will not be reduced to chance: constraints built into life at the most fundamental level guarantee that life is going to follow the same evolutionary pathways to achieve limbs, respiration, vision, balance, an immune system, indeed all the remarkable features we associate with living things across the great spectrum of life.[518]

---

[517] Farrell, John. Why Teleology Isn't Dead

[518] Idem.

The Final Cause and the Explanatory Method

7.1 Indeed, there is a purpose in nature

Let me introduce the explanatory method of science with the following reflection, which demonstrates to scientists that, indeed there is purpose in nature: "Protect me from violent tests and preserve and shelter me in the strongly fortified fortress of Thy Covenant and Testament."[519]

This picture corresponds to the Khotyn Fortress, "Under the rule of Stephen the Great of Moldavia the fortress was greatly expanded. Under his leadership, new 5–6-meter (16–20 ft) wide and 40 meters (130 ft) high walls were built."[520] [521]

God protects us in many other ways. The following quote by 'Abdu'l-Bahá, inspired the author to generate the ideas below:

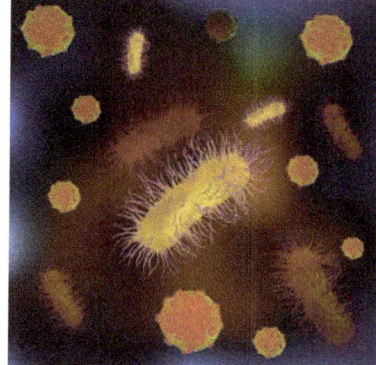

"But true felicity for the animal consists in passing from the animal world into the human realm, like the microscopic beings that, through the air and the water, enter into the body of man, are assimilated, and replace that which has been consumed in his body. This is the greatest honour and felicity for the animal world, and no greater honour can be conceived for it."[522]

"Scientists discovered a deluge of microbes[523] in the air, discovering that a single square meter of the planet's surface could be bathed with hundreds of millions of viruses—and tens of millions of bacteria—in a single day."[524]

Abdu'l-Bahá said: "The more microscopic animals exist in the soil, the better the plants will grow."[525]

The ignorance of the consequences of decisions made during the planting, growing, and harvesting phases concerning the effects on the topsoil layer of the intensive use of herbicides, fungicides, pesticides, chemical fertilizers, and heavy machinery during many cycles.

---

[519] Bahá'í Prayers.

[520] Web. April 2021 <https://en.wikipedia.org/wiki/Khotyn_Fortress>.

[521] Photo 168306108 © Mykola Ivashchenko | Dreamstime.com

[522] 'Abdu'l-Bahá, Some Answered Questions.

[523] Illustration 137135136 © Ustyna Shevhcuk | Dreamstime.com

[524] Mindy Weisberger, Billions of Viruses Are Falling to Earth Right Now (But That Isn't Why You Have the Flu).

[525] 'Abdu'l-Bahá, Additional Tablets, Extracts and Talks.

Additionally, what could happen to the worms, fungus, insects, and microorganisms that were in abundance in the soil, when there is no cover crop for months?[526]

The following is a concrete example if we assume that the soils of the Central Valley of California are almost sterile:

Ecosystem productivity *commonly increases asymptotically with plant species diversity*, and determining the mechanisms responsible for this well-known pattern is essential to predict *potential changes in ecosystem productivity with ongoing species loss*. Previous studies attributed the asymptotic diversity–productivity pattern to plant competition and differential resource use (e.g., niche complementarity). Using an analytical model and a series of experiments, we demonstrate theoretically and empirically that host-specific soil microbes can be major determinants of the diversity–productivity relationship in grasslands. In the presence of soil microbes, *plant disease* decreased with increasing diversity, and productivity increased nearly 500%, primarily because of the *strong effect of density- dependent disease on productivity at low diversity*. Correspondingly, *disease was higher in plants* grown in conspecific-trained soils than heterospecific-trained soils (demonstrating host-specificity), and productivity increased and host-specific *disease* decreased with increasing community diversity, suggesting that *disease was the primary cause of reduced productivity in species poor treatments. In sterilized, microbe-free soils, the increase in productivity with increasing plant species number was markedly lower than the increase measured in the presence of soil microbes*, suggesting that niche complementarity was a weaker determinant of the diversity– productivity relationship.

Our results demonstrate that soil microbes play an integral role as determinants of the diversity–productivity relationship.[527]

Microbes fall on the cuticle of leaves[528]. "In higher plants, a cuticle covers the outer epidermal surface of most above-ground tissues, such as leaves, fruit, and floral organs. The cuticle is well known for its *functions as a diffusion barrier* limiting water and solute transport across the apoplast and for its protection of the plant *against chemical and mechanical damage, as well as pest and*

---

[526] Photo 18958496 © Razvanjp | Dreamstime.com

[527] Schnitzer, Stefan A. et al., Soil microbes drive the classic plant diversity-productivity pattern. Emphasis added.

[528] Photo 15530693 © Eugene Shapovalov | Dreamstime.com

pathogen attack."[529]

"Hundreds of millions of viruses—and tens of millions of bacteria fall in one square meter in a single day"[530], on centipedes' exoskeleton[531] and on the bark of trees.[532] They fall on creeks, rivers, lakes, and on animals' skin and feathers.[533]

"If you were to tear off and spread out the average adult's skin, it would cover approximately 22 square feet (2 square meters)."[534] [535]

Neutrophils are a type of white blood cell. In fact, most of the white blood cells that lead the immune system's response are neutrophils. There are four other types of white blood cells. Neutrophils are the most plentiful type, making up 55 to 70 percent of your white blood cells.[536]

---

[529] Markus Riederer, Biology of the plant cuticle. Emphasis added.

[530] Mindy Weisberger, Billions of Viruses Are Falling to Earth Right Now (But That Isn't Why You Have the Flu).

[531] Photo 3336393 / Cicada Shell © Jun Ji | Dreamstime.com

[532] Photo 57557858 © Karunakaran Parameswaran Pillai | Dreamstime.com

[533] Photo 77614153 © Megatalia20 | Dreamstime.com

[534] Remy Melina, How Much Does Your Skin Weigh? January 13, 2011

[535] Photo 105697569 © Mihail39 | Dreamstime.com

[536] Susan York Morris, Understanding Neutrophils: Function, Counts, and More.

The Final Cause and the Explanatory Method

The nasal mucosa plays an important role in mediating immune responses to allergens and infectious particles which enter the nose. It helps prevent allergens and infections from invading the nasal cavity and spreading to other body structures, for example the lungs. The mucus secreted by and which lines the mucosa provides a physical barrier against invasion by pathogens (harmful microorganisms). It is sticky and t nasal cavity.[537] [538]

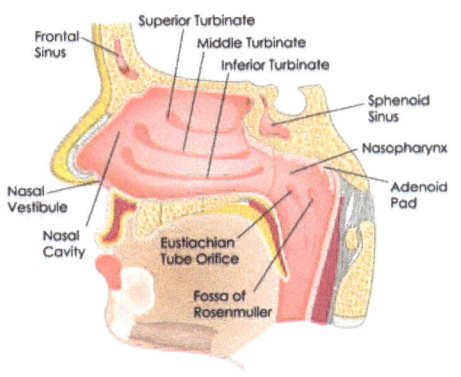

Mucus is also home to phages, viruses that infect and kill bacteria. They can be found wherever bacteria reside, but Barr and his colleagues noticed that there were even more phages in mucus than in mucus-free areas just millimeters away. The saliva surrounding human gums, for example, had about five phages to every bacterial cell, while the ratio at the mucosal surface of the gum itself was closer to 40 to 1. "That spurred the question," Barr says. "What are these phages doing? Are they protecting the host?"[539]

"The bronchi branch into smaller bronchi that evolve into alveoli. Bronchi tubes parallel many blood vessels that perform the respiratory gas exchange. A human adult probably has about 300 million bronchi that could cover an area of around 180m^2. The surface area of a pair of lungs is equal to that of a tennis court.[540]

On the right are lungs whose owner decided to smoke the phages with cigarettes and sterilize his bronchi and alveoli by drinking alcohol, weakening his immune system, making them more vulnerable to harmful microorganisms entering the lungs. On the left healthy lungs, keeping the phages alive and capable of effectively controlling those harmful microorganisms that come into daily contact with the lungs.[541]

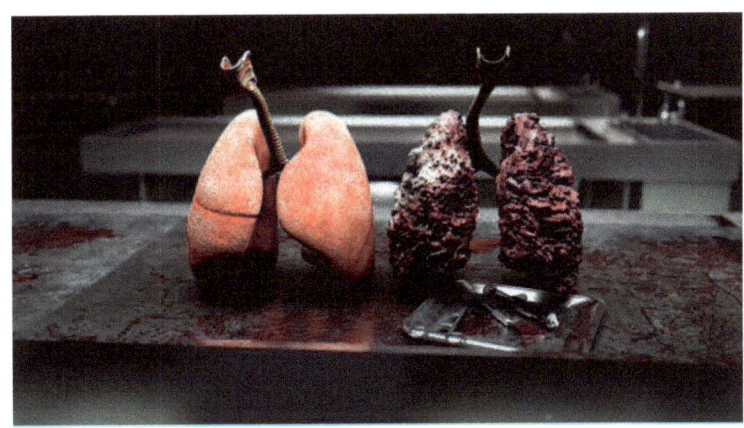

---

[537] Virtual Medical Center, Anatomy and Physiology of the Nasal Cavity (Inner Nose) and Mucosa.

[538] Illustration 31606461 / Anatomical Drawing © Snapgalleria | Dreamstime.com

[539] Beth Skwarecki, Friendly Viruses Protect Us Against Bacteria.

[540] The surface area of a pair of lungs is equal to that of a tennis court https://vedanadosah.cvtisr.sk/en/

[541] Photo 85299525 © Pavel Chagochkin | Dreamstime.com

Cell organelle's membranes also protect: They have doors (pores) for well-selected guests and work as barriers for those not welcome in the fortress.[542] Lysosome: "A lysosome is a membrane-bound cell organelle that contains digestive enzymes. ... They break down excess or worn-out cell parts. They may be used to destroy invading viruses and bacteria. If the cell is damaged beyond repair, lysosomes can help it to self-destruct in a process called programmed cell death, or apoptosis."[543]

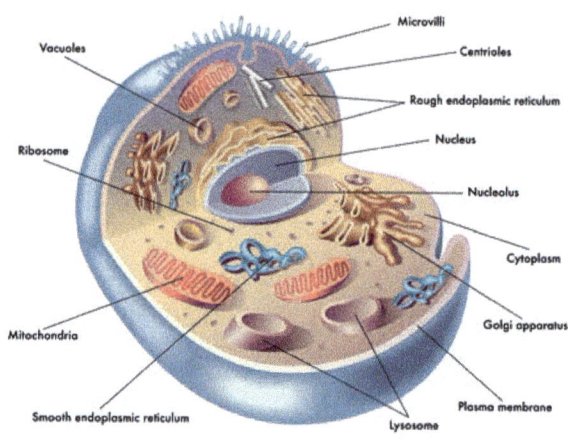

Of course, there is more to say about the defense system of plants, animals, and humans. This highly complex system of barriers did not emerge in all kingdoms by chance; its concatenation is not an accident. There is a purpose, a mission for the organism's survival, and it is called protection.

7.2 Philosophical Argument

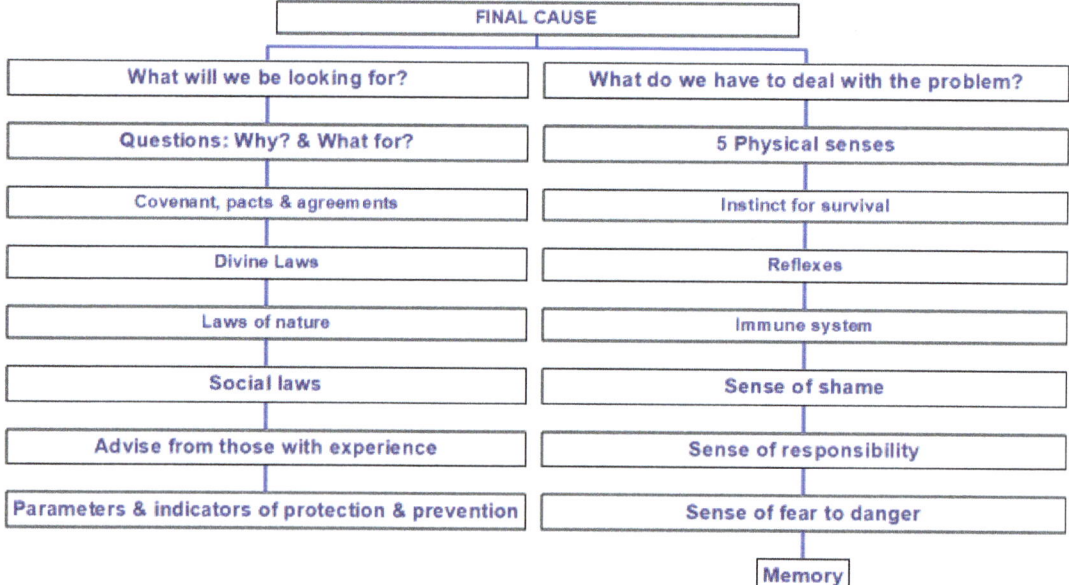

Let us advance with the example of the chair and the "Final Cause." The purpose of a chair's construction is to determine its ideal standard and end. It must answer the questions: *Why?* and *For what?*

---

[542] Illustration © Rob3000 | Dreamstime.com

[543] National Human Genome Research Institute (NHGRI), Lysosome.

The Final Cause and the Explanatory Method

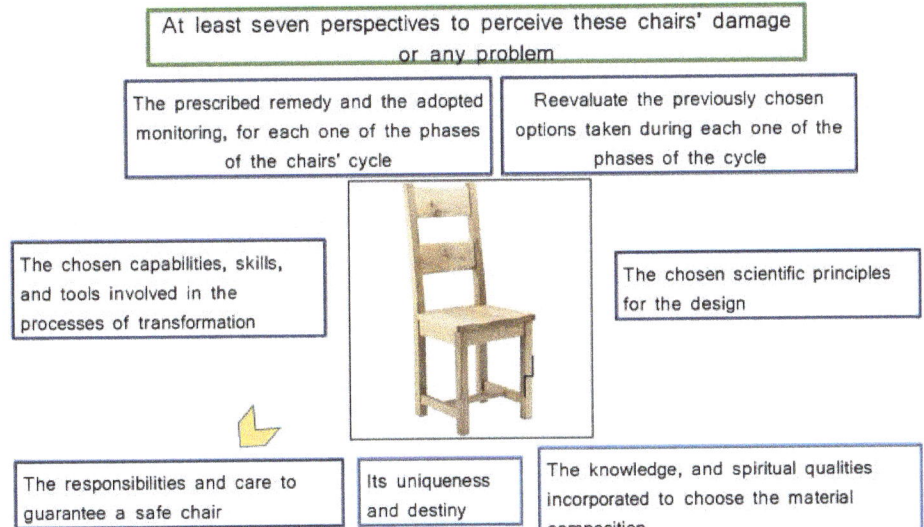

Let us review the quote from 'Abdu'l-Bahá, which serves as the foundation for this search. In this quote, he mentions the Final Cause:

> Essential pre-existence is an existence which is not preceded by a cause; essential origination is preceded by a cause. Temporal pre-existence has no beginning; temporal origination has both a beginning and an end. For the existence of each and every thing depends upon four causes: the efficient cause, the material cause, the formal cause, and *the final cause*. So this chair has a creator who is a carpenter, a matter which is wood, a form which is that of a chair, *and a purpose which is to serve as a seat*. Therefore, this chair is essentially originated, for it is preceded by, and its existence is conditioned upon, a cause. This is called essential or intrinsic origination.[544]

The author supposes that 'Abdu'l-Bahá is telling us that the *"purpose which is to serve as a seat "* is related to the Final Cause. Answering the questions Why? and, For what? It also must provide safety to whoever rests in it. The garden's weather inclemency is not the location for a living room sofa; its place is to be indoor.

The essence of the Laws of God, as the final cause, guided the author to the right track to articulate the set of notions and the corresponding set of human faculties associated with the Explanatory method of science. It will help us answer the questions: "Why?" and "For what do I exist?"

Philosophers have linked the final cause with concepts such as end, aim, purpose, and ideal (Walter Brugger, Diccionario de Filosofía; The University of Vermont).

The Final Cause is associated with the questions: "Why?" and "For what?". It is difficult to relate the questions "Why?" and "For what?" to concepts other than the law. Let us carefully study the following extract of the dialogue 'Crito' / Or the Duty of a Citizen to understand the relationship between the final cause and laws. After Socrates condemnation to death, his friend Crito visits him and says:

> **Crito**. So let it be, then. But answer me this, Socrates, are you not anxious for me and other friends, lest, if you should escape from hence, informers should give us trouble, as

---

[544] 'Abdu'l-Bahá, Some Answered Questions, pp.155 -56. Emphasis added.

having secretly carried you off, and so we should be compelled either to lose all our property, or a very large sum, or to suffer something else besides this? For, if you fear anything of the kind, dismiss your fears; for we are justified in running the risk to save you—and, if need be, even a greater risk than this. But be persuaded by me, and do not refuse.

Socrates. I am anxious about this, Crito, and about many other things.[545] Further down, we find:

Socrates. My dear Crito, your zeal would be very commendable were it united with right principle; otherwise, by how much the more earnest it is, by so much is it the more sad. We must consider, therefore, whether this plan should be adopted or not. For I not now only, but always, am a person who will obey nothing within me but reason, according as it appears to me on mature deliberation to be best. And the reasons which I formerly professed I cannot now reject, because this misfortune has befallen me; but they appear to me in much the same light, and I respect and honor them as before; so that if we are unable to adduce any better at the present time, be assured that I shall not give in to you, even though the power of the multitude should endeavor to terrify us like children, by threatening more than it does now, bonds and death, and confiscation of property. How, therefore, may we consider the matter most conveniently? First of all, if we recur to the argument which you used about opinions, whether on former occasions it was rightly resolved or not, that we ought to pay attention to some opinions, and to others not; or whether, before it was necessary that I should die, it was rightly resolved; but now it has become clear that it was said idly for argument's sake, though in reality it was merely jest and trifling. I desire then, Crito, to consider, in common with you, whether it will appear to me in a different light, now that I am in this condition, or the same, and whether we shall give it up or yield to it. It was said, I think, on former occasions, by those who were thought to speak seriously, as I just now observed, that of the opinions which men entertain some should be very highly esteemed and others not. By the gods! Crito, does not this appear to you to be well said? For you, in all human probability, are out of all danger of dying to-morrow, and the present calamity will not lead your judgment astray. Consider, then; does it not appear to you to have been rightly settled that we ought not to respect all the opinions of men, but some we should, and others not? Nor yet the opinions of all men, but of some we should, and of others not? What say you? Is not this rightly resolved?

Crito. It is.

Socrates. Therefore we should respect the good, but not the bad?

Crito. Yes.

Socrates. And are not the good those of the wise, and the bad those of the foolish?

Crito. How can it be otherwise?[546]

Further down, we find:

Socrates. Observe, then, what follows. By departing hence without the leave of the city,

---

[545] Plato, Crito: or, the Duty of a Citizen, p. 4.

[546] Ibid. p. 6.

are we not doing evil to some, and that to those to whom we ought least of all to do it, or not? And do we abide by what we agreed on as being just, or do we not?

Crito. I am unable to answer your question, Socrates; for I do not understand it.

Socrates. Then, consider it thus. If, while we were preparing to run away, or by whatever name we should call it, the laws and commonwealth should come, and, presenting themselves before us, should say, 'Tell me, Socrates, what do you purpose doing? Do you design any thing else by this proceeding in which you are engaged than to destroy us, the laws, and the whole city, so far as you are able? Or do you think it possible for that city any longer to subsist, and not be subverted, in which judgments that are passed have no force, but are set aside and destroyed by private persons?'— what should we say, Crito, to these and similar remonstrances? For any one, especially an orator, would have much to say on the violation of the law, which enjoins that judgments passed shall be enforced. Shall we say to them that the city has done us an injustice, and not passed a right sentence? Shall we say this, or what else?

Crito. This, by Jupiter! Socrates.[547]

Socrates. What, then, if the laws should say, 'Socrates, was it not agreed between us that you should abide by the judgments which the city should pronounce?' And if we should wonder at their speaking thus, perhaps they would say, 'Wonder not, Socrates, at what we say, but answer, since you are accustomed to make use of questions and answers. For, come, what charge have you against us and the city, that you attempt to destroy us? Did we not first give you being? and did not your father, through us, take your mother to wife and beget you? Say, then, do you find fault with those laws among us that relate to marriage as being bad?' I should say, 'I do not find fault with them.' 'Do you with those that relate to your nurture when born, and the education with which you were instructed? Or did not the laws, ordained on this point, enjoin rightly, in requiring your father to instruct you in music and gymnastic exercises?' I should say, rightly. Well, then, since you were born, nurtured, and educated through our means, can you say, first of all, that you are not both our offspring and our slave, as well you as your ancestors? And if this be so, do you think that there are equal rights between us? and whatever we attempt to do to you, do you think you may justly do to us in turn? Or had you not equal rights with your father, or master, if you happened to have one, so as to return what you suffered, neither to retort when found fault with, nor, when stricken, to strike again, nor many other things of the kind; but that with your country and the laws you may do so; so that if we attempt to destroy you, thinking it to be just, you also should endeavor, so far as you are able, in return, to destroy us, the laws, and your country; and in doing this will you say that you act justly—you who, in reality, make virtue your chief object? ... .[548]

The argument continues highlighting the ascendancy of the laws and the obedience they require from the citizens. Socrates also emphasizes the importance of compliance to the Republic, even after it has mistakenly condemned him to death.

Socrates perceived his material existence as something temporary and was willing to sacrifice it to illuminate the world.

---

[547] Ibid. p. 11.

[548] Ibid. p. 12.

7.2.1 Divine Laws are the Breath of Life unto All Created Things

Having appreciated the importance of laws, a question that immediately arises is: are they inherent to all things? Western civilization has taught us that social rules and religious regulations are external or extrinsic to human beings. But this is not the case. Consider the following quotes, which refer to Divine Laws:

> The laws and ordinances that constitute the major theme of this Book, Bahá'u'lláh, moreover, has specifically characterized as *'the breath of life unto all created things'*, as 'the mightiest stronghold', as the 'fruits' of His 'Tree', as 'the highest means for the maintenance of order in the world and the security of its peoples', as 'the lamps of His wisdom and loving-providence', as 'the sweet-smelling savour of His garment', and the 'keys' of His 'mercy' to His creatures. 'This Book', He Himself testifies, 'is a heaven which We have adorned with the stars of Our commandments and prohibitions'. ... 'Say, O men! Take hold of it with the hand of resignation ... By My life! It hath been sent down in a manner that amazeth the minds of men. Verily, it is My weightiest testimony unto all people, and the proof of the All-Merciful unto all who are in heaven and all who are on earth.[549]

Moreover, in the same Book, Shoghi Effendi in the Introduction of The Most Holly Book quotes Bahá'u'lláh:

> Blessed the palate that savoureth its sweetness, and the perceiving eye that recognizeth that which is treasured therein, and the understanding heart that comprehendeth its allusions and mysteries. By God! Such is the majesty of what hath been revealed therein, and so tremendous the revelation of its veiled allusions that the loins of utterance shake when attempting their description.' And finally: 'In such a manner hath the Kitáb-i-Aqdas been revealed that it attracteth and embraceth all the divinely appointed Dispensations. Blessed those who peruse it! Blessed those who apprehend it! Blessed those who meditate upon it! Blessed those who ponder its meaning! *So vast is its range that it hath encompassed all men ere their recognition of it.* Erelong will its sovereign power, *its pervasive influence and the greatness of its might be manifested on earth.*[550]

Later: "Say: This is the very soul of all Scriptures which hath been breathed into the Pen of the Most High, causing all created beings to be dumbfounded, save only those who have been enraptured by the gentle breezes of *My loving-kindness and the sweet savours of My bounties which have pervaded the whole of creation.*"[551]

The above is the reason why the author decided to choose "the Laws of God" as the essential origination of the Final Cause.

The Arhuaco tribe is one of the indigenous populations of Colombia with their philosophy of life. They consider a violation of social law as against their true nature. Sometimes we feel laws as an imposition upon us as if they were external to ourselves. Indeed, accepting the challenge of attempting to establish harmony between social rules and the Laws of God, whether manifested in nature or the Sacred Books revealed by Him, we start grasping their inherent wisdom, and our love for them increases. We may also find that some regulations are very oppressive, and we must work to remove them.

---

[549] The Kitáb-i-Aqdas, p.16. Emphasis Added.

[550] Ibid. p. 16. Emphasis Added.

[551] Bahá'u'lláh, The Kitáb-i-Aqdas, p. 68. Emphasis Added.

Order in a country, and any community, is dependent on respect for its constitution, social laws, natural laws, and obedience to those in authority.

Submission to the laws and norms which govern all things guarantees the flourishing of their potentialities and powers. For human beings, conscious conformity to the law must be reached willingly, through an understanding of the reasons that underpin it and through "love for God as the motive of obedience to His Laws."[552] The causes of the most deep-rooted problems must be looked for within this recognition; this is the key to establishing the general purposes of any institution, government, or individual.

The information about the final cause can be systematized using natural laws, religious laws and ordinances, social norms, recommendations, risks, consequences, relationships of causality, parameters, and indicators of prevention and protection.

An essential input to enrich the concept of human being in the Final Cause is the understanding of what citizenship means. William Huitt expresses:

> Discussions regarding citizenship have received extensive attention in recent years as humanity moves rapidly into a new era of globalization (e.g., Isin, 2000; King, 2000; Peters, Britton, & Blee, 2008; Roth & Burbules, 2007).
>
> There are at least three issues that should be of concern to educators: identity, loyalty and responsibility, and rights.
>
> In many ways, one's concept of citizenship is an essential element of one's self identity. At the earliest stages of human evolution, the group affiliation that provided a source of one's identity was the family or band, which then evolved into tribe, city state, and empire (McNeill & McNeill, 2003). These changes in affiliation and identity took tens of thousands, then thousands, then hundreds of years. While most people had an identity at only one of these levels at any given time, the concept of world citizen (derived from the Greek word kosmopolitês) had its advocates even in antiquity (Kleingeld, 2006). The geographical explorations of the world in the fifteenth to seventeenth centuries (McNeill, & McNeill, 2003), the rapid changes in technology in the nineteenth and twentieth centuries (Huitt, 2007), combined with an increased diaspora that is expected to continue (Castles & Miller, 2003; OECD, 2008), has created a complexity of affiliation and identity seen only in isolated individuals in the past (Banks, 2007; Grimshaw & Sears, 2008; Marshall, 2009). Unfortunately, recent attempts to make sense of these changes result in contradictory views. One such example is found in the statements that the world is flat (Friedman, 2007) and the world is curved (Smick, 2008). Townsend (2009) argued that this particular contradiction results from a change in focus--either on the whole (flat) or the differentiation of the parts (curved)--and that in the postmodern world, people (especially leaders) need to think and act at both levels and all of those in between. Abrams and Primack (2011) make the case that one's identity should have a relationship to the cosmos as each human being, at least his or her material form, is a direct result of the evolution of the cosmos. This is the perspective taken by Brown (2007) and Christian (2005) in their development of the concept of big history. A major advantage for those advocating a local-global mindset (Bell-Rose & Desan, 2006; Townsend, 2009) is that people are able to understand one level above where they are

---

[552] Shoghi Effendi qtd. in Synopsis and Codifications of the Laws and Ordinances of The Kitáb-i-Aqdas in Bahá'u'lláh, The Kitáb-i-Aqdas, p. 163.

able to act (Perry, 1999; Reimer, Paolitto, & Hersch, 1983). Developing an identity as a global citizen will be easier for those who identify themselves as citizens of the cosmos.

Another issue involves the contradictions of loyalty (Hansen, 2010) and responsibility (Karlberg, 2008). On the one hand, individuals owe certain loyalties and have responsibilities to their communities, as they are the primary contexts with which individuals have daily contact (Shinn & Toohey, 2003). In fact, it is neighborhoods that are the first level of community and serve as a developmental context, especially for children in poverty (Vaden-Kierman, D'Elio, O'Brien, Tarullo, Zill, and Hubbell- McKey, 2010). However, at least in the USA, individuals are becoming more isolated and detached from their neighborhoods and immediate geographical regions (Putnam, 2000). Turkle (2012) suggested this is a direct result of being tethered digitally throughout every waking moment of one's life. With the spread of wireless technology and the use of smartphones and other mobile devices throughout the world (Corasaniti, 2010; International Telecommunications Union, 2011), this phenomenon is likely to increase. While it is advocated that people spend more time "untethered" (Bourg Carter, 2012), perhaps having a sense of loyalty to the cosmos as a cosmic citizen will encourage people to develop their full potentials as a means of being loyal to the cosmos into which they were born and using their competencies in the development of their local communities.

The reality is, however, that in the modern world, the nation-state is the focus of one's loyalties and responsibilities (Koczanowicz, 2010). At the same time, the populations of the nation-state across the world are morphing rapidly as there is an unprecedented number of foreign-born individuals within a specific nation state (OECD, 2009). This is putting tremendous pressure on nation states to both prepare their citizens to interact with and perhaps live and work in other countries while at the same time integrating non-native people into the society (Koczanowicz, 2010). To make matters even more complex, the various relationships are nested (i.e., individual within local community within province, state, or region within nation within international region within world), and there are reciprocal (i.e., back-and- forth) relationships at all levels (Huitt, 2012a). The fluidity of people's movements in, out, and through these various relationships is unprecedented in human history, at least on a global basis, making it very difficult to define one's loyalties and responsibilities.

Another contradiction is the discussion of whose rights should be central to the concept of citizenship: that of the individual (Hall, Coffey, & Williamson, 1999) or that of the community (Stevenson, 2010). While there are excellent rationales provided for both of these as a focus for identifying rights of citizens, there is also an advocacy that the most important consideration is to provide a dynamic balance between the perspectives of individual autonomy and collective benefit (McIntyre-Mills, 2009).

This theme is adopted in the United Nations (1948) Universal Declaration of Human Rights. Pykett (2010) suggested that discussing the tensions between individual freedoms and social order is crucial to developing a sustainable view of citizenship education and guiding social reforms. At a time in history when society is in great flux, a lack of a coherent policy results in jumping back and forth between these two advocacies in a manner that is neither satisfying nor effective. Again, having an identity as a cosmic citizen can impact the development of a concept of human rights from a global perspective.[553]

Bahá'u'lláh referring to citizenship, stated:

> It is incumbent upon every man of insight and understanding to strive to translate that which hath been written into reality and action......That one indeed is a man who, today, dedicateth himself to the service of the entire human race. The Great Being saith: Blessed and happy is he that ariseth to promote the best interests of the peoples and kindreds of the earth. In another passage He hath proclaimed: It is not for him to pride himself who loveth his own country, but rather for him who loveth the whole world. The earth is but one country, and mankind its citizens.[554]

☙❧

A set of human faculties is common to all human beings, and with it, we respond to any menace, injury, danger, or hostility. Besides our immune system, a survival instinct, and reflexes, human beings have some other faculties to protect others and ourselves:

7.2.2 Sense of obligation also called the sense of responsibility

When considering "Why?" and "For what?" it is imperative to recognize not just an individual but also a collective sense of responsibility.

'Abdu'l-Bahá said, "O ye lovers of truth, ye servants of humankind! Out of the flowering of your thoughts and hopes, fragrant emanations have come my way, wherefore an inner *sense of obligation* compelleth me to pen these words."[555]

Shoghi Effendi says: "... Nothing but a fiery ordeal, out of which humanity will emerge, chastened and prepared, can succeed in implanting that *sense of responsibility* which the leaders of a newborn age must arise to shoulder."[556]

If religious and natural laws are inherent to human beings, why is science held to be non-responsible? Mikael Stenmark says:

> Science should be non-responsible in the sense that scientists have not special accountability for the application of science. The ends to which scientific results are to be applied should be determined by society. Within, for instance, the framework of democracy, people are equally allowed to use the teachings and the findings of science to whatever vision of a good human life they endorse, be it a feminist, a Christian, or a Buddhist vision. On the other hand, scientists qua scientists should themselves no take side in these debates. To take a stand on questions about the utility of science is not part of the scientist's task, nor does scientists have any special responsibility that goes beyond those they have as citizens. The only thing scientist should care about in their professional role is finding out the truth, or how nature works.[557]

---

[553] Citizenship. Cosmic-Citizenship.

[554] Bahá'u'lláh, Gleanings from the Writings of Bahá'u'lláh, p. 250.

[555] 'Abdu'l-Bahá, Selections, p. 246. Emphasis Added.

[556] Shoghi Effendi, The World Order of Bahá'u'lláh, p. 46. Emphasis Added.

[557] Mikael Stenmark qtd. in LeRon Shults (ed.), The Evolution of Rationality, p. 51.

Consider the following: Using the qualitative method of science, a scientist may discover that a compound is an excellent painkiller. Do you agree that the role of the scientist is limited to finding out if the new compound is an effective painkiller? What if the medical community realizes later that the painkiller is highly addictive?

7.2.3 The senses of fear and shame

For our protection and that of others, we should be aware of the warning signs of our senses of responsibility, fear, shame, and our survival instinct:

> The Great Being saith: The structure of world stability and order hath been reared upon, and will continue to be sustained by, the twin pillars of *reward and punishment*. And in another connection He hath uttered the following in the eloquent tongue: Justice hath a mighty force at its command. It is none other than *reward and punishment* for the deeds of men. By the power of this force the tabernacle of order is established throughout the world, causing the wicked to restrain their natures for *fear of punishment*.[558]

> The *fear of God* hath ever been a sure defense and a safe stronghold for all the peoples of the world. It is the chief cause of the protection of mankind, and the supreme instrument for its preservation. Indeed, there existeth in man a faculty which deterreth him from, and guardeth him against, whatever is unworthy and unseemly, and which is known as his *sense of shame*. This, however, is confined to but a few; all have not possessed, and do not possess, it. It is incumbent upon the kings and the spiritual leaders of the world to lay fast hold on religion, inasmuch as through it the *fear of God* is instilled in all else but Him.[559]

However, I believe we can pray and ask God, the All Generous, to provide us with the sense of shame as a powerful instrument for our protection and that of others.

It is of the utmost importance to introduce the following argument setting as a foundation, absolute respect for the free will of the mature individual so s/he can assume the outcome of his (her) decision:

Let us now briefly examine alcoholic drinks in the context of the set of human faculties mentioned above to perceive their importance when addressing issues:

> Abdu'l- Bahá explains that the Aqdas prohibits "both light and strong drinks", and He states that the reason for prohibiting the use of alcoholic drinks is because "alcohol leadeth the mind astray and causeth the weakening of the body."[560]

> Further, Abdu'l- Bahá says: "Alcohol consumeth the mind and causeth man to commit acts of absurdity, but this opium, this foul fruit of the infernal tree, and this wicked hashish extinguish the mind, freeze the spirit, petrify the soul, waste the body and leave man frustrated and lost."[561]

Science supports this rationality, and research shows that alcohol is a depressant of the central nervous system, it is not a stimulant. It depresses our immune system[562] and our reflexes.

---

[558] Bahá'u'lláh, Tablets of Bahá'u'lláh. Emphasis Added.

[559] Bahá'u'lláh, Epistle to the Son of the Wolf, p. 27. Emphasis Added.

[560] Bahá'u'lláh, The Kitáb-i-Aqdas, p. 227, Note 144.

[561] Ibid. 239, Note 170.

[562] The National Institute on Alcohol Abuse and Alcoholism says: "A number of reviews in the

Alcohol also depresses our senses of responsibility, shame, and fear[563]. Of course, it depresses all human faculties. The alcoholic beverages industry labels alcohol as causing euphoria; it only lasts while the individual consumes between one to four drinks and then disappears. It is just a false sense of wellbeing. How many chronic illnesses alcohol causes?[564] I leave it to the reader to determine if alcohol disconnects the neuron's synapsis or kills the neurons. What are the consequences of having our memory affected?[565] The liquor industry's ultimate goal is to depress our right to free will, compromising our intelligence, becoming their regular clientele by keeping us addicted.

Suppose that acting mercifully, the administrative order of the town I live in decided to thoroughly educate others and me about why alcoholic beverages are prohibited. However, one day, I decided to drive a car under the influence of alcohol, and I killed somebody. Do I not deserve a severe punishment? How can I ask for mercy? The institution should keep me in jail, following the law, showing its mercifulness to the public.

Abdu'l-Bahá, the Master said:

The object of punishment is not vengeance but the prevention of crime.

Kings must rule with wisdom and justice; prince, peer and peasant alike have equal rights to just treatment, there must be no favour shown to individuals. A judge must be no "respecter of persons", but administer the law with strict impartiality in every case brought before him.

---

literature provide an overview of current knowledge concerning alcohol's effects on the human immune system (Baker and Jerrells 1993; Cook 1995, 1998; Frank and Raicht 1985; Ishak et al. 1991; Johnson and Williams 1986; Kanagasundaram and Leevy 1981; MacGregor and Louria 1997; Mendenhall et al. 1984; Mufti et al. 1989; Palmer 1989; Paronetto 1993; Watson et al. 1986. Web. December 2016 <https://pubs.niaaa.nih.gov/publications/10report/chap04b.pdf >.

[563] There are many life histories of Alcoholics Anonymous members that can serve to develop our awareness of what can happen to our senses of shame, fear and responsibility. There are also many books that dial with this subject such as The Treatment of Shame and Guilt in Alcoholism Counseling By Ronald T. Potter-Efron, Patricia S. Potter-Efron and in From Guilt Through Shame to AA: A Self-Reconciliation Process by Ed Ramsey 1987 Alcoholism Treatment Quarterly 4(2) 87-107.

[564] To get a glimpse of the consequences of alcohol consumption there are several interesting studies by Jürgen Rehm who is the Director, Social and Epidemiological Research Department and Senior Scientist and Head of Group Population Health Research at the Centre for Addictions and Mental Health (CAMH) in Toronto. One of his publications is: Rehm, Jürgen, Jens Klotsche and Jayadeep Patra. Comparative quantification of alcohol exposure as risk factor for global burden of disease. International Journal of Methods in Psychiatric Research. June 2007. Volume 16, Issue 2 66–76. Web. January 2013

[565] "Korsakoff's syndrome (also called Korsakoff's dementia, Korsakoff's psychosis, or amnestic-confabulatory syndrome) is a neurological disorder caused by a lack of thiamine (vitamin B1) in the brain or viral encephalitis. Its onset is linked to chronic alcohol abuse or severe malnutrition, or both." Wikipedia, Korsakoff Syndrome.

The Final Cause and the Explanatory Method

If a person commits a crime against you, you have not the right to forgive him; but the law must punish him in order to prevent a repetition of that same crime by others, as the pain of the individual is unimportant beside the general welfare of the people.[566]

7.2.4 Memory

When considering "Why?" and "For what?" it is very beneficial to listen carefully and remember the advice of those with experience. Also, in our memory, we store what we have learned, for example, after several times of trying to do something, we develop the skill.

In Aristotle, *The Metaphysics*, we find:

> ... it is from *memory* that men derive their experience. For many recollections of the same thing perform the function of a single experience. Indeed, it is thought that experience is more or less similar to knowledge and skill, and that men acquire knowledge and skill through experience. ...

> Now the circumstances in which skill arises are from the many cases of thinking in experience a single general assumption is formed in connection with similar things. For instance, to have the assumption that when Callias is ill with such and such a disease such and such medicine is appropriate and similarly for Socrates and for many others individually is a matter of experience. But the knowledge that for all such people ... when ill with such and such a disease, such and such a medicine is beneficial belongs to skill.[567]

'Abdu'l-Bahá taught:

> There are five outward material powers in man which are the means of perception — that is, five powers whereby man perceives material things. They are sight, which perceives sensible forms; hearing, which perceives audible sounds; smell, which perceives odours; taste, which perceives edible things; and touch, which is distributed throughout the body and which perceives tactile realities. These five powers perceive external objects.

> Man has likewise a number of spiritual powers: the power of imagination, which forms a mental image of things; thought, which reflects upon the realities of things; comprehension, which understands these realities; *and memory, which retains whatever man has imagined, thought, and understood.*[568]

Memory is among the set of faculties related to the Final Cause because memory is a repository of what we comprehend. Memory is similar to other kinds of reservoirs or storing organs in the vegetable, animal, and human kingdoms to preserve nutrients, natural defenses, or to isolate toxic substances acting as differential responses to needs and menaces.

> If we were to claim that all these effects proceed from the powers of the animal nature and the physical senses, then we see plainly and clearly that, with regard to these powers, the animals are superior to man. For example, the sight of animals is much keener than that of man, their hearing is more acute, and likewise, their powers of smell and taste. Briefly, in the powers which man and animal share in common, the animal often has the advantage. *Take the power of memory: If you carry a pigeon from here to a faraway country, and there*

---

[566] 'Abdu'l-Bahá, Paris Talks, p. 152.

[567] Aristotle. The Metaphysics. Emphasis Added.

[568] 'Abdu'l-Bahá, Some Answered Questions, p. 56. Emphasis Added.

set it free, it will remember the way and return home. Take a dog from here to the heart of Asia, set it free, and it will return home without ever losing its way. And so is it with the other powers, such as hearing, sight, smell, taste, and touch.[569]

Somewhere else, we find:

Immunological *memory* is a hallmark of the adaptive immune system. However, the ability to remember and respond more robustly against a second encounter with the same pathogen has been described in organisms lacking T and B cells. Recently, NK cells have been shown to mediate Ag-specific recall responses in several different model systems. Although NK cells do not rearrange the genes encoding their activating receptors, NK cells experience a selective education process during development, undergo a clonal-like expansion during virus infection, generate long- lived progeny (i.e., memory cells), and mediate more efficacious secondary responses against previously encountered pathogens—all characteristics previously ascribed only to T and B cells in mammals. This review describes past findings leading up to these new discoveries, summarizes the evidence for and characteristics of NK cell *memory*, and discusses the attempts and future challenges to identify these *long- lived memory* NK cell populations in humans.[570]

7.2.5 Relationships

The relationships between the individual and the Institutions should be of:
- gratitude, loyalty, reciprocity, mutuality,
- guilt, repentance, punishment, and reward
- and those of protection and fear to disobey.

7.2.6 Parameters and indicators of protection

To monitor and evaluating the situation's protection, equality, prevention, and security, such as:
- those that guarantee that the law is for all,
- those to measure the handling of aggressive behaviors,
- those used to ensure food security and food safety for all
- The proportion of inhabitants vaccinated
- Level of understanding of the wisdom of the laws and ordinances.

In conclusion, one of the phases of the cycle of scientific research is the explanatory phase; and the notions and the set of human faculties associated with the final cause are ideal to answer the questions Why? and What for?

ⅽⰄⰂⰑ

In his search for a single science in relation to its *mission,* Aristotle wrote:

And the science which knows to what *end* each thing must be done is the most authoritative of the sciences, and more authoritative than any ancillary science; and this *end* is the good of that thing, and in general the *supreme good* in the whole of nature. Judged by all the tests we

---

[569] 'Abdu'l-Bahá, Some Answered Questions, p. 48. Emphasis Added.

[570] Sun, Joseph C. et al. Emphasis Added.

have mentioned, then, the name in question falls to the same science; this must be a science that the first principles and causes; for the good, i.e. the *end*, is one of the causes.[571]

Let us consider again what the wise Diotima taught Socrates about love in connection to the words where the added emphasis appears:

> ... So that if a virtuous soul have but a little comeliness, he will be content to love and tend him, and will search out and bring to the birth thoughts which may improve the young, until he is compelled to contemplate and see the beauty of *institutions and laws*, and to understand that the beauty of them *all* is of one family, and that personal beauty is a trifle ...[572]

7.3 Research Steps for the Explanatory Method

If we follow the art of constructing the Single Science, we conclude that each method of science intertwines with the other methods.

The author made an effort to relate this method to the City of Contentment, as revealed in The Seven Valleys. These Valleys "mark the wayfarers' journey from their mortal abode to the heavenly homeland"[573]; in the City of Contentment, we find: The wayfarer, after traversing the high planes of this supernal journey, entereth into the city of contentment. In this valley he feeleth the breezes of divine contentment blowing from the plane of the spirit. He burneth away the veils of want, and with inward and outward eye perceiveth within and without all things the day of "God will satisfy everyone out of His abundance." From sorrow he turneth to bliss, and from grief to joy, and from anguish and dejection to delight and rapture.[574]

When we reflect on the wisdom of the following phrase, we understand that the source of this abundance is not only spiritual but also material:

> First and foremost is the principle that to all the members of the body politic shall be given the greatest achievements of the world of humanity. Each shall have the utmost welfare and well-being. To solve this problem we *must* begin with the farmer; there will we lay a foundation for system and *order* because the peasant class and the agricultural class exceed other classes in the importance of their service.[575]

The following are just general ideas that may serve as guiding steps for the explanatory method of science. The suggested steps should link the problem's facts and symptoms to protection and prevention. Also, let us always remember the connection of the problem with the personal and the collective responsibility of the individual, with the commitment to protect all. Obedience, respect, and protection to the bahá'í institutions are the duty of the individual.

---

[571] The Metaphysics, Emphasis added.

[572] Plato, Symposium. Emphasis added.

[573] Bahá'u'lláh, The Seven Valleys, The Call of the Divine Beloved.

[574] Idem.

[575] 'Abdu'l-Bahá, Foundations of World Unity, p. 39. Emphasis Added.

First: The exploratory step.

Search for changes in the risk prevention safeguards that failed and caused the problem's facts and symptoms. Also, looking for the connection of the problem with the senses of responsibility, fear of punishment, shame, and our memory, and the right to exercise free will and the parameters of protection, equality, prevention, and security. Let us always respect the institutional order, laws, norms, natural laws, religious laws, ordinances, pacts, agreements, and covenants.

"Education of the individual Bahá'í in the Divine law is one of the duties of Spiritual Assemblies."[576]

- What were the monitoring and evaluation results of the risks prevention safeguards adopted during the last cycle?
- What is the strategy's curricular approach to introducing new believers and continuing their education to the Divine law?
- Which decision broke the law, pact, or agreement that caused the problem's facts and symptoms?
- Which symptoms or warning signs show damage to any kingdom?
- To contribute to the control of damaging insects, are there any alarm pheromones to make them run away from the crop?

Second: The formative step, focusing on the general properties of matter and the faculties closely related to the Formative Cause. Concentrate on locating the Cause of the risk situation.

- How harmful is the risk? How many localities, homes, ecosystems, or people may be affected?
- How secure is the location where the problem's facts and symptoms' are present?
- How can a change in the form, arrangement, or modification in the other general properties of matter, lower the risk?
- How will we guarantee, during the following cycles, compliance with the changes agreed?
- What advantages brings an early warning system to control pest on crops?

Third: The qualitative step, focusing on the specific properties of matter

- What appropriate measures to prevent more significant damage and Identifying first responders (expert professional, agency) to address the immediate consequences of the risk?
- What can I do to assume a higher degree of responsibility in caring for others and trust in protecting the rest of the kingdoms?
- If codifying protection on this site, what would be the rank assign to it?

---

[576] The Universal House of Justice, Department of the Secretariat, Issues Related to the Study of the Bahá'í Faith. 8 Feb. 1998.

- What modification in the material composition can prevent harmful consequences?

Fourth: The Root Cause. Before proceeding further, please verify that the root cause of the problem is, in reality, what you are looking for because there may be other reasons. Sometimes negligence, prejudice, lack of self-accountability of faults, transgressions, omissions, wrongdoings, carelessness, failing, oversight, inadvertence, or mere ignorance, are the Cause.

- What would be the destiny of this facility if the preventive risk situation continues as is now?
- Which symptoms or warning signs show disobedience to the previously agreed preventive measures?
- Is it carelessness if we do not comply in protecting the Faith, its Institutions, the bahá'ís, the friends, biodiversity, farmers, and food security for all? What is the educational path to secure our collective destiny?
- Which is the curricular strategy to deepen the understanding of destiny, fate, predestination, and their originating causes: Providence, Grace, and Fear of God?

Fifth: The Explanatory Step. Assure yourself to become knowledgeable of the laws and standard recommended precautions for any potential risk that you may blame in a lawsuit or punish by the Institutions.

- Why and what for, when punishing some crimes, measures should collectively announce to dissuade others from following the same behavior? What is the strategy to secure the protection of all?
- Is included in your budget the reserves for equipment maintenance, utilities, and insurance payment on a timely basis, checking for pests, and repairing leaks?
- What about locating your valuable assets, arms, dangerous tools, drugs, and harmful substances in a safe place?
- Why is it a collective responsibility to have food security for all? Have health insurance for all employees? And take care of farmers because agriculture is the foundation of order, and peasants and everybody else, are supposed to protect the environment and our health?
- What norms and regulations can protect us from future hazards?
- Installed devices to prevent dangerous situations?

Sixth: The Experimental Step
- Which educational experiences contribute to developing further the following faculties related to the Final Cause: the senses of responsibility, fear of punishment, and shame; our memory, and our right to exercise free will?
- Which experiment or test tells us that the situation now is safe to prevent the problem's facts and symptoms?
- Can we change those processes that seem to be riskier? Those tools or equipment that require more protective measures?

Seventh: The Propositional Step
- Do we perceive any possible harmful consequences on the different kingdoms?
- Which parameters or indicators of prevention and protection do we want to discard, improve, or define and formulate anew?

7.4 The explanatory phase of the farmers' situation

To help the reader understand the farmers' example, the author emphasizes the discourse's fundamental notions reasonably linked to each one of the Causes.

Let us then continue with our example of the farmers using the Explanatory Method.

Let us illustrate the final cause continuing with the example of the farmers. They can find the most deep-rooted reasons for their attitudes towards peasant life if they examine:

The *sense of fear of going bankrupt* because of the *crop's vulnerability* once harvested, without having the means to protect it from decomposing, because the buyer does not appear the day he was supposed to pick up the produce. For example:

> "In Akuapem South district, pineapples are a high-value, nontraditional crop grown primarily for export as whole fruits. As described in Conley and Udry (2010), the opening of European pineapple markets to Akuapem farmers in the mid-1990s had a transformative effect on local agriculture. But, as Fold and Gough (2008) describe, unanticipated changes in the European market around 2004 *caused major disruptions* for Ghanaian pineapple growers and *fundamentally altered the terms of their contracts. Verbal agreements were not honored*, and in some cases firms that had begun the process of harvesting pineapples from smallholder farms *neglected to return to pick up the fruit, leaving the farmers with unsellable produce and without payment*. Both farmers and exporting firms *lost their businesses* as a result of the demand shock, leading to a period of intense rationalization in the industry. Farmers interviewed in 2009 *expressed regret for accepting verbal contracts* with the buying firms, and reported that they would *no longer sell without a written and legally binding agreement* (Harou and Walker 2010)."[577]

In an Agriculture Working Group of the Association of Bahá'ís studies, we consulted and agreed, that a Cooperative was an advisable way to legalize a storehouse:

> "Despite having to pay a cooperative membership fee in Ghana, most Ghanaian pineapple farmers join because these groups have greater bargaining power, the *ability to demand written contracts and the financial might to take legal action in response to breach of contract.* Cooperatives are also a vehicle for accessing resources and skills training. In Ghana, 27 percent of cooperative members mentioned the increased likelihood of receiving help from the government or from an NGO as their main reason for joining a cooperative."[578]

> "From our observations across the case study countries, the problem of *holdup* by firms appears to increase in the number of smallholders with whom the firm *contracts*. As firms face a larger pool of prospective suppliers, especially when the *contract product is perishable*, firms appear more likely to speciously reject commodities as not meeting *agreed quality standards*, or simply not show up to *purchase contracted commodities*. In Ghana, firms and their middlemen commonly come to harvest the crop. If they do *not show to harvest,* collect and pay for the crop, the smallholder's only outside option is sale on the local market at a much lower price, roughly half, or *outright loss due to spoilage caused by waiting on the*

---

[577] Christopher B. Barrett, Smallholder Participation in Agricultural Value Chains: Comparative Evidence from Three Continents. p. 17. Emphasis added.

[578] Idem. p. 27. Emphasis added.

contracting firm. Similar problems were observed in India and Nicaragua in horticultural products."[579]

Their understanding of and attitude toward natural laws teaches us that only plants, algae, fungi, and some bacteria, are direct collectors of the sun's energy and all other organisms are indirect consumers.

The social pacts concerning multiple risk protection of peasants and agriculture. For example:

"The European Economic Community was designed to create a common market among its members through the elimination of *barriers* and the establishment of a *common external trade policy*. The *treaty* also provides for a *common trade agricultural policy*, which was established in 1962 to *protect EEC farmers from agricultural imports*. The first reduction in *EEC tariffs* was implemented in January 1959, and by July 1968 *all internal tariffs were removed*. Between 1958 and 1968 trade, among the EEC's members, quadruple in value."[580]

The International Labor Organization says: "In several countries, the fatal accident rate in agriculture is double the average for all other industries. ILO estimates that workers suffer 250 million accidents every year and out of a total of 335,000 fatal workplace accidents worldwide, there are some 170,000 deaths among agricultural workers."[581]

If they cautiously inspect the banking laws and regulations that are affecting loans to farmers.

If they carefully listen to the recommendations of those with credit experience.

Religious teachings, which in this case say:

First and foremost is the principle that to all the members of the body politic shall be given the greatest achievements of the world of humanity. Each shall have the utmost welfare and well-being. To solve this problem we *must* begin with the farmer; there will we lay a foundation for system and *order* because the peasant class and the agricultural class exceed other classes in the importance of their service. In every village there *must* be established a general *storehouse* which will have a number of revenues.[582]

This storehouse[583] is supposed to play an essential role in providing food to the poor and protecting the farmer if s/he loses the harvest—so long as it is not through her (his) own negligence.

The foundation of an orderly society implies food security for all, the protection of farmers, our health, the environment, and its biodiversity.

---

[579] Idem. p. 29. Emphasis added.

[580] Encyclopedia Britannica. Emphasis Added.

[581] ILO Statement to the 56th Commission on the Status of Women. Adoption of international labour standards key to supporting rural women. Emphasis added.

[582] 'Abdu'l-Bahá, Foundations of World Unity, p. 39. Emphasis Added.

[583] The Merriam-Webster dictionary defines storehouse as a building for storing goods (such as provisions); an abundant supply or source; a place, room, or container where something is deposited or stored.

The Final Cause and the Explanatory Method

Let us briefly consider the *risks* involved in being a farmer, *risks* associated with the environment, pest control, insecticides, and price fluctuations. Compare farmers' *risks* to those assumed by the agribusiness transforming cereal into oats or flour, the merchant who sells the product, and the profits each receives. Have you heard about the coffee farmers and their income in third-world countries? Why do most of the world's farmers receive a meager income if one of the basic principles of economics states that *risk* and profit are directly related: more *risk* exposure should lead to higher gain and vice versa?

> It is believed that the link of the supply chain that is the closest to the farm-gate may be the least competitive one; as *cash trapped farmers* in remote areas *lack good market information and encounter relatively few buying agents*[584] (Wilcox and Abbott, 2006). Despite the apparent importance of government support, few producer countries have policies that provide small farmers with a level of playing field. Scale economies in processing, marketing, and distribution as well as market power may lie behind the larger observed margins. A lack of competition along the cocoa supply chain means that farmers capture as little as *0.5 percent of the retail price of cocoa*. Small farmers, contrary to plantations, *are rarely able to by-pass intermediaries*[585] as they do not have basic processing or transportation facilities. In addition, small producers *do not have good access to international price information*[586], which enables local traders to take bigger margins. Finally, farmers *cannot chose the timing of their sale* as they lack access to credit or warehousing facilities, and often have to sell their harvest in advance to cover immediate expenses. High marketing costs such as in-country transportation reduces the share captured by farmers. Farmers living in producing regions far away from any export point, for instance in big and *landlocked countries*, are bound to receive a lower price than farmers close to a sea port."[587]

The situation at the local market is similar:

> It was revealed at the Aruligo meeting that all of the farmers present sold their pineapple as whole fruit at the market.
>
> When asked about processing the fruit into slices none of the farmers present indicated that they do this.
>
> During market investigations it was revealed that there were middlemen who were buying fruit at reduced prices early in the marketing day and then taking these fruits to smaller markets or their own settlements and processing them into slices. The small pineapple were being purchased by non-market vendors for as low as $1 per fruit and then taken to their settlements or commercial areas, cut in half and sold for
>
> $2 per slice. This same size was available at the market later in the day as a whole fruit for around $5. Larger pineapples were being sold as whole fruit for around $10- 15 each.
>
> The sale of pineapple slices at the Honiara central market was only observed at one stall. This stall was selling whole pineapple fruit from Malaita and pineapple slices that were kept in a

---

[584] These are dangerous exposures.

[585] Meaning more vulnerability.

[586] Idem.

[587] Traoré Cocoa and Coffee Value Chains in West and Central Africa: Constraints and Options for Revenue-Raising Diversification. p. 27-28. Emphasis added.

sealed plastic container and sold for $2 per slice; it appeared that the same size fruit used that was cut was available whole for around $5. For this case it is not quite clear why the farmer chose to peel and slice his product only to sell it for less than he was selling his whole fruit for. More investigation into this activity at the Honiara Central Market would be worthwhile to determine the profit margins and if it is a good diversification activity for Aruligo market vendors.

It would appear that this type of practice is most common during the peak pineapple season when the market is full of pineapple and the price drops significantly, however this would have to be validated by further market data collection. One Aruligo farmer informed us that even when the market is very full of pineapple he has never seen anyone through fruit away at the end of the day. Vendors who live far from Honiara will often sleep at the market for as many days as necessary to sell their produce.[588]

To reach an agreement about the storehouse role, we should explore the feasibility of the below-mentioned ideas. As I understand, the storehouse should:

- Guarantee food security for all and minimize farmers' risk creating strategies to mitigate risks.
- sell its products at a price that ensures farmers' survival and prosperity, for not dealing with farmer's bankruptcy,
- with the taxes received to provide food for the needy
- even sharing with the farmer's part of the profit of storing the harvest, after paying expenses,
- Two quotes from Star of the West which bring light to our minds contain the following phrases:
    - "A general store may be built in that village for all the incomes and products to be brought therein. All products and incomes gathered and collected must be put in the general store."[589]
    - "He said the products of the community should be stored in a storehouse, that each man's share should be noted and when the property was sold, each should receive his proportion, and the tax he should pay to the community would be estimated from his share in the property."[590]

These phrases hint that the storehouse should also be an agricultural processing facility to lower the perishability of the harvest and increase the income of farmers.

FUNDAEC's research proposal says:

The establishment and strengthening of support and service micro-enterprises

A rural economy that is *entirely dependent* on the sale of its agricultural and animal products to an external market and the purchase of every article and service from outside may, at best, achieve momentary prosperity under very special conditions. In order to find mechanisms that would help a rural population *break out of the state of total dependence*,

---

[588] Kyle Stice. Aruligo Pineapple Value Chain – Mapping Report. pp. 6-7. Emphasis added.

[589] Alfred E. Lunt, The supreme affliction. A study in Bahá'í economics and socialization.

[590] Mary Hanford Ford, The economic teaching of 'Abdu'l-Bahá. 21 March 1917. Star of the West – 5. p. 13

the Rural University has been involved for the past years in a learning process concerned with small enterprises for the support of production activities, for processing agricultural products, or for those services that can be easily established within the region itself. Although a number of smaller projects have been successful in the past years, the most important endeavor of this process is the establishment in one of the villages of a small agro-industrial center presently consisting of five enterprises. The flow of subproducts from one operation to the other, and the sharing of facilities, are among the factors that make the center an economically viable enterprise, providing the Rural University with a model that promises to be an important structure possibly to be replicated in every group of four to five villages. In the near future, it is hoped that the model and the facilities will also serve as a center for the training of human resources for the promotion of small rural agroindustries, but the experience is as yet not complete and can only be used as the basis of courses and educational programs in two or three years."[591]

Suppose the population and the institutions agree that their mission is to provide a sustainable food source. In that case, they must make the utmost effort to *abide by the norms and laws endorsed* and adjust the existing *legislation* to reflect these new *pacts and agreements*.

To develop a proposed solution, farmers and others interested could use a macro-economic forecasting software to determine the *consequences* of what will happen if the farmer makes more profit than merchants and those who pack or transform the produce.

The estimated marginal profitability must be proportional to the risks avoided by each of the actors involved in the value chain, such as:

- harvest and products related risks,
- accidents, and health-related issues connected to the production, marketing, and selling of the products and by-products,
- and the potential damages to the ecosystem: soil, water, air, plants, and animals.

My *advice* is to apply these changes gradually: As a matter of analysis with the macro-economic forecasting software, the role of the *storehouse* should include how it will handle situations of scarcity and abundance and at what price it will buy the food to provide for the poor. The storehouse should buy from the farmers the food to provide for the poor, at a price fixed by *the laws of supply and demand*, departing from the price agreed in the consultation process between farmers, agro-industry, merchants, academia, and government officials. The price agreed in the consultation process must become a sustainable price, a minimum to be paid to farmers. The rest of the food for the village and surrounding areas will require a consultation process and *pacts* reached in terms of hectares planted, quantities, and quality to be received by the storehouse.

'Abdu'l-Bahá said the following about the revenues of the *storehouse*:

As to the first, the *tenths or tithes*: we will consider a farmer, one of the peasants. We will look into his income. We will find out, for instance, what is his annual revenue and also what are his expenditures. Now, if his income be equal to his expenditures, from such a farmer nothing whatever will be taken. That is, he will not be subjected *to taxation* of any sort, needing as he does all his income. Another farmer may have expenses running up to one thousand dollars we will say, and his income is two thousand dollars. From such an one a

---

[591] Arbab, Farzam et al. Fundaec: Its Principles and its Activities. Emphasis added.

tenth will be required, because he has a surplus. But if his income be ten thousand dollars and his expenses one thousand dollars or his income twenty thousand dollars, he will have to pay as *taxes*, one-fourth. If his income be one hundred thousand dollars and his expenses five thousand, one-third will he have to pay because he has still a surplus since his expenses are five thousand and his income one hundred thousand. If he pays, say, thirty-five thousand dollars, in addition to the expenditure of five thousand he still has sixty thousand left. But if his expenses be ten thousand and his income two hundred thousand then he *must* give an even half because ninety thousand will be in that case the sum remaining. Such a scale as this will determine allotment of *taxes*. All the income from such revenues will go to this general *storehouse*.

Then there must be considered such *emergencies* as follows: a certain farmer whose expenses run up to ten thousand dollars and whose income is only five thousand, he will receive necessary expenses from the *storehouse*. Five thousand dollars will be allotted to him so he will not be in need.[592]

Before continuing with this search and a proposal, they should invite all farmers to a community gathering to establish a collective pact based on the warning of the Master, who said in regard to the question of capital and labour: "The solution of this problem is one of the fundamental principles of His Holiness Bahá'u'lláh. But it must be solved with justice and not with force. If this problem is not solved lovingly it will result in *war*."[593] To *avoid war*, my recommendations are:

First: Let us reflect on the following quote by Shoghi Effendi in *"Directives from the Guardian"* says:

> "As you say, the Writings are not so rich on this subject and many issues at present baffling the minds of the world are not even mentioned. The primary consideration is the spirit that has to permeate our economic life, and this will gradually crystallize itself into definite institutions and principles that will help to bring about the ideal condition foretold by Bahá'u'lláh.
>
> No, Bahá'u'lláh did not bring a complete system of economics to the world. Profit sharing is recommended as a solution to one form of economic problems. There is nothing in the teachings against some kind of capitalism; its present form, though, would require adjustments to be made.
>
> There are practically no technical teachings on economics in the Cause, such as banking, the price system, and others. The Cause is not an economic system, nor its Founders be considered as having been technical economists. The contribution of the Faith to this subject is essentially indirect, as it consists of the application of spiritual principles to our present-day economic system. Bahá'u'lláh has given us a few basic principles which should guide future Bahá'í economists in establishing such institutions which will adjust the economic relationships of the world ... Social inequality is the inevitable outcome of the natural inequality of man. Human beings are different in ability and should, therefore, be different in their social and economic standing. Extremes of wealth and poverty should, however, be abolished ...

---

[592] 'Abdu'l-Bahá, Foundations of World Unity, p. 40. Emphasis added.

[593] 'Abdu'l-Bahá, Compilations, Bahá'í Scriptures, p. 340. Emphasis added.

> The Master has definitely stated that wages should be unequal, simply because that men are unequal in their ability and hence should receive wages that would correspond to their varying capacities and resources.[594]

Second: Always avoid debates and encourage consultation processes. Invite *all the parties* (government representatives, industry, business sectors, farmers, artisans, merchants, and academia) to grasp the outcome when applying the macro-economic forecasting software.

Third: Have at least the same number of women and men participating in the consultation process.

> The emancipation of women, the achievement of full equality between the sexes, is one of the most important, though less acknowledged prerequisites of peace. The denial of such equality perpetrates an injustice against one half of the world's population and promotes in men harmful attitudes and habits that are carried from the family to the workplace, to political life, and ultimately to international relations. There are no grounds, moral, practical, or biological, upon which such denial can be justified. Only as women are welcomed into full partnership in all fields of human endeavor will the moral and psychological climate be created in which international peace can emerge.[595]
>
> However, a women's majority will help *ensure a peaceful result*.

Fourth: Some agreements between the farmers, the agroindustry, and the merchants could include profit sharing.

---

[594] Shoghi Effendi, Directives from the Guardian.

[595] The Universal House of Justice, The Promise of World Peace, p. 6. Emphasis added.

8 The Efficient Cause and the Descriptive and Experimental Methods

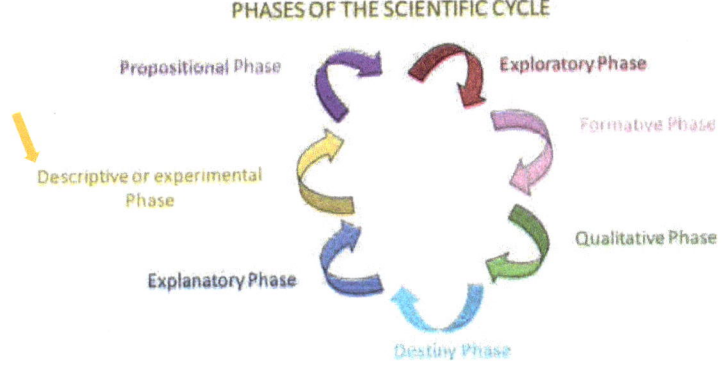

To introduce this chapter, let me tell you that in it you will find a proposal to replace the construction industry as the engine of the economy today.

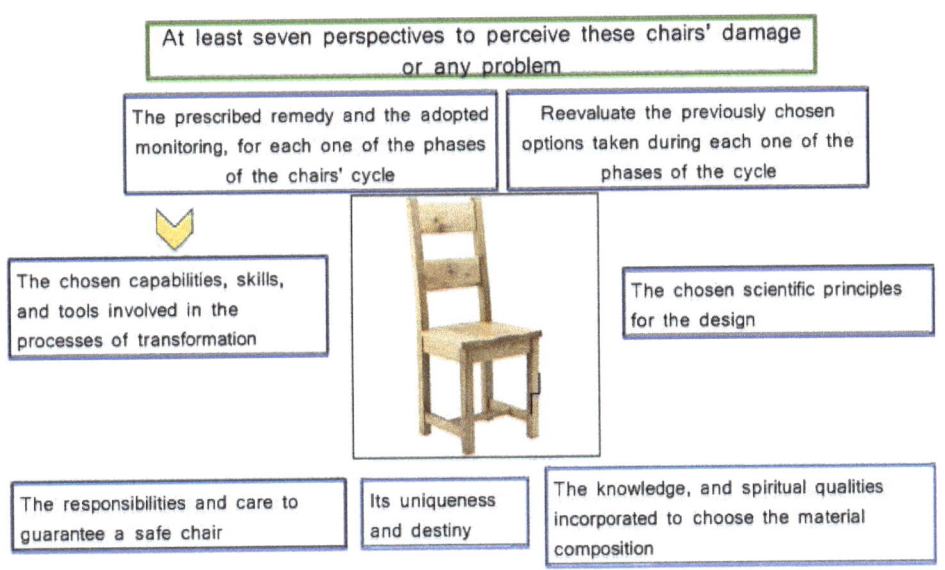

To continue introducing the "Efficient Cause," let us continue with the chair's example. To comprehend *"Who made the chair,"* let us think of the capacities of the carpenter and the methods he employed. We should also consider the tools and equipment the carpenter used to make it, including those not directly incorporated in the chair, such as the sandpaper used to polish it, the rag used to dust it before painting it, and so on. However, more profound is admitting that to make a chair, the potential to develop the arts and tools existed in a latent stage, within the essence of those substances and compounds needed to make a chair. That, happening, even before the making of the first chair.

The Efficient Cause and the Descriptive and Experimental Methods

The essence of the Word of God, as the efficient cause, was the key used by the author to generate the set of notions and the corresponding set of human faculties associated with the Descriptive and Experimental methods of science. It will also help us answer the questions: "Who am I?" and we will be able to direct our efforts to address our misconceptions and wrongdoings.

8.1 Philosophical Argument

In philosophy, the true principle of change is defined as The Efficient Cause. It is the cause of which all actions produce an effect. "The efficient cause is the thing or agent, which brings it about."[596] The Efficient Cause is associated with the question *"With whom?"* or "With what being?" It is related to the faculties of doing, teaching, and learning. We also find it associated with the concept of category, type, or set; in other words, with the capacity to classify notions.

The Efficient Cause is associated with concepts such as movement and the verbs "to be" and "to do." In Plato's Dialogues and The Sophist, it is evident that these are the fundamental aspects of the discussion between Theaetetus and the Stranger:

> **Stranger**: Let us push the question; for if they will admit that any, even the smallest particle of being, is incorporeal, it is enough; they must then say what that nature is which is common to both the corporeal and incorporeal, and which they have in their mind's eye when they say of both of them that they 'are.' Perhaps they may be in a difficulty; and if this is the case, there is a possibility that they may accept a notion of ours respecting the nature of being, having nothing of their own to offer.
>
> **Theaetetus**: What is the notion? Tell me, and we shall soon see.
>
> **Stranger**: My notion would be, that anything which possesses any sort of power to affect another, or to be affected by another, if only for a single moment, however trifling the cause and however slight the effect, has real existence; and I hold that the definition of being is simply power.
>
> **Theaetetus**: They accept your suggestion, having nothing better of their own to offer.[597]

---

[596] Aristotle. The Metaphysics. Emphasis Added.

[597] Plato, Sophist.

The Efficient Cause and the Descriptive and Experimental Methods

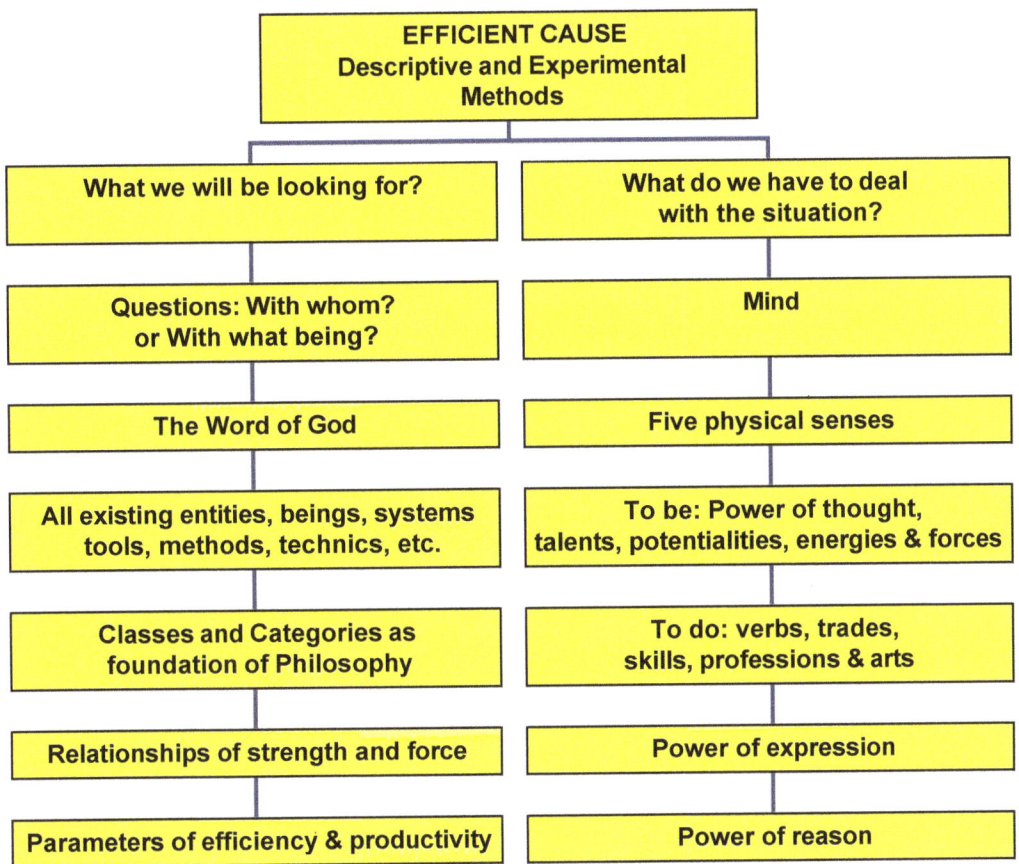

8.1.1 The mind

"As for the mind, it is the power of the human spirit. The spirit is as the lamp, and the mind as the light that shines from it. The spirit is as the tree, and the mind as the fruit. The mind is the perfection of the spirit and a necessary attribute thereof, even as the rays of the sun are an essential requirement of the sun itself."[598]

'Abdu'l-Bahá is reported to have said:

> There is, however, a faculty in man which unfolds to his vision the secrets of existence. It gives him a power whereby he may investigate the reality of every object. It leads man on and on to the luminous station of divine sublimity and frees him from all the fetters of self, causing him to ascend to the pure heaven of sanctity. This is the power of the mind, for the soul is not, of itself, capable of unrolling the mysteries of phenomena; *but the mind can accomplish this and therefore it is a power superior to the soul.*[599]

> "This supreme emblem of God stands first in the order of creation and first in rank, taking precedence over all created things. Witness to it is the Holy Tradition, "Before all else, God created the mind." From the dawn of creation, it was made to be

---

[598] 'Abdu'l-Bahá, Some Answered Questions, p. 55.

[599] 'Abdu'l-Bahá, Divine Philosophy, p. 121. Emphasis added.

8.1.2 "To be" and "to do."

In letters written on his behalf, Shoghi Effendi has explained the significance of the "letters B and E." They constitute the word "Be," which, he states, "means the creative Power of God Who through His command causes *all things to come into being*" and "the power of the Manifestation of God, His great spiritual creative *force*."[601]

The efficiency of an individual, a group, an institution or an enterprise is the result of the relationship between the power of the mind, and the potentialities of the individuals, their skills and capabilities, the equipment, the methodologies, and techniques, used to get the results.

Think of beings as any noun, thing, entity, system, animate or inanimate object, creature, plant, animal, or human being when being asked: *Who? What tool? What is the agent causing this effect?* Here we include all supplies not incorporated in the final product, tools, and equipment that may help in the process of reaping a product or service, and fruits, byproducts, and waste.

"To be" implies concepts such as energy, forces, capacities, potentialities, talents, vocations, arguments, and thoughts.

When thinking about "to do," we should include movements, verbs, methods, processes, activities, procedures, lines of actions, techniques, skills, abilities, arts, technologies, and mechanisms.

> Know that nothing that exists remains in a state of repose—that is, all things are in motion. They are either growing or declining, either coming from non-existence into existence or passing from existence into non-existence. So this flower, this hyacinth, was for a time coming from non-existence into existence and is now passing from existence into non-existence. This is called essential or natural motion, and it can in no wise be dissociated from created things, for it is one of their essential requirements, just as it is an essential requirement of fire to burn.[602]

8.1.3 The Word of God Pervades All Created Things

The Efficient Cause implies an active agent external to the object but acts upon it and generates an effect. We can associate this cause with other concepts such as the forces in nature, energy, heat, work, word, discourse, the Divine argument of the authorized interpreters of the Word of God, and human argument. In the words of Plato:

> **Stranger**- But now that the imitative art has enclosed him, it is clear that we must begin by dividing the art of creation; for imitation is a kind of creation of images, however, as we affirm, and not of real things.
>
> **Theaetetus**- Quite true.
>
> **Stranger**- In the first place, there are two kinds of creation.
>
> **Theaetetus**- What are they?

---

[600] 'Abdu'l-Bahá, The Secret of Divine Civilization, p. 1. Emphasis added.

[601] Bahá'u'lláh, The Kitáb-i-Aqdas: The Most Holy Book. Emphasis added.

[602] 'Abdu'l-Bahá, Some Answered Questions, p. 233.

Stranger- One of them is human and the other divine.

Theaetetus- I do not follow.

Stranger- Every power, as you may remember our saying originally, which causes things to exist, not previously existing, was defined by us as creative.

Theaetetus- I remember.

Stranger- Looking, now, at the world and all the animals and plants, at things which grow upon the earth from seeds and roots, as well as at inanimate substances which are formed within the earth, fusile or non-fusile, shall we say that they come into existence-not having existed previously-by the creation of God, or shall we agree with vulgar opinion about them?[603]

Continuing with Plato:

"**Stranger-** When any one says "A man learns," should you not call this the simplest and least of sentences?

Theaetetus- Yes."[604]

Bahá'u'lláh is even clearer and far more emphatic, expressing:

> Know thou, moreover, that the Word of God -- exalted be His glory -- is higher and far superior to that which the senses can perceive, for it is sanctified from any property or substance. It transcendeth the limitations of known elements and is exalted above all the essential and recognized substances. *It became manifest without any syllable or sound and is none but the Command of God which pervadeth all created things. It hath never been withheld from the world of being.* It is God's all-pervasive grace, from which all grace doth emanate. It is an entity far removed above all that hath *been* and shall *be*.[605]

The author decided to choose "the Word of God" as the essential origination of the Efficient Cause, because Bahá'u'lláh said, *It hath never been withheld from the world of being.*[606]

The arts

The philosopher Socrates considered the purpose of the arts to complement science, a theme we will address below. In Theaetetus, or Of Science, Socrates distinguishes the arts from the sciences by saying that art is the object of science. He clarifies by saying:

Theaetetus- Then, I think that the sciences which I learn from Theodorus-geometry, and those which you just now mentioned-are knowledge; and I would include the art of the cobbler and other craftsmen; these, each and all of, them, are knowledge.

Socrates- Too much, Theaetetus, too much; the nobility and liberality of your nature make you give many and diverse things, when I am asking for one simple thing.

Theaetetus- What do you mean, Socrates?

Socrates- Perhaps nothing. I will endeavour, however, to explain what I believe to be my meaning: When you speak of cobbling, you mean the art or science of making shoes?

---

[603] Plato, Theaetetus.

[604] ibid.

[605] Tablets of Bahá'u'lláh, pp. 140 – 141. Emphasis added.

[606] idem.

Theaetetus- Just so.

Socrates- And when you speak of carpentering, you mean the art of making wooden implements?

Theaetetus- I do.

Socrates- In both cases you define the subject matter of each of the two arts?

Theaetetus- True.

Socrates- But that, Theaetetus, was not the point of my question: we wanted to know not the subjects, nor yet the number of the arts or sciences, for we were not going to count them, but we wanted to know the nature of knowledge in the abstract. Am I not right?

Theaetetus- Perfectly right.[607]

There are many arts other than those traditionally recognized as such. In a passage that seemingly alludes to comedy and caricature, Socrates says:

Stranger- And is there any more artistic or graceful form of jest than imitation?

Stranger- We divided image-making into two sorts; the one likeness-making, the other imaginative or phantastic.

Theaetetus- True.

Stranger- And may we not fairly call the sort of art, which produces an appearance and not an image, phantastic art?

Theaetetus- Most fairly.

Stranger- These then are the two kinds of image making-the art of making likenesses, and phantastic or the art of making appearances?[608]

In other paragraph we find a reference to diverse arts:

Stranger- Let us grant, then, that from the discerning art comes purification, and from purification let there be separated off a part which is concerned with the soul; of this mental purification instruction is a portion, and of instruction education, and of education, that refutation of vain conceit which has been discovered in the present argument; and let this be called by you and me the nobly-descended art of Sophistry.[609]

Perceiving the diverse perspectives from which reality can be seen, leads one to realize how enriching the art of consultation could become.

Art, then, results from the abilities, skills, and capabilities of the individual and the potentialities of the other means employed to reap the fruit.

8.1.4 Relationships

Relations of strength and force help us choose the most meaningful words to describe a situation adequately or persuade others with reasonable arguments. Also, the link of all social actors with the labor force.

---

[607] Plato, Theaetetus.

[608] Plato, Sophist.

[609] ibid.

The Efficient Cause and the Descriptive and Experimental Methods

The relationships between the agent's potential capacity and what is achievable as an existential reality with the methods and tools employed when educating or cultivating. Which evolutionary processes permitted the appearance of volatiles' potentialities in plants?

"Plants are champion synthetic chemists; they take advantage of their anabolic prowess to *produce* volatiles, which they *use* to protect themselves against biotic and abiotic stresses and to provide *information* — and *potentially disinformation* — to mutualists and competitors alike. As transferors of *information*, volatiles have provided plants with solutions to the challenges associated with *being* rooted in the ground and *immobile*."[610]

8.1.5 Power of thought

We can perceive ourselves as mere spectators of a reality built by others or feel that we have the power to construct a better world.

The Master said, "The reality of man is his *thought*, not his material body. The thought force and the animal force are partners. Although man is part of the animal creation, he possesses a power of thought superior to all other created beings."[611]

"Man has likewise a number of spiritual powers: the power of imagination, which forms a mental image of things; *thought, which reflects upon the realities of things*; comprehension, which understands these realities; and memory, which retains whatever man has imagined, *thought*, and understood."[612]

Let us have a reasonable exemplification to understand that the "reality of man is his thought not his material body" in a fascinating reflection made by William Huitt:

Countries around the world are seeking ways to better prepare their children and youth for successful adulthood in the twenty-first century (Smith & Day, 1990; Jakobi & Teltemann, 2011). Unfortunately, there is no easy, readily available solution to this challenge. One reason is that the alternatives that one considers depends on the worldview and/or paradigm one uses to describe a human being and the value of education (Huitt, 2015). Therefore, discussion of alternatives is often an implicit discussion of worldviews and paradigms. For example, if one adopts a secular/materialistic worldview, one sees a human being in strictly materialistic terms whereas if one adopts a cosmic-spiritual worldview one would propose that there is some part of the human being that survives physical death. And if one were to adopt a God-centered worldview, one would likely look to a set of scriptures or traditions to define a human being and a life after earthly existence. Holders of each of these alternative worldviews will develop somewhat different alternatives and establish different criteria for choosing among them.

Likewise, one's paradigm can influence how one interprets reality and organizes facts, concepts, and principles (Huitt, 2011b). For example, if one were to adopt a mechanistic or reductionistic paradigm, one would look at formal education as a separate entity and investigate how the structure and functions of teachers and schools would impact human development (eg, Hattie, 2009; Squires, Huitt, & Segars, 1982). However, if one adopted an

---

[610] Ian T. Baldwin, Plant volatiles. Emphasis added.

[611] 'Abdu'l-Bahá, Paris Talks, p. 24. Emphasis added.

[612] 'Abdu'l-Bahá, Some Answered Questions, p. 56. Emphasis added.

existential/phenomenological paradigm, one would look at human perceptions and interpretations of schooling (Rogers, & Freiberg, 1994). And if one were to adopt an organismic/systems paradigm, one would seek to describe the whole person embedded in multiple layers of context or environment (Huitt, 2012).[613]

8.1.6 Power of prayer

"Just as the phenomenal sun shines upon the material world producing life and growth, likewise, the spiritual or prophetic Sun confers illumination upon the human world of thought and intelligence, and unless it rose upon the horizon of human existence, the kingdom of man would become dark and extinguished."[614] [615]

Bahá'u'lláh, in The Kitáb-i-íqán, revealed:
"Fasting is illumination, prayer is light."[616]

"Should Prayer take the form of action?"

'Abdu'l-Bahá: "Yes: In the Bahá'í Cause arts, sciences and all crafts are (counted as) worship. The man who makes a piece of notepaper to the best of his ability, conscientiously, concentrating all his forces on perfecting it, is giving praise to God. Briefly, all effort and exertion put forth by man from the fullness of his heart is worship, if it is prompted by the highest motives and the will to do service to humanity. This is worship: to serve mankind and to minister to the needs of the people. Service is prayer. A physician ministering to the sick, gently, tenderly, free from prejudice and believing in the solidarity of the human race, he is giving praise."[617]

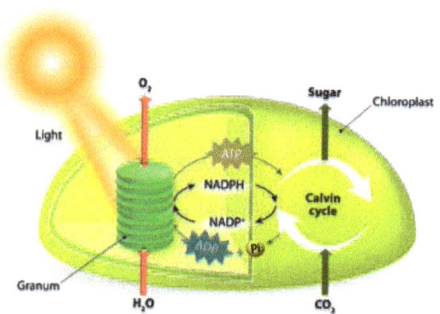

'Abdu'l-Bahá said, "There is nothing sweeter in the world of existence than prayer."[618]

Within the Calvin cycle, sugar is produced and stored in the tissues of plants. All herbivores and humans crave this energy to move and sustain all their metabolic processes.[619]

---

[613] William Huitt, What is a Human Being and Why is Education Necessary.

[614] Idem.

[615] Photo 198527382 © ElenaBelgo | Dreamstime.com

[616] Bahá'u'lláh, The Kitáb-i-íqán.

[617] 'Abdu'l-Bahá, Paris Talks.

[618] 'Abdu'l-Bahá, cited in Star of the West, vol. 8, p. 41. Web. January 2022 <https://www.bahai.org/beliefs/life-spirit/devotion/prayer/>.

[619] Illustration 55911251 © Designua | Dreamstime.com

"A photon of light energy travels until it reaches a molecule of chlorophyll. The photon causes an electron in the chlorophyll to become "excited." The energy given to the electron allows it to break free from an atom of the chlorophyll molecule. Chlorophyll is therefore said to "donate" an electron."[620]

8.1.7 Power of reason

The cause and effect relationships illumined by our reason lead us to understand the description of the results as a logical outcome of the chosen agent, its force, and process, for example, if water when heated in a gas or electric stove, the molecules start moving quicker until the water boils.

Consider the following quotes:

> When we consider the third criterion—traditions—upheld by theologians as the avenue and standard of knowledge, we find this source equally unreliable and unworthy of dependence. For religious traditions are the report and record of understanding and interpretation of the Book. By what means has this understanding, this interpretation been reached? By the analysis of human reason. When we read the Book of God the faculty of comprehension by which we form conclusion is reason. *Reason is mind*. If we are not endowed with perfect *reason*, how can we comprehend the *meanings of the Word of God*? Therefore human reason, as already pointed out, is by its very nature finite and faulty in conclusions. It cannot surround the Reality Itself, the Infinite Word. Inasmuch as the source of traditions and interpretations is human reason, and human reason is faulty, how can we depend upon its findings for real knowledge?[621]

> In the vegetable world, too, there is the power of growth, and that power of growth is the spirit. In the animal world there is the sense of feeling, but in the human world there is an all-embracing power. In all the preceding stages the power of reason is absent, but the soul existeth and revealeth itself. The sense of feeling understandeth not the soul, whereas the *reasoning power of the mind* proveth the existence thereof.[622]

> Bahá'u'lláh has declared that religion must be in accord with science and reason. If it does not correspond with scientific principles and the processes of reason, it is superstition. For God has endowed us with faculties by which we may comprehend the realities of things, contemplate reality itself. If religion is opposed to reason and science, faith is impossible; and when faith and confidence in the divine religion are not manifest in the heart, there can be no spiritual attainment.[623]

8.1.8 Power of expression

The capacity to express our thoughts in a meaningful conversation is achievable by developing the capability to classify ideas. It is possible to achieve this by practicing the grouping criteria on the originating causes of the scientific research methods.

---

[620] Molnar, Charles and Jane Gair. Concepts of Biology

[621] 'Abdu'l-Bahá, Foundations of World Unity, p. 48. Emphasis added.

[622] 'Abdu'l-Bahá, 'Abdu'l Bahá's Tablet to Dr. Forel, p. 2. Emphasis added.

[623] 'Abdu'l-Bahá, The Promulgation of Universal Peace, p. 1063.

The Efficient Cause and the Descriptive and Experimental Methods

Shoghi Effendi, the beloved Guardian, said,

> I feel the necessity of entrusting this highly important and delicate task to a special committee, to be appointed most carefully by the National Spiritual Assembly of America, and consisting of those who by their knowledge of the Cause, their experience in matters of publicity, and particularly by their power of expression and beauty of style will be qualified to produce a befitting statement on the unique history of the Movement as well as its lofty principles.[624]

☙❧

This set of human faculties can help us answer the question: "Who am I?" In conclusion, the cycle of scientific research also has an experimental and descriptive phase, and the notions and the set of human faculties associated with the Efficient Cause are appropriate to be able to answer the questions: "With whom?" and "What is the agent that is causing this effect?"

The concept of development cannot be just industrialization. The Efficient Cause brings new ideas about possible definitions of development of our mind, character, intelligence, soul, and responsibilities:

- What is the methodology to discover the latent potentialities of those residing in our village and enable them to find the potential of other beings in nature?
- How can we succeed in digging the mine rich in gems endowment on ourselves to improving our character, including the unearthing of the moral principles that will serve us as a foundation to make sound choices in life and achieve a better quality of life?
- If "Every age hath its own problem, and every soul its particular aspiration,"[625] what is the master plan for each in our village or neighborhood to discover and channel their particular aspiration?
- "The injury of one shall be considered the injury of all; the comfort of each, the comfort of all; the honor of one, the honor of all."[626] How is everyone in my village or neighborhood reaching this commitment?

Science is the most subversive thing that has ever been devised by man. It is a discipline in which the rules of the game require the undermining of that which already exists, in the sense that new knowledge always necessarily crowds out inferior antecedent knowledge.

This is what the patent system is all about. We reward a man for subverting and undermining that which is already known Man has a tendency to resist changing his mind. The history of the physical sciences is replete with episode after episode in which the discoveries of science, subversive as they were because they undermined existing knowledge, had a hard time achieving acceptability and respectability. Galileo was forced to recant; Bruno was burned at the stake; and so forth. An interesting thing about the physical sciences is that they did achieve acceptance. Certainly in the more economically advanced areas of the Western World, it has become commonplace to do everything possible to accelerate the undermining of existent knowledge about the physical world. The underdeveloped areas of the world

---

[624] Bahá'í Administration, p. 58.

[625] Bahá'u' lláh, Tabernacle of Unity.

[626] 'Abdu'l-Bahá, The Promulgation of Universal Peace.

today still live in a pre-Newtonian universe. They are still resistant to anything subversive, anything requiring change; resistant even to the ideas that would change their basic concepts of the physical world.[627]

Hauser wrote the ideas mentioned-above in the 20th century. Where is the unity of thought among scientists to achieve the masses' acceptance of the more economically advanced areas of the Western World? Where is one initiative of the scientific community in the western world to achieve unity of thought among professors on economics? "There's a joke about economists: if you ask five economists the same question you'll get six different answers. Granted, it's not a very good joke, but it's a fair call. Ours is a complex field, and a growing number of economists are acknowledging that the theory sitting behind mainstream economics is mostly rubbish. As a result, it's very difficult to find consensus on real world events."[628]

I believe the current conception of science is not responding to the true nature of what is a human being.

Problem-solving social issues without a clearer conception of a human being becomes very difficult. I think this is the most significant determinant of ill-health and excellent health anywhere.

- What is the concept of a human being (child, woman, man) among religious leaders?
- What is a human being among merchants? And among government officials at the national, regional, provincial, and local levels?
- What is a child or an adolescent for parents? What is a human being among teachers and professors?
- What is a human being in the military? And among scientists? And among managers and engineers in industries?
- Among drug dealers, human traffickers, and alcoholic beverage producers, what is a human being?
- Why are the masses today tired of the lack of opportunities to get an education?

Are we just mammals? Are we just greedy consumers of material possessions and unappeasable in egotistical pleasures? What does science opine about the statement that we are just sinners? Is it reasonable to fight endless wars to dominate others to impose a set of economic values or an ideology? Out of what schools of thought came the description of the poorly educated masses as the mob, subversive, terrorist used by government officials to control them with the army when they peacefully protest oppression? What is the standard conceptual definition for science today of a human being? For a profound reflection and very enriching discussion about this theme, I want to refer the reader to *The Bahá'í Philosophy of Human Nature* by Ian Kluge; in it, you will find exciting paragraphs:

> At the beginning of *The Blank Slate: The Modern Denial of Human Nature*, cognitive scientist and philosopher Steven Pinker asserts that everyone has a theory of human nature. Everyone has to anticipate the behavior of others, and that means we all need theories about what makes people tick. A tacit theory of human nature—that behavior is caused by thoughts and

---

[627] Philip M. Hauser, Demographer and Census Expert, qtd. in Theodore Berland's The Scientific Life.

[628] Warwick Smith, Tony Abbott achieves the impossible: unity among economists.

feelings—is embedded in the way we think about people Rival theories of human nature are entwined in different ways of life and

different political systems, and have been the source of much conflict over the course of history. ...

The Russian Revolution of 1917 was an attempt at creating a new society by remaking human nature into the "New Soviet Man" (Bauer et al. 157). Communist efforts were based on two principles—that human nature is almost infinitely malleable and that humans are entirely shaped by their natural, social, and, above all, economic environments. There is no innate, pre-determined human nature to be overcome. On 22 June 1941, this materialist and radical environmentalist philosophy of human nature found itself at war with its diametric opposite, German National Socialism, whose philosophy of human nature combined three main principles:

First, it accepted Joseph Arthur, Comte de Gobineau's belief that race is the determining factor in history and that Aryans—white and mostly European—are the superior race.

Second, it taught that the stronger races were in a Darwinian struggle against the numerically superior but weaker races whom it considered ultimately unfit to survive or rule. The concept of "survival of the fittest" was applied to national and international politics, societies, cultures, and, of course, races.

Third, it believed that human nature was genetically determined and that superior gene pools should not be "polluted" by mixing themselves with inferior ones.

The one principle that united Communism and National Socialism was that the value of the individual is determined by his or her usefulness to the state. Individuals have no rights against the state and the supposed welfare of the majority.

On 7 December 1941, a third theory of human nature emerged in the midst of war— one that held that the individual has intrinsic value and, therefore, inherent fundamental rights against the state and society in general. Although the liberal capitalist theory of human nature emerged victorious, it was eventually challenged by yet a different theory of human nature endorsed by politicized radical Islam.

Two of the foundational theorists of radical Islam are Hassan al-Banna, founder of the Muslim Brotherhood, and Sayyid Qutb, the latter of whom advocates for, among other things, a Muslim version of Vladimir Lenin's doctrine of the evolutionary elite to lead the attack on the West. He also calls for isolation from all non-Muslim learning and the establishment of rigorous Sharia law. In his best known book, Milestones, he calls for cease-less violent jihad against all non-Muslims, but especially against the West. Samuel P. Huntington's The Clash of Civilizations and the Remaking of World Order is an in-depth study of this incipient conflict.[629]

In conflict resolution approaches, it is critical to address the concept of a human being. I also found myself surprised when I asked teachers who are they educating?

I believe that we have to elucidate the connection between the Book of Creation and the Book of Revelation[630], in an attitude of respect.

---

[629] Ian Kluge, The Bahá'í Philosophy of Human Nature.

[630] Understood as the compilation of all the Holy Books.

The Efficient Cause and the Descriptive and Experimental Methods

When asked by an astounded atheist, if he were in fact deeply religious, Einstein replied: Yes, you can call it that. Try and penetrate with our limited means the secrets of nature and you will find that, behind all the discernible concatenations, there remains something subtle, intangible, and inexplicable. Veneration for this force beyond anything that we can comprehend is my religion. To that extent I am, in point of fact, religious.[631]

For Einstein, "science without religion is lame, religion without science is blind." He told William Hermanns in an interview that "God is a mystery. But a comprehensible mystery. I have nothing but awe when I observe the laws of nature. There are not laws without a lawgiver, but how does this lawgiver look? Certainly not like a man magnified." He added with a smile "some centuries ago I would have been burned or hanged. Nonetheless, I would have been in good company."[632]

However, we find the assertion of the celebrated physicist Stephen Hawking. When ABC News' Diane Sawyer asked if there was a way to reconcile religion and science, Hawking said, "There is a fundamental difference between religion, which is based on authority, [and] science, which is based on observation and reason. Science will win because it works."[633]

In my opinion, if we humbly find out the scientific reasons for how wise He is, we can demonstrate His authority. Just an example: I can ask a hardworking materialistic individual belonging to a communist or capitalistic ideology if honesty is a spiritual reality and materialists do not believe in such a thing, why is he mad if someone goes into his house to rob? His response: I believe in the survival of the fittest, and that is the way I protect what is mine. Then I can ask, is it ok if the robber is stronger? Another question would be, then permanent civil unrest among neighbors is ok? Another one, do you value honesty? What has been the source of the concept of honesty, science, or religion?

God teaches that spiritual values are part of the endowment of every human being. Still, we must nurture them to reach a better quality of life for the individual and society.

Conclusion: Religion will also win because science and religion work welded and joined in reality. Reality is one.

Once the reader understands and tries this conceptual framework, s/he will perceive no conflict between science and religion. As we try and persevere to see them in peaceful coexistence, we will establish peace in the blue planet when seen from space.

In The Proclamation of Bahá'u'lláh we find: "In such a world society, science and religion, the two most potent forces in human life, will be reconciled, will co-operate, and will harmoniously develop."[634] When a new paradigm emerges by merging basic notions from the realms of science and religion into one conceptual framework to perceive, interpret, and transform reality, it becomes an extraordinary coalescence force to bring about a new civilization's ideals.

---

[631] Max Jammer, Einstein and Religion, pp. 39-40.

[632] Wikipedia, Religious and philosophical views of Albert Einstein. Web. May 2021 <https://en.wikipedia.org/wiki/Religious_and_philosophical_views_of_Albert_Einstein>.

[633] Stephen Hawking qtd. in interview by ABC News' Diane Sawyer.

[634] Shoghi Effendi in the Introduction of a book by Bahá'u'lláh, The Proclamation of Bahá'u'lláh, p. XI.

The Efficient Cause and the Descriptive and Experimental Methods

In his search for a single science, Aristotle wrote:

> "Now for each one *class* of things, as there is one perception, so there is one science, as for instance grammar, being one science, investigates all *articulate* sounds. Hence to investigate all the *species of being qua being*[635] is the work of a science which is generically one, and to investigate the several *species* is the work of the specific parts of the science."[636]

Let us read again what Diotima taught Socrates about the meaning of love in connection to the words where added emphasis shows: "... in the next stage he will consider that the beauty of *the mind is* more honourable than the beauty of the outward form. So that if a virtuous soul has but a little comeliness, he will be content to love and tend him, and will search out and bring to the birth *thoughts* which may improve the young, ..."[637]

8.1.9 Parameters of efficiency

Some examples of efficiency parameters are rotation of inventories profitability, productivity, rotation of accounts receivables, and the productiveness of conversion of nutrients.

8.2 Research Steps for the Experimental and Descriptive Methods

If we follow the art of constructing the Single Science, we conclude that each method of science intertwines with the other methods.

The author made an effort to relate this method to the Valley of Wonderment, as revealed in The Seven Valleys. These Valleys "mark the wayfarers' journey from their mortal abode to the heavenly homeland"[638]; in the Valley of Wonderment, we find:

> O thou who art mentioned in these Tablets! Know thou that he who embarketh upon this journey will marvel at the signs of the power of God and the wondrous evidences of His *handiwork*. Bewilderment will seize him from every side, even as hath been attested by that Essence of immortality from the Concourse on high: "Increase My *wonder* and *amazement* at Thee, O God!" Well hath it been said:
>
> > I knew not what *amazement* was
> > Until I made Thy love my cause.
> > O how amazing would it be
> > If I were not *amazed* by Thee![639]

---

[635] Metaphysics ... is one of the principal works of Aristotle and the first major work of the branch of philosophy with the same name. The principal subject is "being qua being," or being insofar as it is being. It examines what can be asserted about anything that exists just because of its existence and not because of any special qualities it has. Also covered are different kinds of causation, form and matter, the existence of mathematical objects, and a prime-mover God. Wikipedia, Metaphysics.

[636] The Metaphysics. Emphasis added.

[637] Plato, Symposium. Emphasis added.

[638] Bahá'u'lláh, The Seven Valleys, The Call of the Divine Beloved.

[639] Bahá'u'lláh, The Seven Valleys, The Call of the Divine Beloved.

Meditate on what the poet hath written: "*Wonder* not, if my Best-Beloved be closer to me than mine *own self*; *wonder* at this, that I, despite such nearness, should still *be* so far from Him."... Considering what God hath revealed, that "We are closer to *man* than his life-vein," the poet hath, in allusion to this verse, stated that, though the revelation of my Best-Beloved hath so permeated my *being* that He is closer to me than my life-vein, yet, notwithstanding my certitude of its *reality* and my recognition of my station, I am still so far removed from Him. By this he meaneth that his heart, which is the seat of the All-Merciful and the throne wherein abideth the splendor of His revelation, is forgetful of its Creator, hath strayed from His path, hath shut out itself from His glory, and is stained with the defilement of earthly desires.[640]

The following are just general ideas that may serve as guiding steps for the experimental method of science. The suggested steps should link the problem's facts and symptoms to the reasons that support the outcome. Let us always remember the connection of the problem with the potentialities and actualities of all beings in nature, especially the empowerment of the masses of humanity with the methods of science, using as a learning context the latent powers of biodiversity.

First: The exploratory step. Search for modifications in the processes, training of personnel, or equipment that caused the problem's facts and symptoms. Also, looking at the link of the powers of the mind (thought, reason, expression) and the parameters of productivity and efficiency to the problem's facts and symptoms. Let us always keep present that God's Word is Law and is light to illuminate our path. We must also respect laws of thermodynamics, grammar rules, labor law, tools regulations, and grammar rules.

- What were the monitoring and evaluation results of the processes' efficiency carried out during the last cycle?
- Which decision might cause the problem's facts and symptoms' impact on the productivity and efficiency of the system's results?
- Which processes, arts, skill, or capability do we want to research to deliver results more efficiently?
- Besides contributing to the control of insects with physical methods or mechanical devices, which other options do we have to increase the results?
- To contribute to the control of damaging insects, are there any alternatives to rapidly increase the reproduction of beneficial insects?

Second: The formative step, focusing on the general properties of matter and the faculties closely related to the Formative Cause

- How could a modification on the distribution of resources in time and space contribute to increasing efficiency?
- How the diversity of functions of the different organisms (birds, plants, fish, insects, microorganisms, etc.) contribute to the ecosystem's efficiency? For example, microorganisms are very efficient in providing nutrients to plants:
- "Mycorrhizal fungi increase the surface absorbing area of roots 100 to 1,000 times, thereby greatly improving the ability of the plant to access soil

---

[640] Bahá'u'lláh, Gleanings from the Writings of Bahá'u'lláh.

resources."[641]

Third: The qualitative step, focusing on the specific properties of matter
- What modification in the quality of the product or the service can bring better-integrated results?

Fourth: The Root Cause step. Before proceeding further, please verify that the root cause of the problem is, in reality, what you are focused on. For example, in some cases, sloth, passivity, and apathy are the reason for inefficiency.
- Without encouraging confession or finger-pointing, how can you guide an individual to overcome such behavior?

Fifth: The Explanatory Step
- Why is perceiving each human being as potentially responsible for the collectivity's well-being an accountable attitude to increase productivity?
- What rehearsal is ideal for determining the response of all of us to a very harmful situation?

Sixth: The Experimental Step
- What kind of waste is reasonable in the activities of production and service that we perform?
- What beings, methods, skills, capabilities, tools, equipment, and machinery bring about excellence in our efficiency and productivity?
- In the context of the responses to the questions of this list: which experiment tells us if the equipment, machinery, arts, skills, and actual capabilities of the personnel and other beings demonstrate their efficiency?

Seventh: The Propositional Step
- Which regularity is best to program equipment maintenance?
- Which educational experiences contribute to developing artistic skills, the powers of reason and expression, and willingness to serve His Cause?
- Which parameters or indicators of productivity and efficiency do we want to discard, improve, or define and formulate anew?

---

[641] Mycorrhizal Online LLC.

The Efficient Cause and the Descriptive and Experimental Methods

8.3 The descriptive and experimental phase of the farmers' situation

To help the reader understand the farmers' example, the author emphasizes the discourse's fundamental notions reasonably linked to each one of the Causes.

Let us then continue with our example of the farmers using Descriptive and Experimental Methods—and without intending that this be an exhaustive analysis—it is necessary to examine:

1. The eating *customs* of humans and the resources *used* to feed livestock
2. The *caloric, nutritional, industrial, and health potential* of the species planted; and the *potentiality* of those species that *exist* in the region but not being planted
3. The *potentialities* of the soil and the soil management *techniques*
4. The *techniques* employed in planting, cultivation, harvesting, packing and transporting, and the *results* reaped from these

Let us reflect on the sustainability of agribusiness:

> After cars, the *food system uses more fossil fuel* than any other sector of the economy — 19 percent. And while the experts disagree about the exact amount, the way we feed ourselves *contributes more greenhouse gases to the atmosphere than anything else we do* — as much as 37 percent, according to one study. Whenever farmers clear land for crops and till the soil, large quantities of *carbon are released into the air*. But the 20th-century industrialization of agriculture has increased the amount of greenhouse gases *emitted by the food system* by an order of magnitude; chemical fertilizers (made from natural gas), pesticides (made from petroleum), farm machinery, modern food processing and packaging and transportation have together transformed a system that in 1940 *produced 2.3 calories of food energy for every calorie of fossil-fuel energy it used* into one that now *takes 10 calories of fossil-fuel energy to produce a single calorie of modern supermarket food*. Put another way, when we *eat* from the *industrial-food system*, we *are eating oil and spewing greenhouse gases*. This state of affairs appears all the more absurd when you recall that every *calorie we eat is ultimately the product of photosynthesis* — a *process* based on *making food energy from sunshine*.[642]

Another source says, "Modern agriculture and the food system as a whole have developed a strong dependence on fossil energy; *7.3 units of (primarily) fossil energy are consumed for every unit of food energy produced*."[643]

We must find *productive systems* where human labor is more intensive and profitable for farmers opting for the biodiversity's potential.

The three sisters' polyculture[644] of maize, squash, and beans is an excellent example of the research for systems. In it, we discover the symbiotic potential of plants of different *taxonomical families* becoming much more *efficient*. The corn stalk serves as a support for the climbing bean plant. The bean plant fixes nitrogen in the soil that benefits itself and the maize and the squash. The shade provided by the squash's big leaves slows the evaporation process, which benefits the three sisters.

---

[642] Pollan, The Food Issue, Farmer in Chief. Emphasis added.

[643] University of Michigan, Center for Sustainable Systems. Emphasis added.

[644] Postma, Complementarity in root architecture for nutrient uptake in ancient maize/bean and maize/bean/squash polycultures.

"Search for alternative systems of production in small farms

... Here, a few words are included about the *development* of a specific *learning process* which, next to the ... effort to create a formal *educational system* for rural areas, has absorbed the greatest share of the resources of FUNDAEC for about twelve years. Its most visible success has been the *development* of appropriate *systems of production* and *technologies* for norte del Cauca, but its value also lies in the creation of a *working methodology* and a *conceptual framework* that is gradually being disseminated to other regions.

The first two years of this *learning process* were dedicated to collaborative *efforts* with some of the *farmers* of the region in small *experiments* mostly concerned with the physical arrangement of crops on the farm and the distribution of the time of the *farmers* among various *tasks* in a diversified plan of *production* in small modules. The *fruits* of this *experience* were not so much new and more appropriate *technologies* but the knowledge that was *generated* as the FUNDAEC professionals immersed themselves in the life of the people of the region. It was strongly felt at the time that the usual attempts to carry out extensive studies and surveys somehow missed the point that the first stages of a *meaningful* participatory *development process* must create strong bonds of solidarity and friendship among the professionals from outside and the villagers, and that nothing can achieve this goal more effectively than facing the difficulties of life on a daily basis together as *co-workers* in search of new and *practical* solutions to problems that are analyzed jointly and approached with care.

The knowledge *generated* and socialized during this period is not entirely new, but it helped everyone to understand in a more explicit way the many details of the *rationality* that has traditionally governed *agricultural and animal production* on the small farms of the region. Although the purpose of the Rural University is to develop, if necessary, entirely new *systems of production*, it was clearly understood that some of the elements of this *rationality* were to be preserved in future developments and even strengthened. The criteria that were subsequently formulated on the basis of these observations to guide the search for alternative *systems of production* are not very different from what is being discussed today by many groups in the world *working* with different rural populations. The *systems* to be developed would have to improve the *production of food* at the level of each farm and obtain a better nutritional balance for the family. They were to *utilize* more *efficiently* the *resources* of the *farmers*. In contrast to "modernized" monoculture, diversity of *species* was to be promoted so that risk could be minimized. Great care was to be taken to conserve and improve the quality of the *natural resources* of the farm especially through the improvement of the management of the deteriorating soils of the region The *systems* had to be arranged in such a way that the *work* of the family members could be regulated avoiding periods of excess and deficit, and parallel to this, the flow of food *crops* and of money throughout the year had to be also regulated. The use of costly inputs was to be decreased although *systems* totally free from the use of *chemical products* were not envisioned for the near future. Finally it was hoped that the alternative *systems* would not contribute further to the individualistic attitudes that were already being propagated in the country at an alarming rate but that they would be consistent with the spirit of a community built not on competition but on the principle of cooperation.

The intensive *work* carried out with the farmers during this period also showed that, at least in norte del Cauca, the *process* of the disintegration of traditional economy had advanced too far and that it was impossible to bring abrupt changes to the *system* of a farm the owners of which had already made the necessary adjustments and lived from the very low *production* of the farm (managed with minimum input) and supplementary *income* from the occasional outside work of the family members. Moreover, the *kind* of changes that were necessary to increase income from the farm appreciably implied a rather high level of *investment* if from the beginning the entire *system* was to be changed. The answer seemed to lie in focusing attention on smaller areas of the farm at a given time and gradually increase the area *cultivated* according the requirements of new *systems*. It was in the process of implementing this possibility that a new vision of the search for alternative *systems* as sums of *subsystems* rather than aggregates of single elements was formulated. The word *subsystem* is used here somewhat differently from the usual references to *crops, animals*, or even the *family* as *subsystems* of a *farming system*; a *subsystem* in the vocabulary of FUNDAEC refers to a small part of the farm usually between one and five thousand square meters with a definite program for the management of a diversity of *plant and animal species* arranged in space and time following the criteria that have been set forth for the entire *system*. Each *subsystem* involves intensive use of the land including the utilization of fences for *production*. Work with each *subsystem occupies* only a portion of the *farmer*'s time, and by itself is economically *profitable*. Individual *farmer* families can start changing their *production technology* by adopting one single *subsystem* and gradually incorporate more until they have found a suitable *system* for their own specific conditions and aspirations.

During the decade that followed, researchers of the Rural University and an increasing number of participating *farmers* of norte del Cauca dedicated a great deal of *effort* to the *development* of appropriate *subsystems* for three zones of the region divided according to soil and climatic conditions. The *result* is now a set of some 15 well tested *subsystems* that are far superior both to the traditional *systems* and the modern monoculture that unfortunately continues to be propagated by many programs. Encouraged by this success, the Rural University has focused its attention in the past three years on two important *sets* of objectives. First, it has become increasingly involved in two other ecological regions where, in collaboration with other institutions, it is trying to repeat its *experience systematically*, make the necessary adjustments to its *methodology*, and learn to share it with others. The second set of activities have to do with the search for formal village and regional structures: small village plots dedicated to *experimentation*, community *learning* farms, village *technical* committees to manage some of the *experiments*, community funds managed by a committee of the villagers, and possibly a regional fund to handle increasing sums for *investment* and credit to individuals, groups, and community organizations. Both of these new *efforts* are now in progress but it is as yet too early to evaluate *results* and make definite *statements* about the final form of the corresponding *processes* and structures. It should be *stated*, however, that the success of these efforts is essential if the Rural University is to create a permanent basis for a participatory *process* of *technological* transformation.

Examined in the context of *courses* and *educational* programs, the *activities* of this *learning process* can be described in terms of three programs for the *development* of *human resources*. The first is directed towards *farmers* and proposes to *develop* their *capacity* to be excellent *producers* within new and more appropriate *systems* of *agricultural and animal production* as well as effective participants in *research* and in the *application* and the *socialization* of knowledge. The second program is directed towards the field *workers* of other *development* organizations, both official and private, in order to increase their *capacity* to offer *technical* assistance to the *farmers*. The third program involves contributing to a series of *seminar-courses* at the graduate level offered by CELATER for professionals who are already *working* with rural *populations* not as disseminators of *technological* packages, but facilitators of the *generation* of knowledge for change. Parallel to these *educational activities* and intertwined with them, the Rural University continues to carry out the necessary research to create new *subsystems* to increase the diversity of *species*, to find improved *procedures* for the management of the natural *resources* especially the soil within each *subsystem*, and also to examine on entire farms the necessary adaptations that have to be made in order to establish total *systems* that are economically viable and at the same time meet the criteria that have guided the *activities* of this learning *process*."[645]

Another excellent experience that is worthy of mention because it applies to a different ecosystem is Farms Igi Qko: The Tree Farms at Sapobai:

"Baker supervised an *experimental* station in Sapoba while *working* as *forest conservator* on one hundred different concessions ranging from nine to two hundred square miles. The *logging* by British firms was heavy but good management helped to alleviate stress on the forest. More detrimental, Baker believed, was the *clearing* o the forest by indigenous *farmers*, whose networks of small tracts already honeycombed the forest. Yams, their staple, quickly depleted the soil and the *farmers* then *moved* on to a new area of virgin forest where the *process* began again.

> *Population* pressure was beginning to accelerate the impact of this traditional *farming system*.

The solution Baker devised and *implemented* on a small scale was called Farms, Igi Oko. Fifty-two of my forest *workers* needed land in which to grow their *crops*, so instead of allowing them to select their own sites at haphazard wherever they might be inclined in the virgin forest, I persuaded them to agree to a site adjoining the young *plantations*. We chose an area of twenty-six acres of inferior *bush* which in years gone by had been *cultivated* but now contained only inferior growth. This they cleared in their spare time, and we divided it into half-acre plots, *oil palms* being *planted* first to make the boundaries of each allotment. In between their *food crops* they planted an economic selection of the most valuable *trees* available in the nurseries . . . In among the *mahoganies*, as nurse trees, they *planted* soil improvers such as *Ricinodendron africana* and *Pentclethera macrophylla*, both of which provide also useful timber while acting as nurse trees to the mahoganies. Besides the mahoganies we planted African walnut, *Lovoa Klaineana*, which, although not a true walnut, has that characteristic black streak which gives it the appearance of English walnut and has suggested its use in the *making* of gunstocks. We found also places for *Obeche, Triplochitin species*, a soft white wood *used for making ply-wood*. It has a huge cylindrical bole, and before the war was finding a good market in Hamburg for the *manufacture* of packing cases.

In addition to the indigenous *trees*, we *experimented* with Burma teak, *Tectora grandis*. I used to get the Forestry Department in Burma to supply us with quantities of seed, which was shipped to Lagos. For ground cover and as a soil improver and nurse tree we planted *Cassia siamea*, a welcome exotic. All these were planted by the forest cultivators in between

---

[645] Arbab, Farzam et al. Fundaec: Its Principles and its Activities. Emphasis added.

their food crops at a distance of six feet by twelve feet. When I inspected their allotments at the end of the season there were on the average 300 trees flourishing, and one allotment had 366. Among the farm crops grown for food were *corn, yams, gourds, okra, peppers, ildogie, beans and ground-nuts*, the latter are one of the best nitrogenous *crops* and great soil—improvers. Each year a new farm is allotted to successful *cultivators*, so that in time they will create considerable areas of new forest for the growing benefit of their country. I should add that as an incentive to obtain the best *results* I gave a small bonus to those *farmers who* succeeded in establishing not less than 500 trees to the acre. That number is sufficient to take possession of the land and provide the requisite silvicultural conditions to *produce* the best timber."[646]

It is also necessary to examine the potential damage caused by plagues and diseases, and the techniques farmers use to control them, and the tools and equipment employed, and the ideal procedures to do it more efficiently.

With the same forecasting macro-economic software, examine the consequences of reversing the intentional displacement of rural populations, as captured in the following quote:

> For economic development to succeed in Africa in the next 50 years, African agriculture will have to change beyond recognition. *Production* will have to have increased massively, but also *labor productivity*, requiring a vast reduction in the proportion of the population engaged in agriculture and a large move out of rural areas. The paper questions how this can be squared with a continuing commitment to smallholder agriculture as the main route for growth in African agriculture and for poverty reduction. We question the evidence base for an exclusive focus on smallholders, and argue for a much more open-minded approach to different modes of *production*. To allow alternative modes and scale of *production* to emerge, new institutional and policy frameworks are required. A rush to establish 'mega-farms' with government discretionary allocation of vast tracts of land is unlikely to be the answer. Allowing a more dynamic agriculture to develop will require clear institutional frameworks, and not just a narrow focus on smallholders.[647]

- Also, it is necessary to examine the techniques and methodologies employed by commercial establishments, private or governmental development agencies, and those promoted by the educational system.
- Rain and residual water management, including disposal of organic matter.

In *UN report: one-third of world's food wasted annually, at great economic, environmental cost*, we learn:

> The waste of some 1.3 billion tons of food each year is causing economic losses of $750 billion and significant damage to the environment, according to a United Nations report launched today.
>
> The report, Food *Wastage* Footprint: Impacts on Natural Resources, is the first study to analyze the impacts of global food *wastage* from an environmental perspective, looking specifically at its consequences for the climate, water and land use, and biodiversity.

---

[646] Richard St Barbe Baker Igi Qko: The Tree Farms at Sapobai, Nigeria circa 1927; quoted in The Spirit of Agriculture edited by Paul Hanley, pp. 170-171. Emphasis added.

[647] Collier and Dercon, African Agriculture in 50 Years: Smallholders in a Rapidly Changing World? Emphasis added.

One of the key findings of the report is that food that is *produced* but not *eaten* each year guzzles up a volume of water equivalent to the annual flow of Russia's Volga River and is responsible for adding *3.3 billion tonnes of greenhouse gases* to the planet's atmosphere. Similarly, 1.4 billion hectares of land – *28 per cent of the world's agricultural area* – *is used annually to produce food that is lost or wasted.*

Beyond the environmental impacts, food *wastage costs some $750 billion* annually to *food producers.*

"All of us – farmers and fishers; food processors and supermarkets; local and national governments; individual consumers – must make changes at every link of the human food chain to prevent food *wastage* from happening in the first place, and *re-use* or recycle it when we can't," said the Director-General of the Food and Agriculture Organization (FAO) José Graziano da Silva.

"We simply cannot allow one-third of all the food we produce to go to *waste* or be lost because of inappropriate *practices*, when 870 million people go hungry every day." (FAO, Food and Agriculture Organization, 2020)[648]

These factors will help the population to define who they should consult to reach the change in their conception of the physical world and the potential of other beings (animate and inanimate) that they would like to utilize to achieve their objectives. The population also has to make an effort to determine the parameters and indicators that may help them evaluate the efficiency of the community's products and services.

ɑbɢp

The following is a proposal to replace the construction industry as the engine of the economy today:

First and foremost is the principle that to all the members of the body politic shall be given the greatest achievements of the world of humanity. Each shall have the utmost welfare and wellbeing. To solve this problem we must begin with the farmer; *there will we lay a foundation for system* and order *because the peasant class and the agricultural class exceed other classes in the importance of their service*. In every village there must be established a general storehouse which will have a number of revenues.[649]

The challenge ahead for scientists is vast, there are close to 300,000 species of plants, and there are scores of different chemical compounds in a leaf of a plant, each one of them with different potentialities. "It has been estimated that well over 300,000 secondary metabolites exist, and it is thought that their primary function is to increase the likelihood of an organism's survival by repelling or attracting other organisms."[650] Each chemical compound has its potential for agriculture, nutrition, health, or industrial uses. To estimate the possible combinations of 300,000 secondary metabolites, it is necessary to use a calculator such as http://stattrek.com/online- calculator/combinations-permutations.aspx. The results for a set of

---

[648] UN News. UN report: one-third of world's food wasted annually, at great economic, environmental cost. Emphasis added.

[649] 'Abdu'l-Bahá, Foundations of World Unity, p. 39. Emphasis Added.

[650] Brahmkshatriya, Priyanka P. Brahmkshatriya and Pathik S. Brahmkshatriya, Terpenes: Chemistry, Biological Role, and Therapeutic Applications.

2 compounds is 44,999,850,000 possible combinations; for *a set of 4 compounds* is 3.3749325004125E^{20}; and for *a set of 6 compounds* is 1.01244937595624E^{30}.

How many different *uses* can we derive from the whole plant and its parts? How many beneficial *species* of insects, bacteria, viruses, and fungus do we know?

- "Fungal Diversity Revisited: 2.2 to 3.8 Million Species. With 120,000 currently accepted species, it appears that at best just 8%, and in the worst case scenario just 3%, are named so far."[651]
- "Over one million species of insects."[652]
  - "Earth May Be Home to a Trillion Species of Microbes" According to a new estimate, there are about one trillion species of microbes on Earth, and 99.999 percent of them have yet to be discovered.[653] 8% of our DNA is of viral origin.
- The composition of the human gut microbiota is linked to health and disease, but knowledge of individual microbial species is needed to decipher their biological roles. Despite extensive culturing and sequencing efforts, the complete bacterial repertoire of the human gut microbiota remains undefined. Here we identify 1,952 uncultured candidate bacterial species by reconstructing 92,143 metagenome-assembled genomes from 11,850 human gut microbiomes. These uncultured genomes substantially expand the known species repertoire of the collective human gut microbiota, with a 281% increase in phylogenetic diversity.[654]

When studying the Descriptive and Experimental methods of science and thinking about agriculture as a way to "lay a foundation for *system*,"[655] one has to examine the *education and labor systems*. An *educational system* that revolves around agriculture should focus on monitoring and evaluating the *teaching of the methods of sciences* to the *youth* of the world, especially the *daughters and sons of the farmers*, so they can discover the *potential of biodiversity and its results in terms of generating employment*. Once empowered with science methods, they can address their challenges and difficulties and the collective needs of their villages, neighborhoods, and the world's.

To come up with a proposal, having agriculture as a foundation for the system, we may use a macro-economy forecasting software to evaluate the productivity of the investment required in the educational system, when compared to the construction sector, considered today as the engine of the economy.

The discovery of the *potential of nature's biodiversity* by the *world's youth* will be the suggested *learning* context to *apply scientific methods*. The objective of having the *potential of biodiversity* as a *learning* context for the "methods of science" is that it has a more *significant potential* than the construction sector in the *generation of employment*.

---

[651] David L. Hawksworth and Robert Lücking, Fungal Diversity Revisited: 2.2 to 3.8 Million Species.

[652] Camilo, Mora, et al. How Many Species on Earth and in the Ocean?

[653] Nicholas Bakalar, Earth May Be Home to a Trillion Species of Microbes.

[654] Alexandre Almeida, et al., A new genomic blueprint of the human gut microbiota.

[655] 'Abdu'l-Bahá, Foundations of World Unity, p. 39. Emphasis added.

Because I dedicate this paper to the relentlessly persecuted Bahá'ís of Iran, I suggest this endeavor to support the Bahá'í Institute of Higher Education, and I would certainly love to be part of it.

Another potentiality:

The Báb hath revealed that the people of Bahá must *develop* the science of medicine to such a high degree that they will *heal* illnesses by means of foods. The basic *reason* for this is that if, in some component substance of the human body, an imbalance should occur, altering its correct, relative proportion to the whole, this fact will inevitably *result* in the onset of disease. If, for example, the starch component should be unduly *augmented*, or the sugar component *decreased*, an illness will *take* control. It is the *function of a skilled physician to determine* which constituent of his patient's body hath suffered *diminution*, which hath been *augmented*. Once he hath discovered this, he must *prescribe* a food *containing* the diminished element in considerable amounts, to *reestablish* the body's essential equilibrium. The patient, once his constitution *is* again in balance, will *be* rid of his disease.[656]

Alan C. Hamilton, in *Medicinal plants, conservation and livelihoods*, published in 2004, says:

Plants in traditional medicine

It is estimated that 70–80% of people worldwide rely chiefly on traditional, largely herbal, medicine to meet their primary healthcare needs (Farnsworth and Soejarto 1991; Pei 2001). The global demand for herbal medicine is not only large, but growing (Srivastava 2000). The market for Ayurvedic medicines is estimated to be expanding at 20% annually in India (Subrat 2002), while the quantity of medicinal plants obtained from just one province of China (Yunnan) has grown by 10 times in the last 10 years (Pei 2002b). Factors contributing to the growth in demand for traditional medicine include the increasing human population and the frequently inadequate provision of Western (allopathic) medicine in developing countries.

Plants in herbal medicine and botanicals

Herbal medicine is becoming ever more fashionable in richer countries, a market sector which has grown at 10–20% annually in Europe and North America over recent years (ten Kate and Laird 1999). In addition, there are many related botanical products sold as health foods, food supplements, herbal teas, and for various other purposes related to health and personal care. The extent to which herbal preparations are prescribed within conventional medicine varies greatly between countries, for instance being much higher in Germany than in the UK or USA.

Global use and value of medicinal species

In terms of the number of species individually targeted, the use of plants as medicines represents by far the biggest human use of the natural world. Plants provide the predominant ingredients of medicines in most medical traditions. There is no reliable figure for the total number of medicinal plants on Earth, and numbers and percentages for countries and regions vary greatly (Table 2; Schippmann et al. 2002). Estimates for the numbers of species used medicinally include: 35 000– 70 000 or 53 000 worldwide (Farnsworth and Soejarto 1991; Schippmann et al. 2002); 10 000–11 250 in China (He and Gu

---

[656] 'Abdu'l Bahá, Selections from the Writings of 'Abdu'l Bahá. Emphasis added.

1997; Xiao and Yong 1998; Pei 2002a); 7500 in India (Shiva 1996); 2237 in Mexico (Toledo 1995); and 2572 traditionally by North American Indians (Moerman 1998). The great majority of species of medicinal plants are used only in Folk Medicine. Traditional Scholarly Medical Systems employ relatively few: 500–600 commonly in Traditional Chinese Medicine (but 6000 overall) (Pei 2001); 1430 in Mongolian Medicine (Pei 2002b); 1106–3600 in Tibetan Medicine (Pei 2001, 2002b); 1250–1400 in Ayurveda (Dev 1999); 342 in Unani; and 328 in Siddha (Shiva 1996). The number of plant species that provide ingredients for drugs used in Western Medicine is even fewer. It was calculated for an article published in 1991 that there were 121 drugs in current use in the USA derived from plants, with 95 species acting as sources (more than one drug is obtained from some species) (Farnsworth and Soejarto 1991). Despite the small number of source species, drugs derived from plants are of immense importance in terms of numbers of patients treated. It is reported that about 25% of all prescriptions dispensed from community pharmacies in the USA between 1959 and 1973 contained one or more ingredients derived from higher plants (Farnsworth and Soejarto 1991).

The value of medicinal plants to human livelihoods is essentially infinite. They obviously make fundamental contributions to human health, and: "Is not health dearer than wealth?". A study of the 25 best-selling pharmaceutical drugs in 1997 found that 11 of them (42%) were either biologicals, natural products or entities derived from natural products, with a total value of US$ 17.5 billion (Laird and ten Kate 2002). The total sales' value of drugs (such as Taxol) derived from just one plant species (Taxus baccata) was US$ 2.3 billion in 2000 (Laird and ten Kate 2002). The world market for herbal remedies in 1999 was calculated to be worth US$ 19.4 billion, with Europe in the lead (US$ 6.7 billion), followed by Asia (US$ 5.1 billion), North America (US$ 4.0 billion), Japan (US$ 2.2 billion), and then the rest of the world (US$ 1.4 billion) (Laird and Pierce 2002).[657]

Another study published in 2018 by Fatemeh Jamshidi-Kia, Zahra Lorigooini, and Hossein Amini-Khoe, in *Medicinal plants: Past history and future perspective*, says:

> Undoubtedly, the demand for plant-derived products has increased across the world. In the Middle East, Latin America, Africa and Asia more than 85 percent of the populations predominantly rely on traditional medicine, especially on herbal medicines, for their health care needs. About 100 million people in the European Union and in some countries as high as 90% of the population, still use traditional, complementary or herbal medicines. The herbal medicine has an increasing big market. In 2012, the whole sales of Chinese herbal medicines reached more than US$83 billion which was 20% more than the market in 2011. It has been suggested that the whole market for all herbal supplements will reach more than US$115 billion by 2020, which in Asia-Pacific is the fastest and in Europe is the largest growing markets. These demands are predominantly driven by women subjects by growing emphasis on concerns on the adverse effects of synthetic drugs. Therefore, investing in human resource training can be the main source of research development in order to move from production to the production of crops. The importance of research in the field of medicinal plants is felt more than ever. Some medicinal plants are the sources of adjuvant therapy in the health systems worldwide, not only to treat diseases but also to prevent them and maintain health. Despite the extensive

---

[657] Alan C. Hamilton, Medicinal plants, conservation and livelihoods.

experiences in use of medicinal plants in traditional medicine, scientific study and identification of active plant compounds and their effects can lead to the discovery of new therapeutic benefits and the production of nature-based products in the future. To achieve this purpose, extensive research is fundamentally important to control the quality of raw drugs and the formulation to justify their use in the modern medicine system; subsequently, animal studies and clinical trials are required to use the benefits of these plants. In addition, in the development of medicine from medicinal plants, among other things, a practical plan should be developed to preserve these resources.[658]

Mariana Mazzucato, in *The Value of Everything. Making and taking in the global economy*, makes the following statement:

In modern capitalism, value-extraction is rewarded more highly than value-creation: the productive process that drives a healthy economy and society. From companies driven solely to maximise shareholder value to astronomically high prices of medicines justified through big pharma's 'value pricing', we misidentify taking with making, and have lost sight of what value really means.[659]

The oil industry is an example of value extraction; merchants are transferrers of value, while agriculture, education, industry, health, tourism, and artisanship are value- creators. What is the value creation of a hundred thousand jobs, lasting for centuries after the oil well becomes extinct, when discovering the potential for pest control of one more beneficial insect or microorganism? Or the creative value of scientifically validating an affordable remedy for hypoglycemia or diabetes. The World Health Organization says: "About 422 million people worldwide have diabetes, the majority living in low-and middle-income countries, and 1.5 million deaths are directly attributed to diabetes each year."[660]

I want to encourage the reader to watch *Mushrooms as Medicine* with Paul Stamets at Exponential Medicine.[661]

We should also monitor *employment* conditions when the principles of physics (*energy consumption per calorie produced*) help correct the agricultural system's viability. Monitoring and evaluating food waste regularly, perhaps is one of the most critical research to be done.

The International Labor Organization (ILO) a United Nations agency, says: "About 40 percent of the world's three-billion strong labor force, some 1.2 billion workers, are employed in agriculture as self-employed farmers, unpaid family workers, and hired workers. The ILO puts the number of 'waged' or hired workers at 450 million."[662]

One can easily calculate that the number of self-employed farmers and unpaid family workers is 750 million. Therefore, 25 percent of the world's workforce ignored in the calculations of *GDP* (Gross Domestic Product) and *unemployment rates, which are critical parameters for*

---

[658] Fatemeh Jamshidi-Kia, Zahra Lorigooini, Hossein Amini-Khoe, in Medicinal plants: Past history and future perspective.

[659] Mariana Mazzucato, The Value of Everything. Making and taking in the global economy.

[660] World Health Organization, Diabetes.

[661] Web. August 2021 <https://www.youtube.com/watch?v=7agK0nkiZpA>.

[662] International Labor Organization (ILO). ILO: Global Farm Worker Issues.

governments to make decisions. It is not fair to make decisions, only taking into account the 450 million who are agricultural workers earning a salary and benefits. However my previous stament "earning a salary and benefits" is too optimistic, because it is evident that among the hired workers there is underpayment, since many small farmers do not earn a living income. The contribution of these self- employed farmers and unpaid family workers should be taken into account in the *GDP and the unemployment rates* for every region.

They are just part of the informal economy:

> Under this new definition, the informal economy is comprised of all forms of 'informal employment'— that is, employment without labour or social protection— both inside and outside informal enterprises, including both self-employment in small unregistered enterprises and wage employment in unprotected jobs. ...
>
> As part of economic restructuring and liberalization, there has been a fair amount of *deregulation*, particularly of financial and labour markets. *Deregulation* of labour markets is associated with the rise of informalization or 'flexible' labour markets. ...
>
> Some countries include informal employment in agriculture in their estimates. This significantly increases the proportion of informal employment: from 83 per cent of non-agricultural employment to 93 percent of total employment in India; from 55 to 62 per cent in Mexico; and from 28 to 34 per cent in South Africa.[663]

Should we ask the Ministry of Labor or its equivalent in every country if s/he includes farming work done by men and women in their farms in the Gross Domestic Product (GDP)?

Finally, but not less important, is to find out if *self-employed* farmers and unpaid family workers count in the periodic government *report* about *unemployment* in rural areas.

---

[663] Chen, Rethinking the Informal Economy: Linkages with the Formal Economy and the Formal Regulatory Environment. Emphasis added.

9 The Cause of Maturity and the Propositional Method

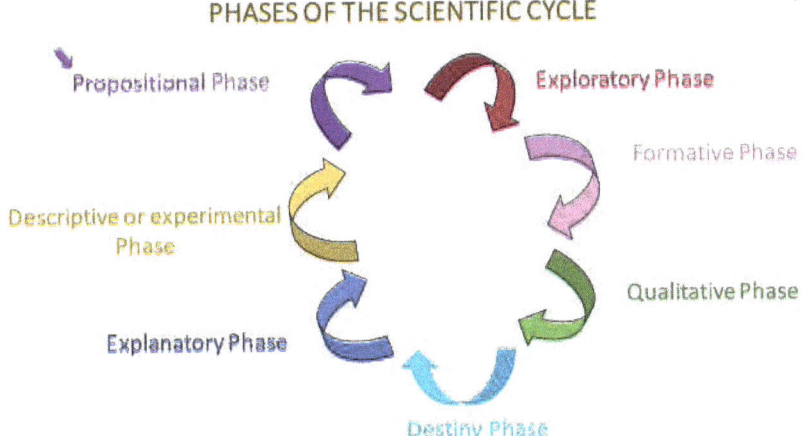

9.1 Philosophical Argument

Let us advance with the chair's example to introduce the Maturity Cause in its relationship to the propositional method.

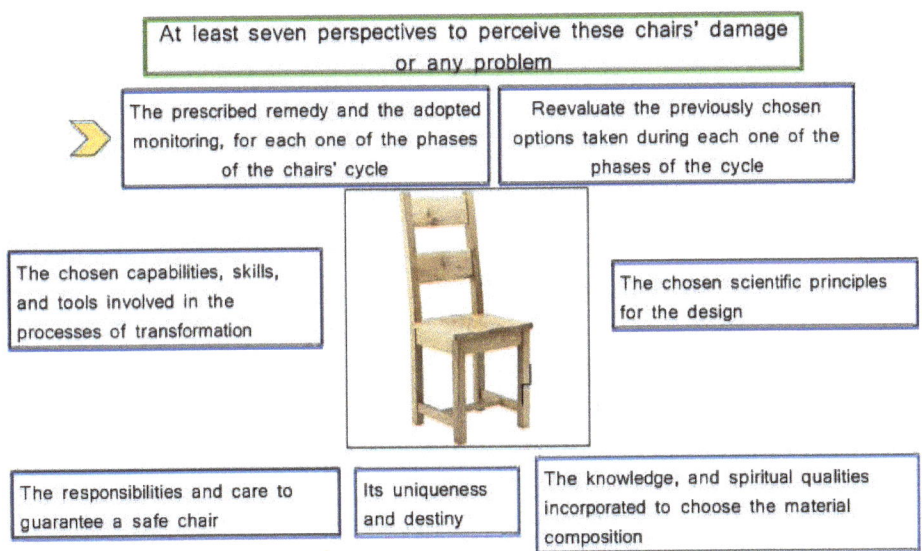

Consider now the options that we have after evaluating if we want to repair a chair. If the option is to fix it, then we may have assessed the remaining expected useful life of the chair. If the decision is to trash it, we may consider recycling its parts; if this is the option, then which parts should be recycled naturally, for example, by composting, and which parts are recyclable in the industrial process? The approach of the book *Cradle to Cradle: Remaking the Way We Make Things*, by William McDonough and Michael Braungart, seems to me to be the one that best fulfills the exercise of stewardship in the management of the environment.

The Cause of Maturity and the Propositional Method

If the legs, made out of special wood, show recurrent cracks, we could go as far as finding out what happened in the cycle of growing, harvesting (plant kingdom), and storing the wood (mineral kingdom). But what if the option is to understand what happened that several chairs broke, so we do not make the same mistakes when building new ones? Then, we should look into all the decisions made during each phase of the prior cycle related to the chair's specific damage.

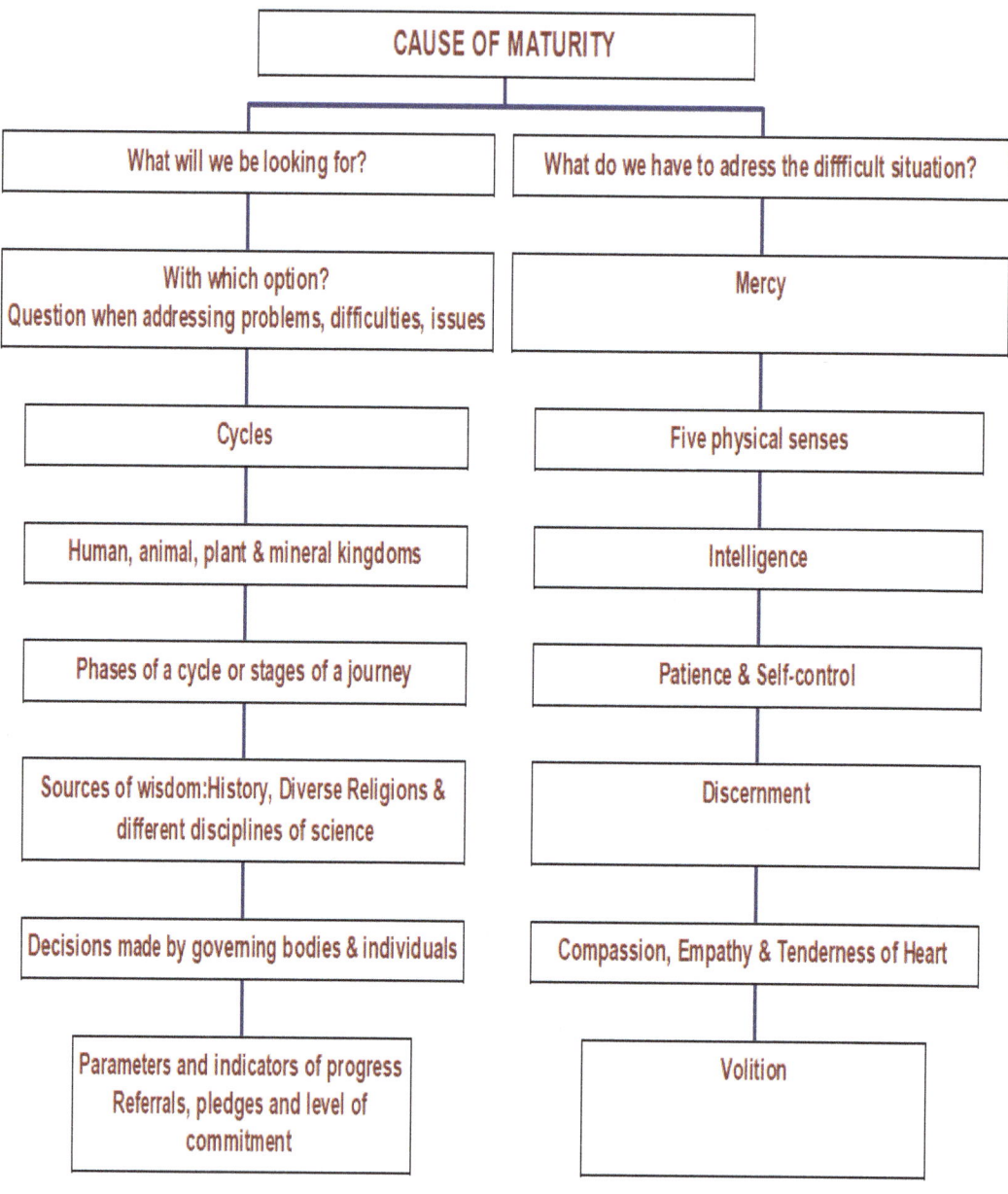

The Cause of Maturity can be systematized and improved using the following categories to organize its most relevant information:

9.1.1 Intelligence

We have been endowed with this most excellent gift to demonstrate our capacity to address aspirations, needs, problems, issues, difficulties, and calamities.

9.1.2 Discernment, self-control, and volition

Search alternatives using our power of discernment by calmly examining all wisdom sources and exercising our free will.

9.1.3 Human kingdom

Institutions, government, parents, and everybody else are responsible for their decisions. These responsibilities include all mature human beings, not necessarily only those in a position of authority since he (she) is exercising his (her) volition.

9.1.4 Kingdoms in nature

The effects of our decisions on the mineral, plant, animal, and human kingdoms should be carefully assessed.

9.1.5 Stages and stations

When considering which options seem wise requires exploring the stages of the journey, and the phases of a cycle, finding where the problem is originating.

9.1.6 Wisdom

Assume decisions based on the sources of wisdom and mercy, tenderness of heart, and empathy.

9.1.7 Mechanism of control

Check the current control mechanisms when going from one stage or phase to the next, suggest new ones, and encourage self-accountability.

9.1.8 Parameters and indicators of progress

Such as respect for the individual to exercise his (her) free will, level of commitment to the chosen alternative, number of pledges, and referrals.

We can perceive that the rational soul, the intelligence, and the human spirit refer to the same being: a free one, with duties and rights, and can decide whether to be responsible since all the fundamental causes discussed above are subject to laws. All mature individuals are equal before the Laws, but in the context of the options taken by each, it is clear that each human being is unique:

- in his (her) potential;
- the distinctive characteristics of how the individual expresses his (her) qualities,

weaknesses, and limitations.
- the degree of excellence which s/he makes manifest his (her) art, profession, or trade
- in his (her) level of love toward others and of obedience to the norms;
- in his (her) noble aspirations

9.2 The quintessence of knowledge

The quintessence of knowledge, O my Lord, proclaimeth its powerlessness to know Thee, and perplexity, in its very soul, confesseth its bewilderment in the face of the revelations of Thy sovereign might, and remembrance, in its inmost spirit, acknowledgeth its forgetfulness and effacement before the manifestations of Thy signs and the evidences of Thy praise. What, then, can this poor creature hope to achieve, and to what cord must this wretched soul cling?[664] It behoveth him who is a wayfarer in the path of God and a wanderer in His way to detach himself from all who are in the heavens and on the earth. He must renounce all save God, that perchance the portals of mercy may be unlocked before his face and the breezes of providence may waft over him. And when he hath inscribed upon his soul that which We have vouchsafed unto him of the quintessence of inner meaning and explanation, he will fathom all the secrets of these allusions, and God shall bestow upon his heart a divine tranquillity and cause him to be of them that are at peace with themselves. In like manner wilt thou comprehend the meaning of all the ambiguous verses that have been sent down concerning the question thou didst ask of this Servant Who abideth upon the seat of abasement, Who walketh upon the earth as an exile with none to befriend, comfort, aid, or assist Him, Who hath placed His whole trust in God, and Who proclaimeth at all times: 'Verily we are God's, and to Him shall we return'.[665]

ɑƀϙρ

In conclusion, the cycle of scientific research has also the propositional phase.

The notions and the set of human faculties associated in the Maturity Cause are appropriate to answer the question: "With which option?"

Because each cause has its effects, then as we practice the Maturity Cause, it will show its results:

- In our understanding of the notions mentioned above (cycle, phases, stages, kingdoms) In the development of our faculties (patience, discernment, self-control, free will)
- In the uses of our powers (intelligence, realizing the existence of two egos, the heart's role)
- In the building of relationships assessing the impact of our decisions on all kingdoms
- In our understanding of the laws of the cycle related to the problem (For example, illness cycle, domestic violence cycle) and our determination and courage to break the cycle

---

[664] Bahá'u'lláh, Prayers and Meditations by Bahá'u'lláh, p. 173.

[665] Bahá'u'lláh, Gems of Divine Mysteries, p. 25.

- In the polishing of virtues (detachment, mercy, empathy, compassion, and tenderness of heart)
- We acknowledge our humble learning mode capacity to define parameters and design indicators to monitor and evaluate the solution to any problem.

This approach will be critical in reaping the fruits of a curriculum that teaches the science of the love of God.

9.3 Research steps for the Propositional method

If we follow the art of constructing the Single Science, we conclude that each method of science intertwines with the other methods.

The author tried to relate this method to the Valley of True Poverty and Absolute Nothingness, as revealed in The Seven Valleys. These Valleys "mark the wayfarers' journey from their mortal abode to the heavenly homeland"[666]; in it, we find:

> In one sense, these letters refer to the states of holiness. The first meaneth "Free thyself from the promptings of self, then approach thy Lord." The second meaneth "Purify thyself from all save Him, that thou mayest offer up thy life for His sake." The third meaneth "Draw back from the threshold of the one true God if thou art still possessed of earthly attributes." The fourth meaneth "Render thanks unto thy Lord on His earth, that He may bless thee in His heaven, albeit in the realm of His unity His heaven is the same as His earth." The fifth meaneth "Remove from thine eyes the veils of limitation, that thou mayest learn that which thou knewest not of the stations of holiness."[667]

The following are just general ideas that may serve as guiding steps for the propositional method. The suggested steps should identify the problem, its history, describe its facts and symptoms, and connect to the faculties related to the Cause of Maturity, such as intelligence, human power, discernment, self-control, tenderness of heart, mercy, and sympathy. Also, looking at the link to God's Power and the parameters of progress in exercising stewardship to manage the trust.

First: The exploratory step

- Which decision should be adopted to eliminate the cause of the problem's facts and symptoms' and its effects in the kingdoms?
- With which frequency are we going to monitor and evaluate the impact of our activities in all kingdoms?
- Have we regularly attending classes to further our understanding of the meaning of stewardship?
- With which recurrency will I monitor and evaluate with humbleness the efforts made to be wise, merciful, and willing to empower others with science methods? What about becoming increasingly aware of emotions such as tenderness of heart, empathy, and mercy when consulting with women?
- Besides traps, barriers, bats, spiders, birds, beneficial insects, alarm pheromones,

---

[666] Bahá'u'lláh, The Seven Valleys, The Call of the Divine Beloved.

[667] Idem.

and biodiversity for controlling harmful insects population, what seems to be the best strategy to handle it?
- Which will be the outcome of a village, neighborhood, or even a whole society committed to a high degree of recycling everything discarded?

Second: The formative step, focusing on the general properties of matter and the faculties closely related to the Formative Cause
- With which regularity are we going to monitor and evaluate the impact of the modifications of the arrangement?
- With which periodicity will I monitor and evaluate the efforts made to be truthful, courteous, and kind?

Third: The qualitative step, focusing on the specific properties of matter

In The Promulgation of Universal Peace, by 'Abdu'l-Bahá, we find: "The fundamental basis of the community is agriculture, tillage of the soil. All must be producers."
- How often should we keep this in mind when deciding to improve the quality of the food produced nearby and the sustenance of the village and the neighborhood?
- With which frequency are we going to monitor and evaluate the quality or efficacy of our product?
- With which regularity will I monitor and evaluate knowledge acquisition and polish my moral values and virtues such as generosity and honesty to reach prosperity?

Fourth: The Root Cause

"O Son of Being!

Bring thyself to account each day ere thou art summoned to a reckoning; for death, unheralded, shall come upon thee and thou shalt be called to give account for thy deeds."[668]

"Set before thine eyes God's unerring Balance and, as one standing in His Presence, weigh in that Balance thine actions every day, every moment of thy life. Bring thyself to account ere thou art summoned to a reckoning, on the Day when no man shall have strength to stand for fear of God, the Day when the hearts of the heedless ones shall be made to tremble."[669]

According to history, the self-transformation of cultures takes centuries, in some cases millennia, requiring an innovative approach. In, *Cultures of Resistance: The Struggle Against Domestic Violence in Arab Societies*, we learn:

Domestic violence is a global phenomenon that affects an estimated 30% of all females worldwide. The health impacts of physical or psychological violence within the home include depression and other mental health issues, reproductive problems, injuries, and even death. While this problem is common to almost all societies, it is not the same everywhere. The focus of this chapter is Arab countries. Understanding domestic violence demands an understanding of the sociocultural context in which it occurs. The most important sociocultural factors in the Arab region are the culture of religion, specifically religious family law, and the politics of the state. We present a comparative framework that emphasizes the

---

[668] Bahá'u'lláh, The Hidden Words.

[669] Bahá'u'lláh, Gleanings from the Writings of Bahá'u'lláh.

interplay among family law, state power, intrafamily violence, and women's rights. Given cultural and political resistance to combatting domestic violence, the field of healthcare can be the most effective site for obtaining more and better data and for developing national policies to redress this problem.[670]

How often will we follow up the objectives, principles, notions, values, methodologies of the approach, and the definition of the parameters and the indicators' design to evaluate the situation's evolution?

Fifth: The Explanatory Step

Let us always keep present that laws regulate all decisions.

In Foundations of World Unity, The Master taught: "To solve this problem we *must* begin with the farmer; there will we lay a foundation for system and *order* because the peasant class and the agricultural class exceed other classes in the importance of their service.[671]

- Why is it required to monitor and evaluate regularly the protection of farmers, crops, the ecosystem, and food security for all?
- How often will we monitor and evaluate the protective measures, the alarm systems, and the equipment coverage for devices lost, stolen, damaged, and the warranty to cover defects?

Sixth: The Experimental Step

In Foundations of World Unity, The Master taught: "To solve this problem we must begin with the farmer; there will we lay a foundation for *system* and order because the *peasant class and the agricultural class* exceed other *classes* in the importance of their *service*.[672]

- If we assume using the potential of biodiversity as a context for empowering the youth with science methods: How often do we want to evaluate the contribution of the educational process to the generation of employment?
- What methods will we use to monitor and evaluate the training procedures and the processes and systems used to achieve results?

Seventh: The Propositional Step

- With which frequency will we monitor and evaluate the consequences of our stewardship management on the different kingdoms' cycles? And the parameters and indicators used?
- Which is the proposed decision?
- Which is the expected result in the kingdoms previously affected?
- Which parameters or indicators of healing, rehabilitation, and regeneration will we use from now on?

---

[670] Lisa Hajjar, et al., Cultures of Resistance: The Struggle Against Domestic Violence in Arab Societies.

[671] 'Abdu'l-Bahá, Foundations of World Unity. Emphasis added.

[672] Idem.

9.4 The farmers' situation and the propositional phase

Let us finish with our example of the farmers with the propositional method of science.

9.4.1 Increasing small farmers' income

How to compete with the current situation of Smallholder Participation in Agricultural Value Chains and establish fair trade initiatives?

Let me suggest a path to address the problem of increasing small farmers' income that requires evaluation with macroeconomic forecasting software as part of the consultation process with farmers, agroindustry, merchants, academics, and governmental officers.

Early in my studies, I discovered how progressive NGOs are, and recently found the following:

> "True Price is a movement of consumers, businesses, and institutions that take action in incorporating environmental and social costs in prices. We envision a world where all products are sold for a true price to enable a sustainable global economy."[673]

To work within the formal market sector, farmers must comply with the stringent quality standards and regular volume requirements of formal buyers as well as be willing to accept that prices may be below those in informal markets.

In a study by Neven, Odera, Reardon, and Wang (2009) that surveyed 115 farmers in Kenya (49 supplying the supermarket channel and 66 supplying traditional channels), the authors undertook primary data analysis of supermarket contracts and prices. They found that the mean price offered by the supermarket was significantly lower than traditional market prices — thus undermining frequent remarks about benefits to the farmer based on stability in the prices of supermarket chains.

There are cases in India where farmers have had to take on relatively large loans, based on contractual agreements. Later, they were later unable to repay when poor rains led to crop failure, or when market prices collapsed. This was highlighted in 2006, when crop failures in the cotton markets led to a series of suicides as farmers chose to die rather than hand over their farms to the debt collectors. Yet, crop insurance schemes associated with sales agreements are rare.[674]

Unlike contracting that focuses on supply coordination, fair trade is based on cooperation. The schemes usually provide farmers with a minimum floor price for their goods and a premium price for highest quality goods."

In the same way that much of the early Green Revolution literature (Feder and O'Mara 1981) focused on limited small farmer uptake of improved seeds, fertilizer, irrigation, and other components of "modern" production systems, a large share of the emerging literature on modern value chains has been concerned with smallholder participation in Agricultural Value

---

[673] True Price.

[674] Shaun Ferris et al. Linking Smallholder Farmers to Markets and the Implications for Extension and Advisory Services. Quoted from Wikipedia: Farmers' Suicides in India

Chains and with whether these same value chains might be leaving many poorer farmers behind.[675]

I would like for the reader to reflect on the bellow diagram[676]. I hope it may serve as an essential contribution to consult with all those interested in reaching food security for all, addressing climate change, and improving the income of small farmers. To achieve that, we must take into account what was said at the very beginning of this book when mentioned the quote, "The more microscopic animals exist in the soil, the better the plants will grow."[677]

In the diagram proposed by IDH and True Price about the situation of small famers of cacao in the Ivory Coast, we find:

"84% of the total external costs of cultivation are social costs, 54% are due to underpayment of hired and family workers. The other largest external cost drivers are land use, child labor, forced labor, and lack of social security."

Income (54%): more than half of the external costs during cocoa cultivation result from the underpayment of hired workers and the underearning of family workers.

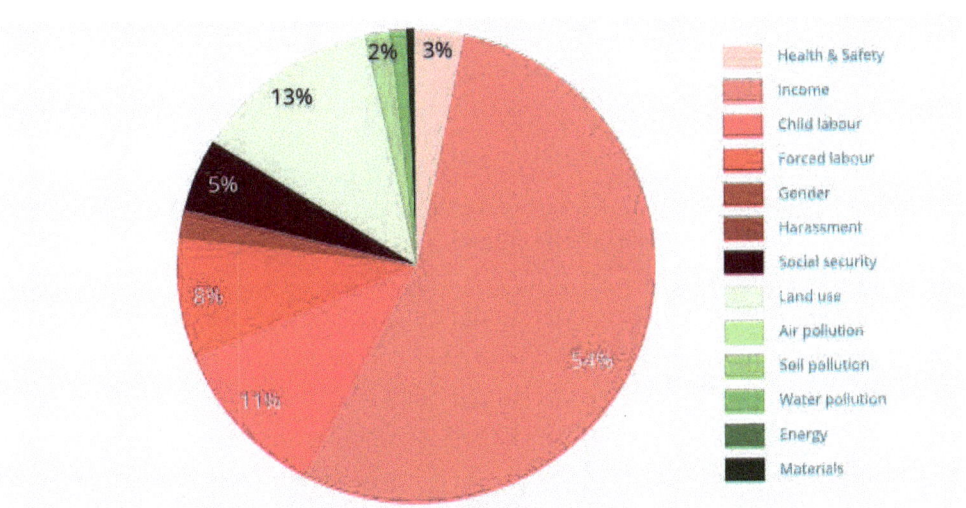

Figure 5 Share of each externality in the total external costs of conventional cocoa cultivation

Underpayment of hired workers and underearning of smallholder farmers constitute by far the largest external cost in cocoa cultivation in Ivory Coast. Hired workers receive an average total wage of €1.6/day, which is only about 20% of the living wage. A farmer household earns on average €3.5/day from the activities on its cocoa farm, which results in a yearly income of around 40% of the living income. On an annual basis, the wage of workers is €477, while the legal minimum wage is €659. The annual living wage for an Ivorian worker, as calculated by True Price, is €2,869. This size of poverty is not only problematic for the livelihoods of

---

[675] Shaun Ferris et al. Linking Smallholder Farmers to Markets and the Implications for Extension and Advisory Services.

[676] IDH and True Price, The True Price of Cocoa from Ivory Coast. Web. December 2022 <https://trueprice.org/wp-content/uploads/2022/07/TP-Cocoa.pdf>.

[677] 'Abdu'l-Bahá, Additional Tablets, Extracts and Talks.

workers and their families, but as well triggers other social issues, such as child and forced labour (Potts, et al., 2014).

It is hard for farmers to pay their workers higher wages, as they themselves do not earn a living income. Raising legal minimum wages, adjusting tax and subsidy structures, increasing farm productivity

Land use (13%): deforestation and other land degradation, caused by the establishment of new farm land and plantations, is the second largest externality;

Child labour (11%): child workers remain common in the Ivorian cocoa industry, often performing hazardous tasks and missing out on education.[678]

The first stage of the suggested strategy is: Prepare ourselves to deal with prejudice against farmers, reach a unity of vision engaging all actors (farmers, agroindustry, merchants, academics, consumers, and governmental officers), and unity of thought about why we reject debates and want to guarantee the participation of women's majority in a consultative dialogue to increase the possibilities of a peaceful outcome.

The second stage of the strategy is: What happens at the macroeconomic level if the government of a country like The Ivory Coast decides to have a storehouse in each village, buying products at the farm gate from smallholder farmers to feed everyone who cannot afford enough food to keep him(her) in good health?

For example, the True Price to pay is increasing the farm-gate price of cacao beans by 91%, explained as follows: 54% to increase income and pay their workers higher wages, 11% for replacing child labor conditioned on sending boys and girls to school, 13% to start building or planting barriers for controlling soil erosion, planting cover crops, and reforesting, 8% to eliminate adult and child forced labor, and 5% to contributing to social security.

    A similar approach to deciding the True Price for the rest of the food staples should be on the table. How many problems will the Ivory Coast solve?

A third stage of the strategy: is an increase to negotiate later on if they become certified as organic, addressing:[679] [680]

- Water pollution: Eutrophocation, acidification, marine ecotoxicity and

---

[678] IDH and True Price, The True Price of Cocoa from Ivory Coast.

[679] Cacao plant. Photo 18947257 © Pindiyath100 | Dreamstime.com

[680] Cup of chocolate. Photo 264042694 © Luis Echeverri Urrea | Dreamstime.com

freshwater ecotoxicity.
- Air pollution Greenhouse gas emissions and other hazardous air pollutants.
- Soil pollution: Terrestrial ecotoxicity and human toxicity.

A fourth stage of the strategy is: "approximately 17.7 million small-scale coffee farmers produce 80 percent of the world's coffee. Brazil, Vietnam, Colombia, Indonesia, and Ethiopia produce 68% of the coffee in the world."[681] If those five countries decide to buy from farmers thru the storehouse in each village the whole coffee harvest at the farm-gate at a price that follows the True-Price principles, what would happen to those countries' economies? And what about the countries importing coffee? Again, the gradual increase in the farm-gate True Price should be consulted and agreed upon with the buyers, agroindustry, retailers, and consumers of coffee and other commodities.

What would happen if Ghana, Indonesia, Nigeria, Cameroon, Brazil, and Ecuador made a pact to follow the same path of the Ivory Coast? "Côte d'Ivoire and Ghana are by far the two largest cocoa growing countries, accounting for over 60 % of global cocoa production, followed by Ecuador with 7 %. In Asia, Indonesia is the largest producer country."[682] A total of 60 countries produce cocoa.

Is it fair to postpone for a **fifth stage** the other problem I mentioned early on: taking into account the *risks* involved in being a farmer, *risks* associated with the environment, pest control, insecticides, and price fluctuations? Compare farmers' *risks* to those assumed by the agribusiness transforming cereal into oats or flour, the merchant who sells the product, and the profits each receives. If one of the basic principles of economics states that *risk* and profit are directly related: more *risk* exposure should lead to higher gain and vice versa, why do most of the world's farmers receive a meager income? Who manages a greater complexity of the three actors (the farmer, the agribusiness, or the retailer)? Who works physically hardest? I believe this matter should be part of the second stage of the agreement.

How many blenders, stoves, pipes, water heaters, electrical wiring, PCs, TVs, and mobile phones will buy the 750 million[683] self-employed farmers and unpaid family workers? What a waste is to ignore the enormous latent potential within nobodies! Without a true-price agreement among all those involved in the food chain, enabling a sustainable global economy will be impossible.

9.4.2 Enabling adolescents to become protagonists of social change

My most profound appreciation to all those who reeducate me in FUNDAEC and the Ruhi Institute. Dr. Farzam Arbab invited me twice to work at FUNDAEC. When working under Edmundo Gutierrez, President of the Rural University, I studied a textbook titled *Service to the Community – Health, an Aspect of Well-being*. Alberto Alzate, Farzam Arbab, and Francia de Valcarcel were coauthors of the text. It was part of FUNDAEC's curriculum for 6th and 7th

---

[681] Eleven Coffees, 25 Top Coffee-Producing Countries in 2022.

[682] Swiss Platform for Sustainable Cocoa.

[683] Page 234 of The Single Science.

graders, which inspired me to continue proposing the empowerment of adolescents following their example.

In the textbook titled *Service to the Community – Health, an Aspect of Well- being* wich was part of FUNDAEC's curriculum for 6th and 7th graders over 40 years ago. It is crucial to notice that the language used in the lessons is a colloquial conversation.

The first three lessons explore among the students of the textbook and the neighbors in the village the current concepts of well-being, education, and health. Also, there is a reflection on the meaning of well-being and education for FUNDAEC.

In each lesson, there is a simple survey to start reading the reality of the village. If there are ten students and each kid visits five families, they collect the facts of 50 households.

Lesson 4: *Intestinal Parasites.* Young kids become protagonists of social change by sharing with the inhabitants of the village their understanding of how we become infected with intestinal parasites, the dimension of the damage caused in our bodies, and different ways to prevent the ill-health determinant.

While sharing their understanding of intestinal parasites with the parents of five homes, they also delivered the survey to find out how many kids were infected in the past with intestinal parasites.

Lessons 5 to 12, are about eight health determinants related to Diarrhea, Typhoid Fever, Environmental Sanitation, Tuberculosis, Respiratory Diseases, Skin Diseases, Eruptive diseases during infancy, and Urinary Infections. Once the student visits the family nine times, sharing his(her) knowledge and collecting information about the village's reality, their commitment increases because learning that theory and practice are linked and fruitful.

Lesson 13: The most empowering event is the participation of the students in the village's consultative process to agree on a whole strategy for dealing with the nine ill health determinants. After figuring out the whole strategy, the villagers examine different scenarios, such as what could be done with $10.000 dollars, $100.000 or just $80.000.

Suppose a Social Worker or graduate in Public Health is present when adolescents show their findings addressing nine ill-health determinants. Will this lead to a fruitful consultation process among the residents of the village or neighborhood? Would adolescents feel that they are protagonists of social change? The gradual empowering of adolescents and the beneficial effects of improving the lives of the village inhabitants is considered a holistic, fascinating outcome.

As part of planning the strategy, Leo suggests developing textbooks to enable adolescents to start learning and practicing research methods gradually. I will mention three ideas to reach the goal:

First: To create a textbook for addressing different addictions, practice the Exploratory method by starting with a simple survey, and gradually learn more about the other critical notions related to the Maturity Cause, for example, when examining addictions, concepts such as chronic[684] or recidivist[685] mean cyclic occurrence. The students will learn how important it is to learn and practice patience, self-control, and forbearance as prerequisites to discern and make sound choices in life.

Second: To develop one textbook for learning to apply the Formative Method when dealing with different types of discrimination: Gender, social class, level of education, minority, nationality, farmers, skin color, religious beliefs, and sexual orientation. Also, prejudice against certain foods, insects, and microorganisms.

About progressively empowering young girls and boys with the Formative Method, Leo suggests beginning to empower adolescents by studying FUNDAEC's *Primary Description Elements: Unit 1 Properties*, where they learn about the General Properties of Matter and virtues closely connected to the Formative Cause.

Documents of research journalism or ethnographic reports will offer the students a great learning context. And they become used to identify them in other readings about ill health determinants, such as the ones addressing different kinds of discrimination.

If living in the United States, the Guardian has clearly expressed that racial discrimination is the most challenging issue for America. If we work in Iran, Pakistan, Central African Republic, Somalia, Sierra Leona, Sudan, Chad, or Afghanistan, where women's human rights are worst, does women's discrimination deserve just one lesson in the textbook? Or should we consider developing an entire text with seven perspectives to read the reality of women's discrimination, collect the facts about this vital theme, start addressing the problem within a village or a neighborhood, and elaborate another text to handle other kinds of discrimination?

The idea is for each adolescent to visit five households for each lesson, sharing the facts about the ill-health determinant, registering the answers to the survey, and bringing their collective findings to a reflection gathering within the neighborhood to a consulting process about the next steps.

The proposed questions for the textbook lessons referring to women's discrimination are:

- Is it for females to wear a hijab or a burka to enable males to control their sexual urges?[686]
- Why must women be accompanied by a male relative when outside the home?[687]
- What is women's health and property ownership status in those countries cultures?[688]

---

[684] [B]eing such by habit and not likely to change. A chronic smoker who has quit—many, many times. Merrian Webster Dictionary.

[685] [A] person who continues to commit crimes even after being caught and punished. Merrian Webster Dictionary.

[686] Mahrukh Arif-Tayyeb, 2021; Timothy R. Jordan et al., 2020 and Concern Worldwide.

[687] Javier Cerrato and Eva Cifre, 2018.

[688] Claire Hawcroft, 2019; Shima Tabatabai and Nasser Simforoosh, 2020; Julie Ray, 2022), Safar Ghaedrahmati and Foad Shahsavari, 2019; Gayle Tzemach Lemmon and Becky Allen, Council on

- Is women's discrimination conducive to war?[689]
- Why do women's witness statements in court carry half the weight of a man's?[690] Why violate women's human rights?
- Mindful of the descriptive and experimental methods:
    - What is the concept of a woman among religious leaders?
    - What is a woman among merchants? And among national, regional, provincial, and local government officials?
    - What is a child girl or an adolescent woman for parents?
    - What is a woman among teachers and professors?
    - What is a woman in the military? And among scientists? And among managers and engineers in industries?
    - And among drug dealers, human traffickers, and alcoholic beverage producers, what is a woman?
- Of all the above, what are superstition, ignorance, imagination, misunderstanding, blind imitations, and dogmatic interpretations, and what conforms to science and religion joined and welded in reality?

Each lesson requires a simple survey for adolescents learning to read their locality's reality.

Third: Produce a textbook to enable adolescents to learn the Qualitative status of a village when examining polluted samples of water, soil, and air, or nutrition deficiencies or excesses of vitamins, proteins, and carbohydrates, or when somebody becomes depressed about being called names. For enabling young girls and boys with the Qualitative Method, Leo has suggested beginning to empower adolescents by studying the exact FUNDAEC text, where they learn about the specific properties of matter and spiritual qualities. Of course, there are other means for adolescents to start reading the truth about ill-health determinants where they live, such as water pollutants, smelling the water for foul odors, or its color.

They will become accustomed to identifying them when recognizing the distinctive features of ill-health determinants within the textbook.

Leo suggests that the selected scientific facts be written in the textbook for adolescents in colloquial language. Both versions will be part of the material to accompany the Junior Youth Groups animators in their learning process.

Leo thinks that utilizing a methodology to identify the notions closely related to the other methods can be done by developing a textbook similar to the FUNDAEC's *Primary Description Elements: Unit 1 Properties*. To write a book addressing this need is desirable to study chapter 10: The Art of Constructing The Single Science.

Foreign Relations, Reforming Women's Property Rights in Afghanistan, 2017).

[689] Universal House of Justice, The Promise of World Peace, 1985; Living with Violence: A National Report on Domestic Abuse in Afghanistan Global Rights: Partners for Justice, March 2008.

[690] Ann Elizabeth Mayer, 2018

9.4.3 Following up on the suggested farmer's strategy

Finishing with our example of the farmers:

We must decide which parameters and indicators will serve to control and evaluate the farmers' situation´s progress and monitor and assess the protection of the ecosystem and our health. The author thinks farmers should make more money than the agro-industry and the merchants. But, any agreement reached between these actors should consider the principles of moderation and gradation. The change cannot be abrupt.

The history of ancient civilizations such as Mesopotamia, Egypt, India, and China provides a good idea of what farmers meant to the ruling classes as a background to understand what they mean to the current world leaders. I want to encourage those interested in addressing this issue to study what happened after discovering agriculture when it became the new normal. Please refer to *The Study of History* by Arnold Toynbee, searching to answer the following questions:

- How much were farmers taxed?
- Who provided the army for the rulers to conquer new territories? Who bore the burden of the construction of infrastructure?
- What caused those civilizations to collapse?

Ian Johnson in Leaving the Land: China's Great Uprooting: Moving 250 Million Into Cities, says: *"China is pushing ahead with a sweeping plan to move 250 million rural residents into newly constructed towns and cities over the next dozen years — a transformative event that could set off a new wave of growth or saddle the country with problems for generations to come."*[691]

We must evaluate the outcome for farmers and the community at large with a macro-economic forecasting software to decide which administrative order and which authorities are strategic contacts for the success of this proposal's implementation.[692]

We must regularly monitor the level of hope of the farmers and their families about the viability of their farming businesses and the reasons not to migrate towards the misery belts around big cities.

Every man of *discernment*, while walking upon the earth, feeleth indeed abashed, inasmuch as he is fully aware that the thing which is the source of his prosperity, his wealth, his might, his exaltation, his advancement and power is, as ordained by God, the very earth which is trodden beneath the feet of all men. There can be no doubt that whoever is cognizant of this truth, is cleansed and sanctified from all pride, arrogance, and vainglory. Whatever hath been said hath come from God. Unto this, He, verily, hath borne, and beareth now, witness, and He, in truth, is the All-Knowing, the All-Informed.[693]

---

[691] Ian Johnson, Leaving the Land: China's Great Uprooting: Moving 250 Million Into Cities.

[692] The author encourages to those interested in the size of the corruption and greed that dives deep into the food production underworld to watch "Rotten" the Netflix TV series.

[693] Bahá'u'lláh, Epistle to the Son of the Wolf, p. 43. Emphasis added.

10 The Art of constructing The Single Science

The author believes that this art will be instrumental in developing a universal language. 'Abdu'l-Bahá, in *Paris Talks*, said:

> One of the great steps towards *universal peace* would be the establishment of a universal language. Bahá'u'lláh commands that the servants of humanity should meet together, and either choose a language which now exists, or form a new one. This was revealed in the Kitáb-i-Aqdas forty years ago. It is there pointed out that the question of diversity of tongues is a very difficult one. There are more than eight hundred languages in the world, and no person could acquire them all.
>
> The races of mankind are not isolated as in former days. Now, in order to be in close relationship with all countries it is necessary to be able to speak their tongues.
>
> A universal language would make intercourse possible with every nation. Thus it would be needful to know two languages only, the mother tongue and the universal speech. The latter would enable a man to communicate with any and every man in the world![694]

When we study the quotes in the Writings related to any notion, such as virtue, human, plant, and being, we must pay attention to the words in the paragraph because in there is the key to our understanding. The quotes provide us with the words that best explain their meaning and energize the potentialities within us. Bahá'u'lláh has manifested that the potential of the Writings is immense:

> It is evident unto thee that the Birds of Heaven and Doves of Eternity speak a twofold language. One language, the outward language, is devoid of allusions, is unconcealed and unveiled; that it may be a guiding lamp and a beaconing light whereby wayfarers may attain the heights of holiness, and seekers may advance into the realm of eternal reunion. Such are the unveiled traditions and the evident verses already mentioned. The other language is veiled and concealed, so that whatever lieth hidden in the heart of the malevolent may be made manifest and their innermost being be disclosed. Thus hath Ṣádiq, son of Muḥammad, spoken: "God verily will test them and sift them." This is the divine standard, this is the Touchstone of God, wherewith He proveth His servants. None apprehendeth the meaning of these utterances except them whose hearts are assured, whose souls have found favor with God, and whose minds are detached from all else but Him. In such utterances, the literal meaning, as generally understood by the people, is not what hath been intended. Thus it is recorded:
>
>> "Every knowledge hath seventy meanings, of which one only is known amongst the people. And when the Qá'im shall arise, He shall reveal unto men all that which remaineth." He also saith: "We speak one word, and by it we intend one and seventy meanings; each one of these meanings we can explain."[695]

Following the grouping criteria, let us proceed to reflect on some examples related to building The Single Science in the context of each Cause:

---

[694] 'Abdu'l Bahá, Paris Talks. Emphasis added.

[695] Bahá'u'lláh, The Kitáb-i-Íqán.

The Art of constructing The Single Science

10.1 The Art of Constructing the Cause of Maturity

The table below shows the methodology followed in building the "Single Science" by inter-weaving the causes horizontally and vertically. It is essential to understand the placement of the following acceptations in one of the Rows (1 to 6) to comprehend how the Cause of Maturity interweaves with the other causes.

	THE CAUSE OF MATURITY AND THE EXPLORATORY AND PROPOSITIONAL METHODS OF SCIENCE	
1	**Question**: With which option? when addressing problems, difficulties, issues, and needs	**CAUSE OF MATURITY**
2	**Essence**: Mercy. **Sources of Wisdom**: What history or experience, the diverse disciplines of science, and world religions say about the addressed issue[696] **Moral principles** **Kingdoms, Cycles, phases, journeys, and stages**	**MATERIAL CAUSE**
3	**Laws** related to the assumed decision's inherent responsibilities, considering the effects in the other Kingdoms and the Human Kingdom. **Goal**: Stewardship in the management of the trust	**FINAL CAUSE**
4	**Powers**: Intelligence, patience, self-control, discernment, compassion, empathy, volition, and tenderness of heart **Decisions made at the individual and collective level**: Governmental Institutions, Administration of private enterprises, NGOs, families, or individuals. The human choices within the stages of a plan or the phases of a cycle **The spontaneous differential response to the stimuli received** by all kingdoms **The effects of the decisions made by humans in all kingdoms** **The developmental stage** of the individual, plant, or animal **Monitoring and evaluating** the strategy's stages and phases	**EFFICIENT CAUSE**
5	Challenging **situations** in which the human kingdom chooses to respond or, in the case of the mineral, plant, and animal kingdoms, the spontaneous differential response to the stimuli received. **Relationships**: are those that interconnect the sources of wisdom	**FORMATIVE CAUSE**
6	**Parameters and indicators** of progress: Such as level of respect for the individual to exercise his (her) free will, level of commitment, number of pledges, and number of recommendations or referrals	**CAUSE OF MATURITY**

The reader learned that the grouping criteria is the foundation of philosophy and has studied the conceptual framework for the Cause of Maturity. Please proceed to check the placement of the definition of each of the following words within the Maturity Cause:

---

[696] Because of their value.

Membrane: "A cell membrane surrounds and protects the contents of a cell. It controls which substances can enter and exit the cell. The membrane also gives a cell its shape and enables the cell to attach to other cells, forming tissues."[697]

The main reason it is advisable to situate it in row 1 is at the intersection of the Cause of Maturity and The Cause of Maturity because the membrane involuntarily controls which substances can enter and exit the cell answering the question: With which option? The author believes that the membranes' function is an expression of the Will of God.

We could also place it in row 3, the intersection of The Cause of Maturity and The Final Cause, because the membrane spontaneously surrounds and protects the contents of a cell as a symbolic expression of stewardship in managing the trust. We could also emplace it in row 4, the intersection of The Cause of Maturity and The Efficient Cause, because *control* is an action. However, we could likewise situate it in row 5, the junction of The Cause of Maturity and The Formative Cause; because the membrane gives a cell its *shape* and enables the cell to *attach* to other cells, *forming* tissues.

Plasma membrane: "The limiting surface of the cytoplasm of a eukaryotic cell is the plasma membrane. It consists of a phospholipid bilayer with a variety of embedded molecules that act as channels and pumps, selectively moving particular molecules into and out of the cell. Surface molecules on the plasma membrane allow specific recognition of each particular cell type. Phospholipids have a head group, which is attracted to water, and a tail group, which is made up of a long hydrocarbon chain repelled by water. Phospholipids are the primary constituent of the lipid bilayers of cells."[698]

It is advisable to situate it in row 1, the intersection of The Cause of Maturity and The Cause of Maturity because its function is *selectively* moving particular molecules into and out of the cell, spontaneously answering the question: With which option? We could likewise situate it in row 2, the intersection of The Cause of Maturity and The Material Cause, if looking into the composition of the plasma membrane consisting of a *phospholipid* bilayer with a variety of embedded *molecules*. We could also emplace it in row 4, the intersection of The Cause of Maturity and The Efficient Cause, if we focus on the "embedded molecules that *act* as channels and pumps, which function as selectively *moving* particular molecules into and out of the cell.

Molt: "To cast off the outer covering. Birds molt old feathers once or twice a year. Reptiles molt old skin, and arthropods cast off the entire cuticle. Mammals also molt hair, but the term shed is usually used in this case."[699]

It is advisable to situate it in row 1, the intersection of The Cause of Maturity and The Cause of Maturity, because birds regularly cast off their old feathers.

Reptiles molt their old skin, and arthropods cast off their rigid exoskeleton as needed to grow larger.

---

[697] "Membrane." <https://quizlet.com>.

[698] "plasma membrane." Eugene M. Mccarthy, Online Biology Dictionary.

[699] "molt." Eugene M. Mccarthy, Online Biology Dictionary.

Additionally, we could situate it in row 4, the intersection of The Cause of Maturity and The Efficient Cause, because molting is a process, and feathers, skin, and hair are the leftover.

Carbon cycle: "The carbon cycle comprises two primary processes, photosynthesis, and respiration. Photosynthesis produces oxygen and glucose from carbon dioxide and water. Respiration reverses this by creating carbon dioxide and water from glucose and oxygen."[700]

It is logical to position it in row 1, the intersection of The Cause of Maturity and The Cause of Maturity because these two primary processes reverse each other results as part of a *continual cycle*. Besides, we could situate it in row 2, the intersection of The Cause of Maturity and The Material Cause, *composed* of two primary processes. We could also position it in row 4, the junction of The Cause of Maturity and The Efficient Cause, if we look at the processes and their results.

Lytic cycle: "Bacteriophage is a virus that parasitizes a bacterium by infecting it and reproducing inside it."[701] "The lytic cycle results in the destruction of the infected cell and its membrane. ... The lytic cycle, which is also commonly referred to as the "reproductive cycle" of the bacteriaphage, is a six-stage cycle. The six stages are attachment, penetration, transcription, biosynthesis, maturation, and lysis."[702]

The main reason it is advisable to situate it in row 1, the intersection of The Cause of Maturity and The Cause of Maturity, is that the organism's *problem* is the existence of a bacterium. We can position it in row 2, the intersection of The Cause of Maturity and The Material Cause, because that infected cell becomes part of the mineral kingdom. We can also place it in row 3, the intersection of The Cause of Maturity and The Final Cause, because the destruction of the bacteria cell results from its obliterated cell's immune system when infected by the bacteriaphage. In row 6, the intersection of The Cause of Maturity and The Cause of Maturity because once it reaches maturity, the bacteriaphage bursts the cell where they reproduce to expand into other bacteria, starting new lytic cycles.

Pupae: "An intermediate usually quiescent stage of a metamorphic insect (as a bee, moth, or beetle) that occurs between the larva and the imago, is usually enclosed in a cocoon or protective covering, and undergoes internal changes by which larval structures are replaced by those typical of the imago."[703]

The main argument why it is logical to position it in row 1, the intersection of The Cause of Maturity and The Cause of Maturity, is because "metamorphosis is a dramatic change, a transformation from one state to a completely different one."[704] We can situate it also in row 2, the intersection of The Cause of Maturity and The Material

---

[700] "carbon cycle." Eugene M. Mccarthy, Online Biology Dictionary.

[701] Bacteriophage, Oxford Dictionary.

[702] Wikipedia, Lytic Cycle.

[703] "pupae." <https://quizlet.com>.

[704] "metamorphic." <https://www.vocabulary.com/dictionary/>

Cause, because it is a *quiescent* stage of the metamorphic cycle. We can also locate it in row 3, The Cause of Maturity and The Final Cause, because the cocoon is a protective covering. We can position it in row 4, the intersection of The Cause of Maturity and The Efficient Cause, because it is a developmental stage *undergoing* internal *changes*.

Forbearance: "is patient self-control; restraint and tolerance under provocation."[705]

The ideal location is in row 1, the intersection of The Cause of Maturity and The Cause of Maturity because all of the above pertain to a mature individual. We can also situate it in row 2, the intersection of The Cause of Maturity and The Material Cause, because they are attributes of an individual under provocation.

Altruism: "Unselfish behavior[706]; within a biological context, behavior that assists others to survive and reproduce, but that does not benefit the individual engaging in the behavior."[707]

The ideal location is in row 1, the intersection of The Cause of Maturity and The Cause of Maturity because an animal assists others in surviving and reproducing, considering its *existence* as unimportant with absolute detachment. Besides, an unselfish behavior of an insect, bird, or bat is a non-conscious decision. We can also emplace it in row 3, the intersection of The Cause of Maturity and The Final Cause, because the result of the altruistic behavior is to assist others in surviving and reproducing.

The same grouping criteria applied to the faculties mentioned above associated with the Cause of Maturity. What are the benefits of classifying those acceptations in this cause? Answer:

First, let us reflect on the following quote:

Every believer needs to remember that an essential characteristic of this physical world is that we are constantly faced with trials, tribulations, hardships and sufferings and that by overcoming them we achieve our moral and spiritual development; that we must seek to accomplish in the future what we may have failed to do in the past; that this is the way God tests His servants and we should look upon every failure or shortcoming as an opportunity to try again and to acquire a fuller consciousness of the Divine Will and purpose.[708]

Second, a similar situation occurs in the other kingdoms. Let us go back to the concept of evolution mentioned above if we accept that God has created every creature with the potential to express its perfections gradually. Then we could imagine that through the passing of many natural cycles and the corresponding trials and tests, the opportunity for an organism to manifest developing potentialities begins to be plausible.

Third, the author believes that students should learn to perceive and interpret nature's responses to the challenges presented to the organism in each phase of the cycle when

---

[705] https://wikidiff.com/forbearance/tolerance#:~:text=Patient%20self-control%3B%20restraint%20and%20tolerance%20under%20provocation.%0A,The%20ability%20to%20endure%20pain%20or%20hardship%3B%20endurance.

[706] https://plato.stanford.edu/entries/altruism-biological/

[707] "altruism." Eugene M. Mccarthy, Online Biology Dictionary.

[708] From a letter written on behalf of the Universal House of Justice to an individual believer, January 9, 1977. Helen Hornby, Lights of Guidance No. 1226.

tweaking the environmental conditions. This capability opens the doors for all of us to be very observant of the opportunities for the plant, the animal, and human kingdoms to express their inherent perfections progressively. Let us appreciate what gardeners do with tulips' bulbs at the end of the flowering season: "In the spring, after the blossoms have passed their peak, clip off the flower heads and allow the green foliage to die back. This technique lets the plant put all its energy into building a strong bulb for next season."[709]

If we follow the art of constructing the Single Science, we conclude that each method of science intertwines with the other methods.

10.2 The Art of Constructing the Formative Cause

	THE FORMATIVE CAUSE AND THE FORMATIVE METHOD OF SCIENCE:	
1	**Question:** How is the arrangement?	MATURITY CAUSE
2	**Essence:** love and soul as its power. The general properties of matter: form, size, mass, temperature, movement, and position in time and space Hormones, feelings, and emotions	MATERIAL CAUSE
3	The **law** of gravity and **ordinances** related to unity, harmony, and justice **Goals:** Aspirations and desires to reach the expected vision or design	FINAL CAUSE
4	**Powers:** imagination and senses of justice and religion The practical application of the principles related to the general properties of matter: form, size, mass, temperature, movement, and position in time and space in relation to the desired organization, arrangement, and design	EFFICIENT CAUSE
5	Religious and Scientific principles related to the general properties of matter	FORMATIVE CAUSE
6	**Parameters and indicators** of unity, beauty, symmetry, harmony, reciprocity, and justice in reaching the desired vision or design	MATURITY CAUSE

The table above shows the methodology followed in constructing the "Single Science" by inter-weaving the causes horizontally and vertically. For the reader to understand the placement of the following acceptations in one of the Rows (1 to 6), it is crucial to know how the Formative Cause interweaves with the other causes.

---

[709] Encourage Your Tulips to Come Back. February 2020 Web. <https://www.americanmeadows.com/tulips-come-back>.

The reader learned that the grouping criteria is the foundation of Philosophy and has studied the conceptual framework for the Formative Cause. Please proceed to check the placement of the definition of each of the following words within the Formative Cause:

Meristem: "A formative plant tissue usually made up of small cells capable of dividing indefinitely and giving rise to similar cells or to cells that differentiate to produce the definitive tissues and organs."[710]

In row 1, the intersection of The Formative Cause and The Cause of Maturity because it involuntarily generates similar cells or differentiated ones to form the definitive tissues and organs. Also, we can position it in row 4, the intersection of The Formative Cause and The Efficient Cause, because it is describing the meristem's *function* in the formation of a tissue. The main reason why it is advisable to situate it in row 5, the intersection of The Formative Cause and The Formative Cause, is because the meristem is a critical tissue in achieving unity in diversity of cells, tissues, and organs on a plant. Similar and differentiated are referring to the *form* of the cells, and small is referring to *size*, which are general properties of matter

Abscisic acid: "A plant hormone inhibiting growth; helps plants withstand adverse conditions."[711]

The main argument why it is best to position it in row 5, the intersection of The Formative Cause and The Formative Cause, is because it is referring to inhibiting growth (size). We can also place it in row 1, the intersection of The Formative Cause and The Maturity Cause; because it helps, plants withstand adverse conditions.

Auxin: "Any of various hormones or similar substances that promote and regulate the growth and development of plants. Auxins are produced in the meristem of shoot tips and move down the plant, causing various effects."[712]

The main reason it is advisable to situate it in row 5, the intersection of The Formative Cause and The Formative Cause, is because it refers to the promotion and regulation of size. We could also situate it in row 1, the intersection of The Formative Cause and The Cause of Maturity because it involuntarily regulates its growth, or even better expressed is the Will of God why auxin does regulate the growth and development of plants. We can also place it in row 4, the intersection of The Formative Cause and The Efficient Cause, because Auxins are *produced* in the meristem of shoot tips and *move* down the plant, causing various *effects*.

Synchrony: "A state in which things happen, move, or exist at the same time."[713]

In row 2, the intersection of The Formative Cause and The Material Cause, because if *things exist* at the same time, we should be able to perceive their existential properties. We could also situate it in row 4, the intersection of The Formative Cause and The Efficient Cause, because it is in a *state* in which things *move* or *exist* at the same time. The main reason why it is advisable to situate it in row 5, the intersection

---

[710] "meristem." <https://quizlet.com>.

[711] "abscisic." <https://quizlet.com>.

[712] "auxin." <https://quizlet.com>.

[713] "synchrony." Merriam-Webster Dictionary.

of The Formative Cause and The Formative Cause, is because time is a general property of matter and synchrony is when things happen at the same time. We can also place it in row 6, the intersection of The Formative Cause and "The Cause of Maturity" because we can use synchrony as a parameter.

Hormone: "Circulating molecules that serve as signals for particular body processes to occur by interacting with target cells. However ... some specific hormones significantly affect human emotions. These hormones include Estrogen, Progesterone, Testosterone, Norepinephrine and Epinephrine, Serotonin, GABA, Dopamine, Acetylcholine, and Oxytocin."[714]

In row 2, the intersection of The Formative Cause and The Material Cause, because hormones are molecules. The main reason why it is advisable to situate it in row 5, the intersection of The Formative Cause and The Formative Cause, is because some specific *hormones* greatly affect human *emotions.*

Activation energy: "The amount of energy (E_a) required to convert a stable molecule into a reactive one. It is the energy needed to produce the unstable condition in which the energy state of the bonds of the reactants is raised to a level corresponding to the unstable transition state that precedes a chemical reaction."[715]

It is proper to locate it in row 2, the intersection of The Formative Cause and The Material Cause, because a *certain amount* of energy is necessary to convert a stable molecule into a reactive one.

Additionally, we can position it in row 3, the intersection of The Formative Cause and The Final Cause. For the reaction to occur, it *must fulfill* the amount of energy required to produce the unstable condition.

Likewise, we could place it in row 4, the intersection of The Formative Cause and The Efficient Cause, because *activation* energy is the *energy* needed to *produce* the unstable condition that precedes a chemical *reaction.*

Besides, we could also situate it in row 5, the intersection of The Formative Cause and The Formative Cause, because it is referring to the amount (*size*) of the energy required which is a scientific principle.

Synthesis: "Something made by combining different things (such as ideas, styles, etc.). The production of a substance by combining simpler substances through a chemical process."[716]

We can situate it in row 2, the intersection of The Formative Cause and The Material Cause because the combination of *different* things defines synthesis. The main argument why it is best to situate it in row 5, the intersection of The Formative Cause and The Formative Cause, is because we can assimilate a combination to unity in diversity, which is one of the principles of unity. We can also place it in row 4, the intersection of The Formative Cause and The Efficient Cause, because it is the

---

[714] "hormone." Eugene M. Mccarthy, Online Biology Dictionary.

[715] "activation energy." Eugene M. Mccarthy, Online Biology Dictionary.

[716] "synthesis." Merriam-Webster Dictionary.

production of a substance by combining simpler substances through a chemical *process* or *made* by combining different things

It is essential to acknowledge that different thought modes serve us to perceive the accepted definition of the words mentioned above.

The same grouping criteria applied to the faculties mentioned above associated with the Formative Cause.

What are the benefits of being able to classify those acceptations in this cause? Suppose students visualize the concatenation of things linked together by their similarities in a way that produces a particular result or effect within organisms. In that case, it will improve their understanding, as was shown at the beginning of this chapter with the forces that keep things united or separated.

10.3 The Art of Constructing the Material Cause

The table below shows the methodology followed in building the "Single Science" by inter-weaving the causes horizontally and vertically. For the reader to understand the placement of the following acceptations in one of the Rows (1 to 6), it is essential to know how the Material Cause interweaves with the other causes.

	THE MATERIAL CAUSE: AND THE QUALITATIVE METHOD OF SCIENCE	
1	With what?	MATURITY CAUSE
2	Essence: The Spirit, Elements, substances, and raw materials. Specific properties of matter. Distinctive peculiarities of things. Virtues are also called spiritual values. Knowledge, beliefs, and social values	MATERIAL CAUSE
3	Goals and objectives	FINAL CAUSE
4	Powers: knowledge, common faculty, and senses of spirituality and appreciation Specific lines of action or policies to consolidate ethics, knowledge, quality, and efficacy	EFFICIENT CAUSE
5	Relationships of belonging conscience, possession, distinction, and characterization	FORMATIVE CAUSE
6	Parameters and indicators of efficacy and quality in attaining the goals and objectives	MATURITY CAUSE

The reader learned that the grouping criteria is the foundation of Philosophy and has studied the conceptual framework for the Material Cause. Please proceed to check the placement of the definition of each of the following words within the Material Cause:

"Truthfulness is the foundation of all the virtues of the world of humanity. Without truthfulness, progress and success in all of the worlds of God are impossible for a soul. When this holy attribute is established in man, all the divine qualities will also become realized."[717]

It is advisable to locate it in row 2, the intersection of The Material Cause and The Material Cause because it is a *foundation* and is a *virtue.*

We can also place it in row 3, the intersection of The Material Cause and The Final Cause, because it is a spiritual law. The inclusion of the following statements: "in all the worlds of God are impossible" and "all the divine qualities will also become realized" mean the consequences of disobeying and obeying the Law.

Prerequisite: Something that you officially must have or do before you can have or do something else.[718]

If taken as a prior condition is best to place it in row 2, the intersection of The Material Cause and The Material Cause. We can also situate it in row 3, the junction of The Material Cause and The Final Cause, because you *must* have or *must* do before)

Cofactor: "Any molecule or ion required for an enzyme's function."[719]

It is OK to position it in row 2, the intersection of The Material Cause and The Material Cause, because it is a molecule or an ion necessary for an enzyme's function.

We could also place it in row 3, the intersection of The Material Cause and The Final Cause because it is *required (must have)* for an enzyme's function)

Coenzyme: "An organic molecule required for an enzyme's function. Most vitamins are coenzymes."[720] It is correct to situate it in row 2, the intersection of The Material Cause and The Material Cause because it is a molecule.

We could also place it in row 3, the intersection of The Material Cause and The Final Cause, because it is *required (compulsory)* for an enzyme's function.

Vitamin: "Any of a wide variety of chemical substances required by the body's metabolism, but that cannot be synthesized by the body."[721]

It is suitable to place it in row 2, the intersection of The Material Cause and The Material Cause because it is a chemical substance.

We could also place it in row 3, the intersection of The Material Cause and The Final Cause, because it is a *requirement* of the body's metabolism

Calcium: "Silver-white metallic element. ... Vertebrates require relatively large amounts of calcium for the production and maintenance of bone. It is also essential to the function

---

[717] 'Abdu'l-Bahá, Tablets of 'Abdu'l-Bahá v2, p. 459. Emphasis added.

[718] "prerequisite." Merriam-Webster Dictionary.

[719] "cofactor." Eugene M. Mccarthy, Online Biology Dictionary.

[720] "coenzyme." Eugene M. Mccarthy, Online Biology Dictionary.

[721] "vitamin." Eugene M. Mccarthy, Online Biology Dictionary.

of nerves and muscles, and is a necessary cofactor for the enzymes involved in blood clotting and a variety of other bodily processes."[722]

It is acceptable to position it in row 2, the intersection of The Material Cause and The Material Cause, because Calcium is a chemical element, and it is essential to the function of nerves and muscles, and is a necessary cofactor for the enzymes involved in blood clotting.

Additionally, we can place it in row 3, the intersection of The Material Cause and The Final Cause, because it is a *requirement* for the production and maintenance of bone, the function of nerves and muscles, blood clotting, and a variety of other bodily processes).

Amino: "Relating to, being, or containing an amine group."[723]

It is proper to locate it in row 2, the intersection of The Material Cause and The Material Cause, because it *contains* an amine group.

But, also in row 4, the intersection of The Material Cause and The Efficient Cause because having an amine *group.*

Knowledge: "Is a familiarity, awareness, or understanding of someone or something, such as facts, information, descriptions, or skills, which is acquired through experience or education by perceiving, discovering, or learning."[724]

It is ideal for situating it in row 2, the intersection of The Material Cause and The Material Cause, because it becomes a possession *acquired* through experience or education by perceiving, discovering, or learning.

It is essential to acknowledge that different thought modes serve us to perceive the accepted definition of the words mentioned above.

The same grouping criteria applied to the above-mentioned faculties associated with the Material Cause. What are the benefits of classifying those acceptations in this cause? The author believes that if students appreciate the mineral's prosperity when passing to the plant, animal, and human kingdoms, it opens the doors to theirs' and our deeper comprehension of concatenating facts.

---

[722] "calcium." Eugene M. Mccarthy, Online Biology Dictionary.

[723] "amino." Eugene M. Mccarthy, Online Biology Dictionary.

[724] Wikipedia, Knowledge.

10.4 The Art of Constructing the Root Cause

	THE ROOT CAUSE: AND THE UNIQUE METHOD OF SCIENCE	
1	Do only scientists require courage to walk a path nobody ever walked to find a unique potential of something? What is the root cause affecting individual and collective destiny?	MATURITY CAUSE
2	Particular aspiration	FORMATIVE CAUSE
3	Essence: Providence, Fear of God, and Grace Each being has its own particular identity Inner temple of the body	MATERIAL CAUSE
4	Wrongdoings, superstition, ignorance, vain imagination, misunderstanding, blind imitations, and dogmatic interpretations.	ROOT CAUSE
5	High Destiny in search of the appearance of unique hidden perfections. Uniqueness is a law that encompasses all kingdoms	FINAL CAUSE
6	Self-reckoning, The Master delineates a method for the individual to prepare for life beyond. The Universal House of Justice also guides us to reach the goal that applies to the unique circumstances and the unique path of the individual. There are unique methods to be discovered to transform minerals. Pruning and grafting of plants. Domestication and crossing of plants and animals.	EFFICIENT CAUSE
7	The care of an experienced gardener, an individual, or an Institution in recognizing and mending their faults privately	MATURITY CAUSE

    I wonder in amazement about the vastness of the challenge, expecting scientists to discover the uniqueness of 300.000 species of plants, one million species of insects, and 2.2 million species of fungi.[725] What about one trillion species of microorganisms and their non-identical functions in nature? What unique aspirations, skills, and capacities will human beings discover within themselves?

    The reader has studied the conceptual framework for the Root Cause and learned that the grouping criteria is the foundation of Philosophy. Please proceed to check the placement of the definition of each of the following words within the Root Cause:

---

[725] David L. Hawksworth et al., Fungal Diversity Revisited: 2.2 to 3.8 Million Species.

Fundamental: "Serious and very important, affecting the most central and important parts of something."[726] Serving as a basis supporting existence or determining essential structure or function; serving as an original or generating source.[727] Also, because it is very important, affecting the most central and important parts of something.[728]

It is advisable to locate it in row 3, the intersection of The Root Cause and The Material Cause, because it is an essential structure or function.

It also places it in row 4, the junction of The Root Cause and The Root Cause, because "fundamental" serves as an original or generating source.

Basis: "the bottom of something considered as its foundation. The principal component of something. Something on which something else is established or based. An underlying condition or state of affairs."[729]

It is advisable to locate it in row 4, the intersection of The Root Cause and The Root Cause, because "basis" is the bottom of something considered as its foundation. As in "The fundamental basis of the community is agriculture, tillage of the soil."[730]

It also places it in row 3, the intersection of The Root Cause and The Material Cause, because it is the principal component of something.

Oneness: "the quality or state or fact of being one."[731]

Because being the only one, it is advisable to locate it in row 4, the intersection of The Root Cause and The Root Cause, because oneness and uniqueness are key terms used in The Root Cause and the Unique Method of Science. The same applies to the definition of Unicity: "the quality or state of being unique of its kind."[732]

---

[726] Oxford Dictionary.

[727] Merriam Webster Dictionary.

[728] Oxford Dictionary.

[729] Merriam Webster Dictionary.

[730] The Promulgation of Universal Peace.

[731] Merriam Webster Dictionary.

[732] Idem.

10.5 The Art of Constructing the Final Cause

The table below shows the methodology followed in building the "Single Science" by inter-weaving the causes horizontally and vertically. For the reader to understand the placement of the following acceptations in one of the Rows (1 to 6), it is vital to know how the Final Cause interweaves with the other causes.

	THE FINAL CAUSE AND THE EXPLANATORY METHOD OF SCIENCE	
1	Questions: Why? For what?	CAUSE OF MATURITY
2	Essence: the power of law itself, i.e., that which is a condition to a particular aspect of life or nature, and explains its purpose, its mission	MATERIAL CAUSE
3	Social order, laws, and norms; Natural laws; Religious laws and ordinances. Pacts, agreements, and covenants. An individual and collective commitment to ideals Goal: the end, the mission, the purpose	FINAL CAUSE
4	Powers: Memory and senses of responsibility, fear, and shame Advice from those with experience. The consequences of obeying and disobeying.	EFFICIENT CAUSE
5	Situations of risk and danger. The relationship between the individual and the institutional order is mutualistic and conducive to loyalty, fear, protection, guilt, repentance, punishment, and reward.	FORMATIVE CAUSE
6	Parameters and indicators of protection, equality, prevention, and security in the fulfillment of the mission	CAUSE OF MATURITY

The reader has studied the conceptual framework for the Final Cause and learned that the grouping criteria is the foundation of Philosophy. Please proceed to check the placement of the definition of each of the following words within the Final Cause:

Mutualism: "A form of symbiosis in which both participants benefit. For example, a clownfish lives inside a sea anemone and is protected by it. In return, it brings scraps to the anemone and lures larger fish into the anemone's tentacles."[733] It is advisable to situate "mutualism" in row 3, the intersection of The Final Cause and The Final Cause because we can assimilate the alliance between the clownfish and the anemone into an involuntary pact. We could also place it in row 5, the junction of The Final Cause and The Formative Cause, because Mutualism is a *form* of symbiosis.

Infection: "Infection is the process or the state wherein an infectious agent (such as pathogenic microorganisms, viruses, prions, viroids, nematodes, and helminths) invades and

---

[733] "mutualism." Eugene M. Mccarthy, Online Biology Dictionary.

multiplies in the body tissues of the host. It could result in the manifestation of symptoms and disease when the immune response of the host is activated. It may also be palpable when the infection results in the competition for nutrients and metabolism."[734] In row 4, the intersection of The Final Cause and The Efficient Cause, because "Infection is the process or the state wherein an infectious agent" may "result in the manifestation of symptoms and disease." But also in row 5, the intersection of The Final Cause and The Formative Cause. It refers to size when an infectious agent invades and multiplies in the host's body tissues. However, it could also be placed in row 3, the intersection of The Final Cause and The Final Cause, because of the possible infectious consequences affecting the purpose of the body tissues of the host.

Toxin: "A poison, produced by an animal or plant, that elicits the production of an antibody (antitoxin) when introduced into bodily tissue, typically by injection."[735] In row 4, the intersection of The Final Cause and The Efficient Cause is a poison *produced* by an animal or plant. However, it could also be placed in row 6, the intersection of The Final Cause and The Cause of Maturity, because it is an involuntary differential response to elicit the production of an *antitoxin* preventing further damage. But it could also be placed in row 3, the intersection of The Final Cause and The Final Cause, because its *poisonous* effect becomes a situation of risk.

Langerhans cell: "A type of dendritic cell found in the epidermis. As part of the epidermal immune system, these cells act as antigen-presenting cells."[736] In row 2, the intersection of The Final Cause and The Material Cause, because an antigen is "any *substance* foreign to the body that evokes an immune response."[737] It could also be placed in row 4, the intersection of The Final Cause and The Efficient Cause, because it refers to a *type* of dendritic cell. But it could also be placed in row 3, the intersection of The Final Cause and The Final Cause, because they are part of the *immune system*.

Trichome: "A filamentous outgrowth; especially: an epidermal hair structure on a plant. A major function of the trichome is thought to be in plant defense against insects. Chemicals produced in the glandular tip can deter feeding, or the trichome can physically prevent the insect from reaching and feeding on the leaf."[738] In row 2, the intersection of The Final Cause and The Material Cause, because chemicals produced in the glandular tip can deter feeding. However, it could also be placed in row 3, the intersection of The Final Cause and The Final Cause, because it is a plant defense against insects.

Wall: "A high thick masonry structure forming a long rampart or an enclosure chiefly for

---

[734] "infection." Biology Online.

[735] "toxin." Eugene M. Mccarthy, Online Biology Dictionary.

[736] "langerhans cell." Eugene M. Mccarthy, Online Biology Dictionary.

[737] "antigen." Merriam-Webster Dictionary.

[738] "trichome." <https://quizlet.com>.

defense."[739] In row 2, the intersection of The Final Cause and The Material Cause, because it is made out of masonry. However, it also could be placed in row 5, the intersection of The Final Cause and The Formative Cause, because it is a high thick masonry structure forming a long rampart or an enclosure. But, it is more advisable to place it in row 3, the intersection of The Final Cause and The Final Cause, because it is chiefly for the protection.

Cell wall: "In some eukaryotic cells, a rigid capsule enclosing the plasma membrane; in plants, it contains cellulose and lignin; in fungi, chitin; in prokaryotes, a stiff capsule enclosing the cell membrane."[740] "The outermost layer of cells in plants, bacteria, fungi, and many algae that gives shape to the cell and protects it from infection. In plants, the cell wall is made up mostly of cellulose. Most animal cells have a cell membrane rather than a cell wall."[741] In row 2, the intersection of The Final Cause and The Material Cause, because of its composition and specific properties. However, it is best to place it in row 3, the intersection of The Final Cause and The Final Cause, because it is a wall for protection.

Lignin: "A hard material that joins with cellulose to form stiff cell walls in vascular plants; it also cements cells together, providing structural strength to the plant as a whole."[742] In row 2, the intersection of The Final Cause and The Material Cause, because it is a material and is hard. However, it could also be placed in row 3, the intersection of The Final Cause and The Final Cause, because its purpose is to form stiff cell walls for protection.

Exudate: "An exudate is a fluid emitted by an organism through pores or a wound, a process known as exuding."[743] Animal and human exudates pertain to any fluid oozing out from the blood vessels, especially as a result of inflammation. When the infection is present, the discharged fluid may contain white blood cells. In instances wherein there is vascular damage, red blood cells may also escape and be found in the exudate. Plant exudates include viscous materials seeping from interstices or pores.

Examples of exudates include saps, gums, resins, and latex."[744] In row 4, the intersection of The Final Cause and The Efficient Cause is the result of a *process* of inflammation. But it could also be placed in row 3, the intersection of The Final Cause and The Final Cause, because it has a protective function.

Reserve: "A supply of a commodity not needed for immediate use but available if required."[745] In row 2, the intersection of The Final Cause and The Material Cause, because it is a *commodity*. But ideally should be placed in row 3, the intersection of The Final Cause and The Final Cause, because is not needed for immediate use but available if

---

[739] "wall." Merriam-Webster Dictionary.

[740] "cell wall." Eugene M. Mccarthy, Online Biology Dictionary.

[741] "cell wall." <https://www.thefreedictionary.com>.

[742] "lignin." Eugene M. Mccarthy, Online Biology Dictionary.

[743] Wikipedia, Exudate. Web. March 2021 < https://en.wikipedia.org/wiki/Exudate>.

[744] "exudate." Biology Online.

[745] "reserve." <https://quizlet.com>.

required; in other words, it is a stock of a resource.

Stockpile: "A large accumulated stock of goods or materials, especially one held in reserve for use at a time of shortage or another emergency."[746] In row 2, the intersection of The Final Cause and The Material Cause. But, the best first option should be to place it in row 3, the intersection of The Final Cause and The Final Cause, because it is held in reserve for use at a time of *shortage or emergency.*

Cyst: "In an animal or plant, a thin-walled, hollow organ or cavity containing a liquid secretion; a sac, vesicle, or bladder. Medicine: in the body, a membranous sac or cavity of abnormal character containing fluid. ... A tough protective capsule enclosing the larva of a parasitic worm or the resting stage of an organism."[747] In row 3, the intersection of The Final Cause and The Final Cause, because it stores potentially harmful elements

Wax: "Waxes are similar to fats except that waxes are composed of only one long- chain fatty acid bonded to a long-chain alcohol group attached. Plants most noticeably use waxes for a thin protective covering of stems and leaves to prevent water loss. Similarly, animals employ waxes for protective purposes; for instance, earwax in humans prevents foreign material from entering and possibly injuring the ear canal area."[748] In row 2, the intersection of The Final Cause and The Material Cause, because the wax is a compound. But, the first option should be to place it in row 3, the intersection of The Final Cause and The Final Cause, because among the purposes of God's creation is our protection and that of animals and plants. It is essential to acknowledge that different thought modes serve us to perceive the accepted definition of the words mentioned above.

The same grouping criteria applied to the above-mentioned faculties associated with the Final Cause.

What are the benefits of classifying those acceptations in this cause? The author believes that if students perceive the correlation of things linked together by their similarities to produce a particular result or effect within organisms, it will improve their understanding and memory. For example, the ozone layer, a wall, a cell wall, a cell membrane, the exoskeleton, the skin and tissues, the cuticle of a leaf, the bark of a tree, the green cover of earth, and the enamel are all protective barriers.

---

[746] "stockpile." <https://quizlet.com>.

[747] "cyst." <https://quizlet.com>.

[748] "wax." <https://quizlet.com>.

10.6 The Art of Constructing the Efficient Cause

The table below shows the methodology followed in constructing the "Single Science" by inter-weaving the causes horizontally and vertically. For the reader to understand the placement of the following acceptations in one of the Rows (1 to 6), it is vital to appreciate how the efficient Cause interweaves with the other causes.

	THE EFFICIENT CAUSE AND THE DESCRIPTIVE AND EXPERIMENTAL METHODS OF SCIENCE	
1	**Questions:** With what being? With whom? With what type?	MATURITY CAUSE
2	**Essence:** The Word of God. "To be" animated and inanimated beings. **Forces:** the power of mind, energy, capacities, capabilities, potential, talents, vocations, arguments, concepts. **Categories:** genres, clusters, groups. Interacting entities within a system or a subsystem.	MATERIAL CAUSE
3	**Laws** of thermodynamics, labor law, and tools regulations. Grammar rules. The Word of God is Law **Fruits of labor:** results, harvest, services, products, and leftover materials	FINAL CAUSE
4	**Powers:** thought, reason, and expression **To do:** movements: methods, processes, activities, abilities, skills, arts, technologies, mechanisms.	EFFICIENT CAUSE
5	**Relations of:** cause and effect, logic and reason.	FORMATIVE CAUSE
6	**Parameters and indicators** of efficiency and productivity in achieving the results	MATURITY CAUSE

The reader has studied the conceptual framework for the Efficient Cause and learned that the grouping criteria is the foundation of Philosophy. Please proceed to check the placement of the definition of each of the following words within the Efficient Cause:

System: "A regularly interacting or interdependent group of items forming a unified whole. A group of related parts that move or work together. A group of organs that work together to perform an important function of the body."[749] In row 2, the intersection of The Efficient Cause and The Material Cause, because the group is composed of *items*, *parts*, or *organs*. But, it is best to place it in row 4, the intersection of The Efficient Cause and The Efficient Cause, because it is a *group* of *items* that *move* or *work* together.

Humoral immune system: "The portion of the immune system that produces antibodies that circulates in the blood and lymph."[750] It is ideal for situating it in row 4, the

---

[749] "system." Merriam-Webster Dictionary.

[750] "humoral immune system." Eugene M. Mccarthy, Online Biology Dictionary.

intersection of The Efficient Cause and The Efficient Cause since it refers to the immune system's portion that *produces antibodies*. We could also locate it in row 4, the intersection of The Efficient Cause and The Formative Cause, because it is a *portion* of the immune system.

Photosynthesis: "A process carried out in plants, algae, and bacteria, which uses energy from sunlight to convert carbon dioxide and water into glucose and oxygen. Photosynthesis is the source of atmospheric free oxygen and is the essential starting point for the construction of all organic molecules present in living things."[751] The best place to situate it is in row 4, the intersection of The Efficient Cause and The Efficient Cause because it is a *process* that uses energy. We could also station it in row 3, the junction of The Efficient Cause and The Material Cause because Photosynthesis is the *essential* starting point for the construction of all *organic molecules* present in *living things*.

It is essential to acknowledge that different thought modes serve us to perceive the accepted definition of the words mentioned above.

The same grouping criteria apply to the above-mentioned faculties associated with the Efficient Cause. What are the benefits of being able to classify those acceptations in this cause? The author believes that students perceiving the systems' coordination and the processes within an organism or an ecosystem will improve their understanding.

Let us listen to the Master, 'Abdu'l-Bahá:

> It is obvious that all created things are connected one to another by a linkage complete and perfect, even, for example, as are the members of the human body.
>
> Note how all the members and component parts of the human body are connected one to another. In the same way, all the members of this endless universe are linked one to another. The foot and the step, for example, are connected to the ear and the eye; the eye must look ahead before the step is taken. The ear must hear before the eye will carefully observe. And whatever member of the human body is deficient, produceth a deficiency in the other members. The brain is connected with the heart and stomach, the lungs are connected with all the members. So is it with the other members of the body.
>
> And each one of these members hath its own special function. The mind force — whether we call it preexistent or contingent — doth direct and coordinate all the members of the human body, seeing to it that each part or member duly performeth its own special function. If, however, there be some interruption in the power of the mind, all the members will fail to carry out their essential functions, deficiencies will appear in the body and the functioning of its members, and the power will prove ineffective.[752]

---

[751] "photosynthesis." Eugene M. Mccarthy, Online Biology Dictionary.

[752] Selections from the Writings of 'Abdu'l-Bahá.

The Art of constructing The Single Science

11 To Fly, a Bird Needs Two Wings Equally Developed

This chapter is about my first practice of consciously learning about the art of persuasion, with the due respect for the intelligence and free-will of the reader.

And among the teachings of Bahá'u'lláh is the equality of women and men. The world of humanity has two wings—one is women and the other men. Not until both wings are equally developed can the bird fly. Should one wing remain weak, flight is impossible. Not until the world of women becomes equal to the world of men in the acquisition of virtues and perfections, can success and prosperity be attained as they ought to be.[753] [754]

Do we all aspire to peace in the world, food security for all, a job to support the family, and being empowered to address the environmental crisis and any problem?

"The Revelation proclaimed by Bahá'u'lláh, His followers believe, is divine in origin, all-embracing in scope, broad in its outlook, scientific in its method, humanitarian in its principles and dynamic in the influence it exerts on the hearts and minds of men."[755] It is protected in the following manner:

"Religion must stand the analysis of reason. It must agree with scientific fact and proof so that science will sanction religion and religion fortify science. *Both are indissolubly welded*[756] *and joined in reality*. If statements and teachings of religion are found to be unreasonable and contrary to science, they are outcomes of superstition and imagination."[757]

I recognized Aristotle's wisdom on many topics, but I want to express my disagreement with the following quotes from him about women:

In Aristotle's Politics, we find gender roles statements:

"[T]he male, unless constituted in some respect contrary to nature, is by nature more expert at leading than the female, and the elder and complete than the younger and incomplete."

"[T]he relation of male to female is by nature a relation of superior to inferior and ruler to ruled."

---

[753] 'Abdu'l-Bahá, Tablets to The Hague. Emphasis added

[754] Photo 75943415 © Ondřej Prosický | Dreamstime.com

[755] Shoghi Effendi, The World Religion of Bahá'u'lláh: A Summary of Its Aims, Teachings and History to the High Commissioner for Palestine.

[756] Photo 109384172 © Gutaper | Dreamstime.com

[757] 'Abdu'l-Bahá, The Promulgation of Universal Peace. June 9, 1912. Talk at Unitarian Church. Fifteenth Street and Girard Avenue, Philadelphia, Pennsylvania. Notes by Edna McKinney. Emphasis added.

"The slave is wholly lacking the deliberative element; the female has it but it lacks authority; the child has it but it is incomplete. "[758]

In this chapter, the reader will find arguments to demonstrate they are false if women have equal access to education; or, even better, priority if there are not enough resources to educate both genders. You are welcome to study the leading roles of Ṭáhirih, Malala Yousafzai, Catalina I, Zenovia Queen of Palmyra[759], and Bahíyyih Khánum. While the Master traveled for three years to Europe, Canada, and The United States of America, Bahiyyih Khánum was in charge of the affairs of the Bahá'i Faith at the World Center. Later on, in 1922, the Guardian temporarily left the administration of the Cause in her hands in a very challenging moment.

The discovery of agriculture:

There are different theses about who discovered agriculture. The attached picture shows volcanic rock ancestral way of cooking in the Andes Ecuador.[760]

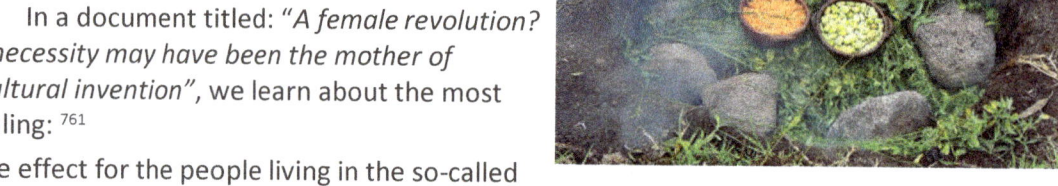

In a document titled: *"A female revolution? How necessity may have been the mother of agricultural invention"*, we learn about the most appealing: [761]

The effect for the people living in the so-called Fertile Crescent was catastrophic, not only had their hunting grounds been drowned by rising sea levels following the Ice Age melt, but now – thanks to this sudden climate change – a severe drought set in and much of their remaining rich and fertile woodland was transformed into barren scrub.

Wild grasses such as wheat were an important part of the staple Natufian diet, but in the now sweltering scrubland they simply withered away. Some experts think this is what may have led Natufian women to experiment with sowing seeds themselves, and deliberately clearing the land to make it suitable for cultivating grasses such as wheat, barley and rye.

In hunter-gathering societies it was generally women who gathered seeds and picked fruits while men went out to hunt game. In the face of starvation, Natufian women are thought to have selected the best seeds they could find, the biggest, sweetest

and most easy to harvest, which they then sewed on specially prepared land as a crop for the following year.

Was it their handiwork – an agricultural insurance policy – that triggered a chain of events

---

[758] Andrea Borghini, Plato and Aristotle on Women: Selected Quotes.

[759] History narrated by 'Abdul-Bahá in The Promulgation of Universal Peace.

[760] Photo 141924104 © Iryna Kurilovych | Dreamstime.

[761] Photo 14857756 © Rinus Baak | Dreamstime.com

that eventually led to the spread of crop farming all over the Middle East, Europe, and northern Africa? Seeds are easy to store and transport. The Natufian women's crop cultivation seems to be the earliest known to history. Evidence of their inventiveness comes from the discovery by modern archaeologists of farming tools, in the form of picks and sickle blades used for harvesting cereal crops. Alongside these ancient farming implements are pestles, mortars and bowls, all essential instruments for gathering and grinding up seeds.[762] [763]

In the Old Testament, there is a promise:

> And it shall come to pass in the last days, that the mountain of the *LORD's house* shall be established in the top of the mountains, and shall be exalted above the hills; and all nations shall flow unto it.
>
> And many people shall go and say, Come ye, and let us go up to the mountain of the LORD, *to the house of the God* of Jacob; and He will teach us of His ways, and we will walk in His paths: for out of Zion shall go forth the Law, and the word of the LORD from Jerusalem.
>
> And He shall judge among the nations, and shall rebuke many people: *and they shall beat their swords into plowshares,* and their *spears into pruning hooks*: nation shall not lift up sword against nation, neither shall they learn war any more.[764]

This promise is not only acknowledged by Jews but also by Christians and Muslims.[765] Baha'u'lláh ratified that promise:

> Whilst in the Prison of 'Akká, We revealed in the Crimson Book that which is conducive to the advancement of mankind and to the reconstruction of the world. The utterances set forth therein by the Pen of the Lord of creation include the following which constitute the fundamental
>
> > Principles for the administration of the affairs of men:
>
> **First:** It is incumbent upon the ministers of the House of Justice to promote the Lesser Peace so that the people of the earth may be relieved from the burden of exorbitant expenditures. This matter is imperative and absolutely essential, inasmuch as hostilities and conflict lie at the root of affliction and calamity.
>
> **Second:** Languages must be reduced to one common language to be taught in all the schools of the world.

---

[762] Independent, A female revolution? How necessity may have been the mother of agricultural invention.

[763] Photo 43772157 © Sjors737 | Dreamstime.com

[764] Isaiah, 2:2 –2:4. Emphasis Added.

[765] ID 42784574 © Saiidghazal. | Dreamstime.com

Third: It behoveth man to adhere tenaciously unto that which will promote fellowship, kindliness and unity.

Fourth: Everyone, whether man or woman, should hand over to a trusted person a portion of what he or she earneth through trade, agriculture or other occupation, for the training and education of children, to be spent for this purpose with the knowledge of the Trustees of the House of Justice.

Fifth: Special regard must be paid to agriculture. Although it hath been mentioned in the fifth place, unquestionably it precedeth the others. Agriculture is highly developed in foreign lands, however in Persia it hath so far been grievously neglected. It is hoped that His Majesty the Shah -- may God assist him by His grace -- will turn his attention to this vital and important matter.[766]

In a statement, Shoghi Effendi mentions that Bahá'u'lláh aims to reconcile conflicting creeds:

The aim of Bahá'u'lláh, the Prophet of this new and great age which humanity has entered... is not to destroy but to fulfil the Revelations of the past, to reconcile rather than accentuate the divergencies of the conflicting creeds which disrupt present-day society.

His purpose, far from belittling the station of the Prophets[767] gone before Him or of whittling down their teachings, is to restate the basic truths which these teachings enshrine in a manner that would conform to the needs, and be in consonance with the capacity, and be applicable to the problems, the ills and perplexities, of the age in which we live.

His mission is to proclaim that the ages of the infancy and of the childhood of the human race are past, that the convulsions associated with the present stage of its adolescence are slowly and painfully preparing it to attain the stage of manhood, and are heralding the approach of that Age of Ages when swords will be beaten into plowshares, when the Kingdom promised by Jesus Christ will have been established, and the peace of the planet definitely and permanently ensured.[768]

In the bahá'í Writings, we find: "Taken in general, women today have a stronger sense of religion than men."[769]

---

[766] Bahá'u'lláh, Tablets of Bahá'u'lláh, pp. 89 – 90. Emphasis Added.

[767] Photo 40201595 © Elena Schweitzer | Dreamstime.com

[768] Shoghi Effendi, Call to the Nations.

[769] 'Abdu'l-Bahá, 'Abdu'l-Bahá in London, p. 104.

The Pew Research Center in, *The Gender Gap in Religion Around the World: Women are generally more religious than men, particularly among Christians*, says: "In the United States, for example, women are more likely than men to say religion is "very important" in their lives (60% vs. 47%).

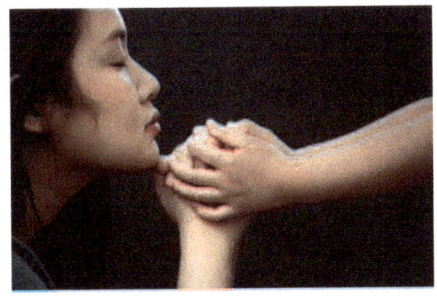

Noting similar gender differences in other countries, mainly in Europe, some social scientists have argued that women are universally more religious than men across all societies, cultures, and faiths."[770] [771]

'Abdu'l-Bahá taught: "The woman's intuition is more correct; she is more receptive and her intelligence is quicker."[772] "Are the following pieces of evidence a corroboration of this quote from the Master?

In all WHO regions, females are less often current drinkers than males, and when women drink, they drink less[773] than men. More than half of the world's female population aged 15 years or older are lifetime abstainers (54.6% or 1.489 billion; for men the figures are: 34.5% or 941 million)."[774]

Because it is a depressant of the central nervous system, alcohol depresses all our faculties, including our sense of shame:

In the United States of America: 37 percent of rapes[775] and sexual assaults will involve offenders under the influence, and that number jumps to 90 percent when the abuses occur on college campuses.

40 percent of child abuse incidents will be connected to alcohol use or abuse, and 70 percent of these abusive individuals (parents or guardians) will suffer from a substance use disorders.

65 percent of intimate partner violence incidents will be carried out by perpetrators who've been drinking. This equates to more than 450,000 such incidents annually.[776]

---

[770] Pew Research Center, The Gender Gap in Religion Around the World. Women are generally more religious than men, particularly among Christian.

[771] Photo 197979664 © Rinchumrus2528 | Dreamstime.com

[772] 'Abdu'l-Bahá,'Abdu'l-Bahá in London, p. 104.

[773] Photo 171286971 / Alcohol Stop © Amazingmikael | Dreamstime.com

[774] WHO, World Health Organization, Global status report on alcohol and health 2018.

[775] Photo 117765605 / Force © Tinnakorn Jorruang | Dreamstime.com

[776] Nena Messina, Shocking Statistics and Facts about Alcohol-Related Crimes.

"*A global study, Crime and Gender. A Study on how Men and Women are Represented in International Crime Statistics*, by Markku Heiskanen and Anni Lietonen, says: "The rate of male suspects (offenders) was higher than the rate of female suspects in all regions   On average 84 per cent of the suspects were male."[777]

---

[777] Markku Heiskanen and Anni Lietonen, , Crime and Gender. A Study on how Men and Women are Represented in International Crime Statistics.

11.1 Employment and biodiversity's potential

'Abdu'l-Bahá stated: "The woman is indeed of the greater importance to the race. She has the greater burden and the greater work.[778] [779]

The palm which carries the fruit is the tree most prized by the date grower. The Arab knows that for a long journey the mare has the longest wind. For her greater strength and fierceness, the lioness is more feared by the hunter than the lion."[780]

Rural women constitute 50 percent of the agricultural labor force in Africa; they are responsible for 80 percent of the food production and 50 percent of the agricultural output. Women reinvest almost 90 percent of their income in their children and household. Since women are the keys to improving household food security and nutritional wellbeing, increasing women's access to financial resources ultimately leads to increased investments in human capital.[781]

Forget about emotional labor. Women living 7000 years ago had to deal with another lopsided workload: farming. Prehistoric women shouldered a major share of the hoeing, digging, and hauling in early agricultural societies, according to a new study. Now, by analyzing the bones of these women, scientists have shown that their upper body strength surpassed even today's elite female athletes. The findings refute popularly held notions that early agrarian women shunned manual labor in favor of domestic work, and they suggest that then—as now—a woman's work was never done.[782] [783]

In, Prehistoric women's manual labor exceeded that of athletes through the first 5500 years of farming in Central Europe. By Alison A. Macintosh, Ron Pinhasi and Jay T. Stock, we learn:

---

[778] Photo 25255633 © Alan Gignoux | Dreamstime.com

[779] Photo 640232 © Eutoch | Dreamstime.com

[780] 'Abdu'l-Bahá, 'Abdu'l Bahá in London, pp. 102–03.

[781] Mpule K. Kwelagobe, Commentary - Investing in Rural Women: Closing the Gender Gap in African Agriculture. Emphasis added.

[782] Michael Price, Strong women did a lot of the heavy lifting in ancient farming societies.

[783] Photo 28910193 © Kamonrutm | Dreamstime.com

Size-standardized humeral and tibial polar second moments of area, cross-sectional shape (Imax/Imin)[784], and interlimb strength proportions are compared between Neolithic, Bronze Age, Iron Age, and Medieval women and living female athletes, as well as recreationally active control subjects as a reference group of low-impact loading. Athletes were included from three sports that load the limbs with differing intensity and directionality: (i) endurance running, high lower limb loading based on ground reaction force and unidirectional loading trajectories; (ii) football (soccer), high lower limb loading based on ground reaction force and multidirectional loading trajectories; and (iii) rowing, higher repetitive upper limb loading based primarily on joint contact forces and unidirectional loading trajectories.

This comparative data set was used to explore the following questions: (i) To what extent can the apparent homogeneity in interpopulation variation in female tibial morphology among early agricultural women be explained by high internal variability? (ii) Were Central European prehistoric farming women more mobile than living sedentary women? (iii) Among prehistoric Central European females, was manual labor a more rigorous behavioral component of agricultural intensification than terrestrial mobility?

The intensification of agriculture is often associated with declining mobility and bone strength through time, although women often exhibit less pronounced trends than men. For example, previous studies of prehistoric Central European agriculturalists (~5300 calibrated years BC to 850 AD) demonstrated a significant reduction in tibial rigidity among men, whereas women were characterized by low tibial rigidity, little temporal change, and high variability. Because of the potential for sex-specific skeletal responses to mechanical loading and a lack of modern comparative data, women's activity in prehistory remains difficult to interpret. This study compares humeral and tibial cross-sectional rigidity, shape, and interlimb loading among prehistoric Central European women agriculturalists and living European women of known behavior (athletes and controls).

Prehistoric female tibial rigidity at all time periods was highly variable, but differed little from living sedentary women on average, and was significantly lower than that of living runners and football players. However, humeral rigidity exceeded that of living athletes for the first ~5500 years of farming, with loading intensity biased heavily toward the upper limb. Interlimb strength proportions among Neolithic, Bronze Age, and Iron Age women were most similar to those of living *semi-elite rowers*.[785]

---

[784] Anthropologists frequently use the shaft bending strength index to infer the physical activity levels of humans living in the past from their lower limb bone remains. Individuals with high Imax/Imin values are inferred to have been very active, whereas individuals with low values are inferred to have been more sedentary. Ian J. Wallace, et al.

[785] Photo 5955199 © Loosli Hans Peter | Dreamstime.com

These results suggest that, in contrast to men, rigorous manual labor was a more important component of prehistoric women's behavior than was terrestrial mobility through thousands of years of European agriculture, at levels far exceeding those of modern women.[786]

In the Bahá'í Writings, we find:

First and foremost is the principle that to all the members of the body politic shall be given the greatest achievements of the world of humanity. Each shall have the utmost welfare and wellbeing. To solve this problem we must begin with the farmer; *there will we lay a foundation for system* and order *because the peasant class and the agricultural class exceed other classes in the importance of their service*. In every village there must be established a general storehouse which will have a number of revenues.[787]

Let us think about the labor and educational systems. I want you to estimate how many employments will generate each one of the following discoveries:

"Not all bugs are bad. Insects get labeled as "pests" when they start causing harm to people or the things we care about, like plants, animals, and buildings. Out of nearly one million known insect species, only about one to three percent are ever considered pests. What about the rest of them? Some insects actually help us by keeping the pests in check[788]", still we know very little about the other 970.000 capacity to prevent damage to crops and their function in nature. When using the Explanatory Method of Science and thinking about the potential of biodiversity to prevent attacks by pests, we find the following challenge:

In an effort to receive advanced warning of destructive pests that could wreak havoc on native plantings, researchers from Europe, the United States and China are growing "sentinel trees" in strategic locations around the world.

The emerald[789] ash borer, introduced to the U.S. from its native range of northeastern Asia, has killed off hundreds of millions of ash trees throughout the country at an estimated cost of nearly $11 billion. The American chestnut, estimated to have numbered between 3-4 billion trees at the turn of the 20th century, is today represented by only a few hundred specimens due to the accidental import of a destructive bark fungus. The spotted lanternfly, first discovered in the U.S. in 2014 and free from natural predators, continues to feed unchecked on 70 plant

---

[786] Alison A. Macintosh, et al. Prehistoric women's manual labor exceeded that of athletes through the first 5500 years of farming in Central Europe. Emphasis added.

[787] 'Abdu'l-Bahá, Foundations of World Unity, p. 39. Emphasis Added.

[788] National Pesticide Information Center, Oregon State University, Beneficial Insects.

[789] Photo 121949131 © Don Bilski | Dreamstime.com

species, including grape vines, fruit trees, ornamental trees and woody trees.[790]

There are other non-pesticide options with the potential to take care of this pest. Using bacteria, fungi, viruses, or soils enriched with an abundance of microorganisms. "According to a new estimate, there are about one trillion species of microbes on Earth, and 99.999 percent of them have yet to be discovered.[791]

Boy discovers microbe that eats plastic[792]

Ph.Ds. have been searching for a solution to the plastic waste problem, and this 16-year-old finds the answer. It's not your average science fair when the 16-year-old winner manages to solve a global waste crisis. But such was the case at last May's Canada-Wide Science Fair in Ottawa, Ontario, where Daniel Burd, a high school student at Waterloo Collegiate Institute, presented his research on microorganisms that can rapidly biodegrade plastic. Daniel had a thought it seems the PhDs hadn't explored: Plastic, one of the most indestructible of manufactured materials, eventually decomposes. It takes 1,000 years but decompose it does, which means there must be microorganisms out there to do the decomposing.[793]

In, A caterpillar that eats and digests plastic in record time, by Federica Bertocchini, we find:

In an accidental discovery, it has been found that larvae of the greater wax moth eat and digest polyethylene plastic bags in record time. Will this find solve our plastic garbage problem?[794] [795]

In cooperation with Paolo Bombelli and Christopher Howe from the Department of Biochemistry at the University of Cambridge, Bertocchini conducted an experiment. The researchers exposed 100 wax moth larvae to a common plastic bag from a British supermarket. And waited to see what happened.

---

[790] Michael d'Estries, Researchers Turn to 'Sentinel Trees' to Warn of Destructive Pests.

[791] Nicholas Bakalar, .Earth May Be Home to a Trillion Species of Microbes.

[792] Photo 163588349 © Péter Gudella | Dreamstime.com

[793] Karl Burkart, Boy discovers microbe that eats plastic.

[794] Photo 94616416 © Verastuchelova | Dreamstime.com

[795] Photo 142456806 © Maryna Lipatova | Dreamstime.com

After only 40 minutes the first holes appeared in the bag. After 12 hours the larvae had eaten 92 milligrams of the plastic. (Fun fact: no bacteria is able to decompose so much material in such a short amount of time).

The Indian mealmoth (Plodia interpunctella) is also able to digest plastics with the help of bacteria, albeit at a much slower rate. Even chemical decomposition methods, like using nitric acid to break down plastic, can take months. Beeswax is just another polymer.

Bertocchini suggests that the wax moth possesses a specific enzyme, which is able to break up chemical bonds that occur both in beeswax and polyethylene molecules.[796]

In the following article of what is known as Styrofoam we read: Biodegradation and Mineralization of Polystyrene by Plastic-Eating Mealworms.

Polystyrene (PS) is generally considered to be durable and resistant to biodegradation. Mealworms (the larvae of Tenebrio molitor Linnaeus) from different sources chew and eat Styrofoam, a common PS product. The Styrofoam was efficiently degraded in the larval gut within a retention time of less than 24 h. Fed with Styrofoam as the sole diet, the larvae lived as well as those fed with a normal diet (bran) over a period of 1 month.[797] [798] [799] [800]

"It has been estimated that well over 300,000 secondary metabolites exist in plants, and it is thought that their primary function is to increase the likelihood of an organism's survival by repelling or attracting other organisms."[801] But the possible combination of these secondary metabolites that exist in plants and animals also has potentialities for industry, health, nutrition, agriculture, climate change, and taking care of the ecosystem's biodiversity.

---

[796] Federica Bertocchini, A caterpillar that eats and digests plastic in record time.

[797] Yu Yang, et al., Biodegradation and Mineralization of Polystyrene by Plastic-Eating Mealworms: Part 1. Chemical and Physical Characterization and Isotopic Test.

[798] Photo 19073914 © Gary Uttley | Dreamstime.com

[799] Photo 210769842 © Profmym | Dreamstime.com

[800] Photo 17263195 © Railman | Dreamstime.com

[801] Brahmkshatriya, Priyanka P. Brahmkshatriya and Pathik S. Brahmkshatriya, Terpenes: Chemistry, Biological Role, and Therapeutic Applications.

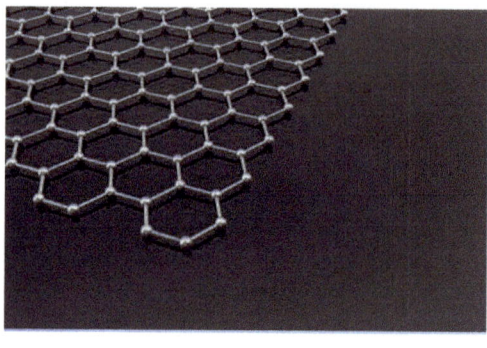
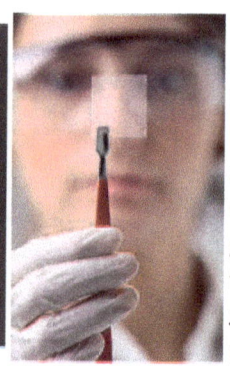

When examining the concept of transmutation in the mineral kingdom using the Root Cause and the Unique Method of science, the author decided to show this example: Someone asked, what is graphene? "Now researchers say they can make graphene with *soybean oil*, potentially making it more commercially viable. Graphene is a form of carbon[802] that exists as a sheet, one atom thick."[803]

It is more than 200 times stronger than steel and conducts electricity better than copper.

About 1% of graphene mixed into plastics could turn them into electrical conductors.

It was produced at the University of Manchester in 2004, winning its discoverers a Nobel Prize in 2010.

Graphene is hoped to have numerous applications, including in electronics, biomedical devices, and water filtration.

"One of the limiting factors in utilizing graphene is the high price compared to other materials,"if we reduce the price, we can promote its application."[804]

Now this new chemical element is produce using soybean oil:

Here we present a single-step, rapid thermal synthesis of uniform and continuous graphene films in an ambient-air environment, using a cheap and renewable form of biomass, soybean oil, as the precursor. This ambient-air process for graphene fabrication is fast, simple, safe, potentially scalable and integration-friendly.

Importantly, it offers the scope to potentially address the critical roadblocks towards large-scale, efficient graphene manufacturing."[805]

---

[802] Two dimensional hexagonal crystal lattice formed atoms carbon thickness one atom. Photo 73498016 © Rostislav Zatonskiy | Dreamstime.com

[803] Photo: showing piece of graphene. 34114299 © Eduard Bonnin Turina | Dreamstime.com

[804] Greg Dunlop, Reporting by the BBC's, Australian scientists use "soybean oil" to create graphene. Emphasis added.

[805] Dong Han Seo, et al., Single-step ambient-air synthesis of graphene from renewable precursors as electrochemical genosensor.

Another example to show the potential of biodiversity as an example for the Experimental Method of Science is the following: **17,000 miles more than a conventional touring tire.**

Yokohama introduced a tire called the dB Super E-Spec, which offered a 20 percent reduction in rolling resistance without the usual corresponding drop in traction. The key: new technology using oil derived from orange peels to displace some of the petroleum typically used in tires. 33% more mileage.[806] [807]

To think about the potential of biodiversity in healing our bodies, let us listen to the following guidance from Bahá'u'lláh: "Treat disease first of all through diet, and refrain from medicine. If you can find what you need for healing in a single herb do not use a compound medicine. Leave off medicine when the health is good, and use it in case of necessity."[808] [809]

The Qualitative Method of Science is vital to achieving the goal of affordable medicine. This method also helps improve our quality of life because the originating cause of this method is the spirit. Some of the notions closely associated with the Material Cause imply developing a spiritual foundation and acquiring valuable knowledge as wellsprings of material well-being in determining prosperity in the quality of life. An honest individual, a faithful husband or wife, and a clean person will have a much better quality of life than the contrary.

In The Promulgation of Universal Peace: Talks delivered by 'Abdu'l-Bahá during his visit to the United States and Canada in 1912, he taught:

"Woman must especially devote her energies and abilities toward the industrial and agricultural sciences, seeking to assist mankind in that which is most needful.

By this means she will demonstrate capability and ensure recognition of equality in the social and economic equation." "The decision-making agencies involved would do well to consider giving first

---

[806] Popular Mechanics,17,000 miles more than a conventional touring tire. <http://www.popularmechanics.com/cars/hybrid-electric/a7593/the-science-behind-yokohamas-orange-oil-tires-8146348/ 33% more mileage>.

[807] Photo 120373429 / Orange Peels © Zahie Werber | Dreamstime.com

[808] Bahá'u'lláh, Tablet to a Physician (Lawh-i-Tibb).

[809] Photo 21356040 © Robyn Mackenzie | Dreamstime.com

priority to the education of women and girls, since it is through educated mothers that the benefits of knowledge can be most effectively and rapidly diffused throughout society."[810] [811]

The challenge to discover the potential of biodiversity to generate employment will take years to produce results. But what about jobs sooner. Starting a small business or a non-profit seems to be a good option, but consult with others before starting anything. There is plenty of reforestation to be done. I suggest consulting experts in recovering riparian areas for protecting creeks and increase the amount of water flowing. Ask their advice to plant *Vetiveria zizanoides* in the mountain's slope to retain organic matter, gradually build terraces, and protect them from erosion because the water is slowly released, instead of coming down at once. The vast root system of *Vetiveria zizanoides* is a promising option for lasting carbon sequestration and expanding the bacteria population in the soil.

Request experts' advice in regenerative agriculture: Specifically, Regenerative Agriculture is a holistic land management practice that leverages the power of photosynthesis in plants to close the carbon cycle, and build soil health, crop resilience and nutrient density. Regenerative agriculture improves soil health, primarily through the practices that increase soil organic matter. This not only aids in increasing soil biota diversity and health, but increases biodiversity both above and below the soil surface, while increasing both water holding capacity and sequestering carbon at greater depths, thus drawing down climate-damaging levels of atmospheric CO_2, and improving soil structure to reverse civilization-threatening human-caused soil loss. Research continues to reveal the damaging effects to soil from tillage, applications of agricultural chemicals and salt based fertilizers, and carbon mining. Regenerative Agriculture reverses this paradigm to build for the future.[812]

11.2 Addressing the environmental crisis

'Abdu'l-Bahá is reported to have said:

> The solution begins with the village, and when the village is reconstructed, then the cities will be also. The idea is this, that in each village will be erected a store-house. In the language of Religion it is called the House of Finance. That is a universal store-house, which is commenced in the village. Its administration is through a committee of the wise ones of the community, and with the approval of that committee all the affairs are directed.[813]

Local farmers' market[814] contributions to reduce carbon print and waste are considerable; less travel implies less waste.

Local farmers' markets are a crucial support for the local economy and employment generation because of the amount of added value that stays within the community: "The

---

[810] 'Abdu'l-Bahá, The Promulgation of Universal Peace.

[811] Photo 136285903 © Roman Volskiy | Dreamstime.com

[812] Regeneration International, Why Regenerative Agriculture?

[813] George O. Latimer, Economic Justice.

[814] Photo 189163690 © Baloncici | Dreamstime.com

waste of some 1.3 billion tons of food each year is causing economic losses of $750 billion and significant damage to the environment, according to a United Nations report launched today.

Adding 3.3 billion tonnes of greenhouse gases to the planet's atmosphere."[815] "This long-distance, large-scale transportation of food consumes large quantities of fossil fuels. It is estimated that we currently put almost 10 kcal of fossil fuel energy into our food system for every 1 kcal of energy we get as food."[816] Is this system sustainable?

National Geographic says "2.9 Trillion pounds of food. That is about 1/3rd of the planet's wasted enough to feed 2 billion people."[819] [820]	The average farm to plate travel distance of the American meal 1500 miles, a head of lettuce 2000 miles.[817] [818]

Fresh Harvest local partner farms in Georgia are all within 100 miles of Atlanta; organic lettuce has a 38-mile commute.[821]

And, in The Promulgation of Universal Peace, by 'Abdu'l-Bahá, we find: "*The fundamental basis of the community is agriculture*, tillage of the soil. All must be producers."[822] If agriculture is the community's fundamental basis, is it of the core activities?

Then, a suggested scientific inquiry: How can the core activities (children's classes, junior youth groups, the sequence of courses, devotional meetings, and reflection gatherings) help alleviate the plight of farmers, especially of the small ones.

Biodiversity is much more successful in pest control management. It is increasingly recognized by experts addressing the environmental crisis, generated in part by using very toxic insecticides, fungicides, and herbicides, many of them of a broad spectrum. For

---

[815] UN report: one-third of world's food wasted annually, at great economic, environmental cost

[816] CUESA. How far do your fruit and vegetables travel?

[817] Idem.

[818] Photo 79704661 © Jerry Coli | Dreamstime.com

[819] National Geographic, One-Third of Food Is Lost or Wasted: What Can Be Done?

[820] Photo 152470792 © Ekaterina Chalysheva | Dreamstime.com

[821] Fresh Harvest, Your direct partnership with local, organic farmers..

[822] 'Abdu'l-Bahá, The Promulgation of Universal Peace.

example, aphids suck the liquids-rich in nutrients from plants: Ladybird[823]: A voracious aphid predator, the harlequin ladybird can eat up to 370 aphids during its larval period and over 5,000 in its adult lifetime. It eats between 10 and 30 per day.[824]

Chrysoperla (=Chrysopa) "The larvae are sometimes called aphid lions, and have been reported to eat between 100 and 600 aphids each, typically feed on soft-bodied insects such as aphids, mealy bugs, thrips, mites, leaf hoppers, whiteflies, caterpillars, other immature insects, and sometimes each other."[825] There are about 85 genera and (differing between sources) 1,300–2,000 species in this widespread group."[826][827][828]

"Many well-studied insect species, such as the ant *Leptothorax acervorum*, the moths Helicoverpa zea and *Agrotis ipsilon*, the bee *Xylocopa sonorina* and the butterfly *Edith's checkerspot*[829] release sex pheromones to attract a mate, and some lepidopterans (moths and butterflies) can detect a potential mate from as far away as 10 km (6.2 mi)."[830] "Entomologists use particular sex-attractant pheromones and aggregation pheromones to lure and trap harmful insects."[831]

Considerable progress is also being made in understanding the important role of MVOCs (Microbial Volatile Organic Compounds). Bacterial and/or fungal MVOCs modulate plant growth and defense, interspecies interaction between plant, bacteria, fungi, and nematodes, play a role as attractants of natural enemies, as bio-control agents and find suitable applications as pest/insect/herbivore management (Leroy et al., 2011; Davis et al., 2013; Weise et al., 2013; D'Alessandro et al., 2014). These progressive studies on MVOCs illustrate their critical roles in multitrophic interactions and their importance in both the ecosystem and sustainable agriculture systems. Microbial volatile organic compounds as plant defense and growth modulators is still in its infancy. Up to now, only 10,000 microbial species described of the

---

[823] Photo 24735922 © Dimijana | Dreamstime.com

[824] Prezi, Harlequin Ladybird.

[825] Erin W. Hodgson Beneficial Insects: Lacewings and Antilions.

[826] Wikipedia, Chrysopidae.

[827] Photo 74185812 © Mario Madrona Barrera | Dreamstime.com

[828] Photo 148044357 © Digitalimagined | Dreamstime.com

[829] Photo 108348741 © Pkzphotos | Dreamstime.com

[830] Wikipedia, Pheromone.

[831] Britannica. Web. March 2021 <https://www.britannica.com/science/pheromone>.

millions of species on Earth and only a 1000 MVOCs released by 400 bacteria and fungi have been described in the literature.[832]

11.3 Food security for all

Food security is one of the most important reasons why people migrate.[833] In a FAO's document, *Women hold the key to building a world free from hunger and poverty*, on December 16, 2016; we learn:

Zero Hunger: No way to get it done without women

Neven Mimica, European Union Commissioner for International Cooperation and Development, told event participants: "It is often said that if you educate a woman, you educate a whole generation. The same is true when we empower women across the board -not only through access to knowledge, but also to resources, to equal opportunities, and by giving them a voice."

In developing countries, women make up 45% of the agricultural labour force, ranging from 20% in Latin America to up to 60% in parts of Africa and Asia.

In developing countries in Africa and Asia and the Pacific, women typically work 12-13 hours more than men per week.

Across all regions women are less likely than men to own or control land, and their plots often are of poorer quality. Less than 20% of the world's landholders are women.

If women farmers had the same access to resources as men, the number of hungry people in the world could be reduced by up to 150 million due to productivity gains.

Women reinvest up to 90% of their earnings into their households - that's money spent on nutrition, food, healthcare, school, and income-generating activities - helping break the cycle of intergenerational poverty.[834]

In the Bahá'í Writings we find the following ordinance: "In every village ther must be established a general storehouse which will have a number of revenues."[835]

The storehouse has seven revenues: Tithes, taxes on animals, property without an heir, all lost objects found whose owners cannot be traced, one third of all treasure-trove, one third of the produce of all mines, and voluntary contributions.

A certain amount must be put aside from the general storehouse for the orphans of the village and a certain sum for the incapacitated. A certain amount must be provided from this

---

[832] Chidananda Nagamangala Kanchiswamy, Chemical diversity of microbial volatiles and their potential for plant growth and productivity.

[833] A young refugee child is sleeping on the shoulder of his father and both look so tired, hungry, and terrified. Photo 196039786 © Mumtaz Ali Magsi | Dreamstime.com

[834] FAO, Women hold the key to building a world free from hunger and poverty.

[835] 'Abdu'l-Bahá, The Promulgation of Universal Peace.

storehouse for those who are needy and incapable of earning a livelihood,[836] [837] [838] and a certain amount for the village's system of education.

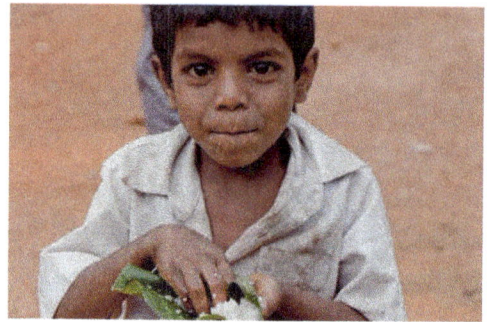

And, a certain amount must be set aside for the administration of public health. If anything is left in the storehouse, that must be transferred to the general treasury of the nation for national expenditures.[839] "

Direct your whole effort toward the happiness of those who are despondent, *bestow food upon the hungry*, clothe the needy, and glorify the humble. Be a helper to every helpless one, and manifest kindness to your fellow creatures in order that ye may attain the good pleasure of God. This is conducive to the illumination of the world of humanity and eternal felicity for yourselves."[840]

Chronic food insecurity

Chronic food Insecurity is a long-term or persistent inability to meet dietary energy requirements (lasting for a significant period of time during the year), FAO defines this as 'undernourishment' and it is the basis for the SDG indicator 2.1.1 published in SOFI. People experiencing moderate food insecurity face uncertainties about their ability to obtain food and have been forced to reduce, at times during the year, the quality and/or quantity of food they consume due to lack of money or other resources. It thus refers to a lack of consistent access to food, which diminishes dietary quality, disrupts normal eating patterns, and can have negative consequences for nutrition, health and wellbeing. People facing severe food insecurity, on the other hand, have likely run out of food, experienced hunger and, at the most extreme, gone for days without eating, putting their health and wellbeing at grave risk (FAO et al., 2019). In 2018 'More than 820 million people in the world were undernourished; […] more than 700 million people were exposed to severe levels of food insecurity' and 'an

---

[836] Photo 68749397 © Tizotov | Dreamstime.com

[837] Photo 188927282 © Mikhail Davidovich | Dreamstime.com

[838] Photo 170843059 © Neeraj Charurvedi | Dreamstime.com

[839] 'Abdu'l Bahá, Additional Tablets, Extracts and Talks. Emphasis Added.

[840] 'Abdu'l-Bahá, The Promulgation of Universal Peace, 5 December 1912. Talk on Day of Departure. On Board Steamship Celtic, New York. Notes by Mariam Haney. Emphasis Added.

additional 1.3 billion people, have experienced food insecurity at moderate levels' (The State of Food Security and Nutrition in the World 2019).[841]

Number of human beings affected in their IQ in Billions

Sadly, the author (Duque, 1999) concluded that around 1.2 billion people were deprived of a significant portion of their potential intellectual capacities; because of their mothers' protein deficiency during pregnancy. (Joseph Enamuthu. Iron 1.6 billion) (Reynaldo Martorell. Iodine 2.1 billion).

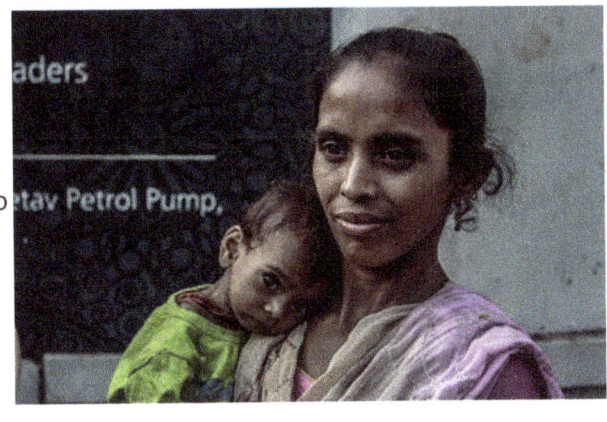

Please stop, and contemplate the image for 30 seconds. Find out if the color of the baby's hair and her features are related to malnourishment.[842]

Dr. Pradilla[843] educated me, telling that in the prenatal period, there are scientific findings that have shown that the organism of the mother and child do not enter into selfish competition for nutrients. The maternal organism sacrifices from its tissues what is necessary for the well-being of the creature.

I believe that this is a concrete manifestation of the wisdom of creation made by a merciful God. However, he told me that the only limitation could be the energy needed to carry out the respective metabolization. Why ask an organism to be forced to reverse its tissues into nutrients? What are the consequences of a second pregnancy and no reserve of energy? The masses of poor women and their immediate descendants are the first to be affected by this injustice.

Does something similar occur in nature? In, *Intelligent Trees - The Documentary*, featuring Suzanne Simard & Peter Wohlleben, we become amazed at what they have found:

"Trees talk, know family ties and care for their young? Is this too fantastic to be true? Scientist Suzanne Simard (The University of British Columbia, Canada) and German forester and author Peter Wohlleben have been investigating and observing the communication between trees over decades. And their findings are most astounding.

Trees are so much more than rows of wood waiting to be turned into furniture, buildings or firewood. They are more than organisms producing oxygen or cleaning the air for us. They are individual beings that have feelings, know friendship have a common language and look after each other. This documentary explores the various ways that trees communicate with each other - from a forester's observations as well as through the microscope of a scientist.

---

[841] Food Security Information Network (FSIN), 2020 Global Report on Food Crises.

[842] Photo 46038610 © Parthkumar Bhatt | Dreamstime.com

[843] Professor of the School of Public Health, Universidad del Valle, Cali, Colombia.

The film centers around the groundbreaking scientific discoveries that Suzanne Simard has been making in the Canadian Wilderness since the 1990s and that seem to be valid for all natural forests around the world![844]

11.4 Our aspiration for peace all over the world

But there is a warning[845] that it is not possible to fix it without women's involvement.

Why is Leo asking if it is time for men to consider, and accept women's superiority, and be willing to learn from them?

"Be anxiously concerned with the needs of the age ye live in, and centre your deliberations on its exigencies and requirements."[846]

Then, what is the need that is so important? It is undeniable that the planet's previous and current rulers have gravely endangered the masses' food security, with few exceptions to find. They have or have had access to information about their options concerning the rural population. This situation's consequences weaken the growing and pauperized masses' necessary intellectual potentials to deal with it successfully and lead them to subjugation and mastery of the well-fed. It is also undeniable that there is an alternative order. The implications of not opting and not acting now will be of greater magnitude and depth in humanity's immediate future! This is the crude reality in which a New Order receives the world![847]

How many elections, of the old-world order, will take to accomplish the suggestions mentioned above? After the M-19 attack to Palace of Justice of Colombia and the dead of 12 magistrates[848] of the Supreme Court of Colombia in 1.985, the government open the door to receive proposals to modify Colombia's Constitution. After consulting with The Universal House of Justice, the National Spiritual Assembly of the Bahá'ís of Colombia summited to the "Asamblea Nacional Contituyente" (National Constituent Assembly) a series of proposals. A team of three bahá'ís, including the author[849], personally visited associations of small and big farmers and the elected members of the National Constituent Assembly to formally present the proposals in "Los Bahá'ís y la Constituyente."

Article 65 of Colombia's Constitution says:

> "*Food production will enjoy the special protection of the State*. For this purpose, it will prioritize the comprehensive development of agricultural, livestock, fisheries, forestry, and

---

[844] Suzanne Simard and Peter Wohlleben, Intelligent Trees - The Documentary.

[845] Photo 165964660 © Steve Allen | Dreamstime.com

[846] Bahá'u'lláh, Gleanings from the Writings of Bahá'u'lláh. Emphasis added.

[847] Duque, Leonardo, En la Encrucijada: Una Perspectiva Nueva: En Honor a las Mujeres del Mundo.

[848] There is a documentary of the events by Pablo Bohorquez showing that some magistrates were killed by the Colombian army.

[849] Juan Carlos Leon and his wife María Mercedes, were the other two members.

agro-industrial activities and the construction of physical infrastructure and adequacy of land. Similarly, the State will promote research and transfer of technology to produce food and raw materials of agricultural origin, to increase productivity."

Shortly after the approval (1991) of the New Constitution by the Congress of Colombia, Congress legalized a massive importation of food commodities (Perfetti, 2017). The openness to globalized markets caused an estimated 1.6 million jobs lost in Colombia's rural areas between 1991 and 2000 (Manuel Alvaro Ramírez Rojas et al., 2003). Approx. 6.4 million rural inhabitants were displaced, mainly not by the guerrillas, not by the narco-traffickers, and not by the paramilitary forces, but by the Congress itself. Between 1990 and 2003, the population growth of all states and municipalities[850] capitals was 33%, 7.6 million. Within the same period, the rural population decreased -0.7%, -72 thousand citizens. In 1990, "the harvested area reached 4.8 million hectares, an area that gradually declined until 2000 with 3.9 million and the already known effects on the level of employment."[851] Colombia's conflict started in 1964; finally today, 23rd March, 2022, there is a glimmering of hope.

The following is the need to be addressed:

> Coffee is grown in more than 70 countries, but nearly 70 percent of the world's coffee is produced by just four of them – Brazil, Vietnam, Colombia, and Indonesia. Latin America is the largest regional producer with a 59 percent share, followed by Asia and Oceania (30%), and Africa (11%).
>
> For countries that produce it, coffee exports generate a significant proportion of national income. They are a vital source of the foreign exchange earnings that governments rely on to improve health, education, infrastructure, and other social services. For instance, Honduras relies on coffee for nearly a quarter of its export earnings, and in Nicaragua over 15 percent of the labour market is employed in coffee production. In Ethiopia, 15 million smallholders, nearly a fifth of the population, depend on coffee for their livelihood, and it generates around a third of the country's total export earnings. In Uganda, 1.7 million households grow coffee, and it contributes to nearly 20 per cent of the country's export earnings.[852]
>
> "The yield of the coffee tree *peaks after 5 to 7 years*. The fruits are left unpicked until they reach the ideal stage of ripeness, *usually after about seven months*"[853] [854] [855]

---

[850] DANE. Estimaciones de población 1985 - 2005 y proyecciones de población 2005 - 2020 total municipal por área. Capitales de Departamentos, Distrito Capital y cabeceras de Municipios, compare to the rest of the area.

[851] Manuel Alvaro Ramírez Rojas et al. Mercado de Trabajo y Condiciones del Empleo en Colombia: Los efectos de la globalización. Translated by the author

[852] Fairtrade Foundation, About Coffee.

[853] Traoré Cocoa and Coffee Value Chains in West and Central Africa: Constraints and Options for Revenue-Raising Diversification. p. 50. Emphasis added.

[854] Photo 48619435 © Andrei Fisenko | Dreamstime.com

[855] Photo 12706938 © Brian Flaigmore | Dreamstime.com

After interviewing Roberto Velez Vallejo, CEO of Colombia's National Federation of Coffee Growers, Bloomberg News says that a cup of coffee at Starbucks costs $3.50, and the Colombian coffee grower receives $0.03.[856] Farmers are receiving less than 1%.

In Selections from the Writings of 'Abdu'l-Bahá, we find: "Although our vision must be world-embracing, the initial stage of economic reconstruction is at the local level, beginning with agricultural reform. 'Abdu'l-Bahá said the solution begins with the village, and when the village is reconstructed, then will the cities be also".[857]

The Universal House of Justice, in a letter to the Bahá'ís of Iran, 2 March 2013, said:

> The deepening environmental crisis, driven by a system that condones the pillage of natural resources to satisfy an insatiable thirst for more, suggests how entirely inadequate is the present conception of humanity's relationship with nature; the deterioration of the home environment, with the accompanying rise in the systematic exploitation of women and children worldwide, makes clear how pervasive are the misbegotten notions that define relations within the family unit; the persistence of despotism, on the one hand, and the increasing disregard for authority, on the other, reveal how unsatisfactory to a maturing humanity is the current relationship between the individual and the institutions of society; the concentration of material wealth in the hands of a minority of the world's population gives an indication of how fundamentally ill-conceived are relationships among the many sectors of what is now an emerging global community.[858]

The Universal House of Justice in *The Promise of World Peace*, says:

> Lack of resources limits the ability of many nations to fulfil this necessity, imposing a certain ordering of priorities. The decision-making agencies involved would do well to consider giving first priority to the education of women and girls, since it is through educated mothers that the benefits of knowledge can be most

---

[856] Alix Steel, Why Colombia Coffee Farmers Are 'Desperate'.

[857] 'Abdu'l-Bahá, Selections from the Writings of 'Abdu'l-Bahá. p. 279

[858] The Universal House of Justice, Non-Involvement in Partisan Politics.

effectively and rapidly diffused throughout society. In keeping with the requirements of the times, consideration should also be given to teaching the concept of world citizenship as part of the standard education of every child.[859]

Transparency International[860] says:

There is a growing consensus, and a large body of evidence from many regions of the world, that countries with more women involved in government or parliament are less prone to corruption.

For example, a study in the early 2000s by the World Bank of 150 countries in Europe, Africa, and Asia suggested that women are more trustworthy and less prone to corruption, a finding later corroborated by additional research."[861]

Let us now consider the impact of prejudice in governance. From 'Abdu'l-Bahá, who passed away in 1921 not long after the end of World War I, we learn:

O ye lovers of truth, ye servants of humankind! Out of the flowering of your thoughts and hopes, fragrant emanations have come my way, wherefore an inner sense of obligation compelleth me to pen these words.

Ye observe how the world is divided against itself, how many a land is red with blood and its very dust is caked with human gore. The fires of conflict have blazed so high that never in early times, not in the Middle Ages, not in recent centuries hath there ever been such a hideous war, a war that is even as millstones, taking for grain the skulls of men. Nay, even worse, for flourishing countries have been reduced to rubble, cities have been leveled with the ground, and many a once prosperous village hath been turned into ruin. Fathers have lost their sons, and sons their fathers. Mothers have wept away their hearts over dead children.

Children have been orphaned, women left to wander, vagrants without a home. From every aspect, humankind hath sunken low. Loud are the piercing cries of fatherless children; loud the mothers' anguished voices, reaching to the skies.

And the breeding ground of all these tragedies is prejudice: prejudice of race and nation, of religion, of political opinion; and the *root cause* of prejudice is blind imitation of the past — imitation in religion, in racial attitudes, in national bias, in politics. So long as this aping of the

---

[859] The Universal House of Justice, The Promise of World Peace.

[860] Photo 120986128 © Jarretera | Dreamstime.com & ID 187349926 © Awargula | Dreamstime.com

[861] Farzana Nawaz, Are Women Less Corrupt than Men? And Other Gender/Corruption Questions

past persisteth, just so long will the foundations of the social order be blown to the four winds, just so long will humanity be continually exposed to direst peril.[862]

Let us examine the situation of farmers and prejudice within the context of the Root Cause and the Unique Method of science: There is a correlation between the various kinds of *discrimination*: dark skin, women, poorly educated, poorly dressed, and of course, being part of the farmers' social class; and its *cumulative* effect on *how much* society values what peasants produce.

United States' Civil War. What happened in the United States when slavery in plantations prevailed as the leading cause of the civil war (1861-1865)?

For 110 years, the numbers stood as gospel: 618,222 men died in the Civil War, 360,222 from the North, and 258,000 from the South — by far the greatest toll of any war in American history.

But new research shows that the numbers were far too low.

By combing through newly digitized census data from the 19th century, J. David Hacker, a demographic historian from Binghamton University in New York, has recalculated the death toll and increased it by more than 20 percent — to 750,000.[863] Slavery on plantations was the main reason for the war.

Was Rwanda's destiny similar?

The Belgian colonial occupation had a much more lasting effect in Rwanda. The most lasting effect was how the colonial authorities racialized the differences between Hutu, Twa, and Tutsi.

Rwanda gained independence from Belgium in 1962, but the post-colonial period was marred by ethnically motivated violence. This violence culminated into the 1994 Rwandan genocide in which more than 800.000 Tutsi people were killed, including thousands of Hutu people who were either part of the opposition or who had refused to take part in the killings.[864]

Hutus were farmers and Tutsis cattle herders. "Upon first entering the region, German colonizers rationalized subjugation of Rwanda's large Hutu population under the flawed assumption that the Tutsi were more Caucasian and thus more fit to rule."[865]

After the genocide, women of both sides met. They consulted and decided to adopt the orphans of the other tribe — Hutu women raising Tutsy orphans, Tutsy women adopting Hutu orphans (Out of Madness, A Matriarchy by Kimberlee Acquaro and Peter Landesman).

What was the fate of those regimens stubbornly supporting oppressive systems of governance?

The emancipation of women, the achievement of full equality between the sexes, is one of the most important, though less acknowledged prerequisites of peace. The denial of such equality perpetrates an injustice against one-half of the world's population and promotes in

---

[862] Selections from the Writings of 'Abdu'l-Bahá. Emphasis added.

[863] Hacker, J. David. New Estimate Raises Civil War Death Toll.

[864] South African History Online (SAHO)

[865] What Impact Did the Belgian Presence in Rwanda Have to Spark Further Conflict? By Stephen Skok.

men harmful attitudes and habits that are carried from the family to the workplace, to political life, and ultimately to international relations. There are no grounds, moral, practical, or biological, upon which such denial can be justified. Only as women are welcomed into full partnership in all fields of human endeavour will the moral and psychological climate be created in which international peace can emerge.[866]

"The woman has greater moral courage than the man; she has also special gifts which enable her to govern in moments of danger and crisis."[867] In, *Women against arrack. Organizing for change: India*[868] [869], we learn about how they addressed a crisis:

Because of a militant movement spearheaded by women in 1992, the sale and consumption of almost all types of potable alcohol is a criminal offense in Andhra Pradesh state in India. Previously, the government had supported a Varuna Vahini (Flood of Liquor) policy which saw the aggressive marketing of the alcoholic beverage "arrack" in rural areas. This policy resulted in revenue from the excise duty on liquor amounting to more than 10% of the state's annual budget in 1991-92. Political parties were also funded by underhanded financial contributions from the liquor industry. While the wisdom of a government-imposed prohibition is still being questioned, the political impact of the women's struggle is plain. The struggle began as individual groups of women tried to keep arrack out of their villages. The women joined forces because of their common experience of the destruction of their lives by arrack and because the government-financed literacy campaign was run by nongovernmental organizations who included in the literacy primers stories about women suffering at the hands of alcoholic husbands.[870] The women were not opposed to alcohol itself but to the new packaging of the cheap beverage which made it readily available in remote locations where people had to trek miles for water. The women were furious that their husbands were spending all of their wages on the drink and were tired of being harassed by drunken husbands. The women focused on arrack suppliers and traders rather than on consumers, so they received the passive support of many men. The women restricted their activities to their own villages and enjoyed the support of the nongovernmental organizations, although the struggle was led by local women themselves. Now the struggle has passed from poor, rural women to middle class, urban women and men who espouse Gandhian ideals and demand

---

[866] The Universal House of Justice. The Promise of World Peace.

[867] 'Abdu'l-Bahá. 'Abdu'l-Bahá in London.

[868] Photo 205630666 © Zz3701 | Dreamstime.com

[869] Right, ID 37202500 © Rafał Cichawa |dreamstime.com

[870] The author understands that Paulo Freire, author of the Pedagogy of the Oppressed methodology, was critical in achieving this goal.

total prohibition. Whether or not the women in the villages will be able to deal with the consequences of state-wide total prohibition, they have managed to shake the foundations of political power.[871]

When 'Abdu'l-Bahá was asked about the solution to the *economic* problem, he said: "The solution of this problem is one of the fundamental principles of His Holiness Bahá'u'lláh. But it must be solved with *justice* and not with force. If this problem is not solved *lovingly* it will result in war.[872]

My recommendations to achieve peace are:

- Always avoid debates and encourage consultation processes.
- Invite *all the parties*: representatives of the government, industry, business sectors, farmers, artisans, merchants, and academia.
- Have at least the same number of both genders participating in the consultation process. More women involved will *ensure a peaceful outcome*.

In The Promulgation of Universal Peace by 'Abdu'l-Bahá, we find:

Children are educated by the women. The mother bears the troubles and anxieties of rearing the child, undergoes the ordeal of its birth and training. Therefore, it is most difficult for mothers to send to the battlefield those upon whom they have lavished such love and care. Consider a son reared and trained twenty years by a devoted mother. What sleepless nights and restless, anxious days she has spent! Having brought him through dangers and difficulties to the age of maturity, how agonizing then to sacrifice him upon the battlefield!

Therefore, the mothers will not sanction war nor be satisfied with it. So it will come to pass that when women participate fully and equally in the affairs of the world, when they enter confidently and capably the great arena of laws and politics, war will cease; for woman will be the obstacle and hindrance to it. This is true and without doubt.[873]

In the *Bible*, we find: And He shall judge among the nations, and shall rebuke many

people: *and they shall beat their swords*[874] *into plowshares*,[875] and their *spears into pruning hooks*: nation shall not lift up sword against nation, neither shall they learn war any more.[876]

---

[871] Anu Joseph, Women against arrack. Organizing for change: India.

[872] 'Abdu'l-Bahá, Compilations, Bahá'í Scriptures. p. 340. Emphasis added.

[873] 'Abdu'l-Bahá, The Promulgation of Universal Peace.

[874] ID 122438425 © Aleksey Satyrenko | Dreamstime.com

[875] Photo 41610509 © Woravit Vijitpanya | Dreamstime.com

[876] Isaiah, 2:2 –2:4. Emphasis Added.

As regards the Constitution of the House of Justice, Bahá'u'lláh addresses the men. He says:

'O ye men of the House of Justice!'

But when its members are to be elected, the right which belongs to women, so far as their *voting and their voice* is concerned, is indisputable. When the women attain to the ultimate degree of progress, then, according to the exigency of the time and place and their great capacity, they shall obtain extraordinary privileges. Be ye confident on these accounts. His Holiness Bahá'u'lláh has greatly strengthened the Cause of women, and the rights and privileges of women is one of the greatest principles of 'Abdu'l-Bahá. Rest ye assured! Ere long the days shall come when the men addressing the women, shall say: 'Blessed are ye! Blessed are ye! Verily ye are worthy of every gift. Verily ye deserve to adorn your heads with the crown of everlasting glory, because in sciences and arts, in virtues and perfections ye shall become equal to man, and as regards *tenderness of heart and the abundance of Mercy and sympathy* ye are superior'.[877]

My suggestion is: the abundance of Mercy, tenderness, and sympathy demonstrated by women is the reason why they should be welcome to participate in more significant number until they reach a majority, in the administrative level of all human affairs, even at the international level, except in the Universal House of Justice as ordained by Bahá'u'lláh.

To clarify the reason for such exception, Shoghi Effendi, in a letter written on his behalf to an individual believer (July 28, 1936), provided the following authoritative elaboration of this theme:

As regards your question concerning the membership of the Universal House of Justice: there is a Tablet from 'Abdu'l-Bahá in which He definitely states that the membership of the Universal House is confined to men, and that the wisdom of it will be fully revealed and appreciated in the future. In the local as well as the national Houses of Justice, however, women have the full right of membership. It is, therefore, only to the International House that they cannot be elected. The Bahá'ís should accept this statement of the Master in a spirit of deep faith, confident that there is a divine guidance and wisdom behind it which will be gradually unfolded to the eyes of the world.[878]

In a letter, The Universal House of Justice expresses the following:

With regard to the status of women, the important point for Bahá'ís to remember is that in face of the categorical pronouncements in Bahá'í Scripture establishing the equality of men and women, the ineligibility of women for membership of the Universal House of Justice does not constitute evidence of the superiority of men over women. It must also be borne in mind that women are not excluded from any other international institution of the Faith. They are found among the ranks of the Hands of the Cause. They serve as members of the International Teaching Centre and as Continental Counselors. And, there is nothing in the Text to preclude the participation of women in such future international bodies as the Supreme Tribunal.[879]

---

[877] 'Abdu'l-Bahá, Paris Talks, p. 794. Emphasis Added.

[878] Shoghi Effendi, qtd. in Women. Compiled by the Research Department of the Universal House of Justice.

[879] The Universal House of Justice, 31 May 1988 – The National Spiritual Assembly of the Bahá'ís

For us, males also mean to address those faculties that we have in less abundance than females: tenderness of heart, sympathy, and mercy, and the others mentioned above. We need to learn, early in life, to listen to those emotions when females describe the situation of a child, the environment, or the world. We have to learn to read in those feelings the dimensions of the problematic situation. That enables both sides to make better choices when consulting about the solution to the needs and tests. 'Abdu'l-Bahá has stated that:

> "The world in the past has been ruled by force, and man has dominated over woman by reason of his more forceful and aggressive qualities both of body and mind. But the balance[880] is already shifting; force is losing its dominance, and mental alertness, intuition, and the spiritual qualities of love and service, in which woman is strong, are gaining ascendancy."[881]

> "Taken in general, women today have a stronger sense of religion than men. The woman's intuition is more correct; she is more receptive and her intelligence is quicker. The day is coming when woman will claim her superiority to man."[882]

Do we all aspire to world peace, food security for all, a job to support the family, and empowerment to address the environmental crisis?

Is it time for men to consider and accept women's superiority and be willing to learn from them?

In a Tablet of 'Abdu'l-Bahá, we learn:

> Know thou that the distinction between male and female is an exigency of the physical world and hath no connection with the spirit; for the spirit and the world of the spirit are sanctified above such exigencies, and wholly beyond the reach of such changes as befall the physical body in the contingent world. In former ages, men enjoyed ascendancy over women because bodily might reigned supreme and the spirit was subject to its dominion. In this radiant age, however, since the power of the spirit hath transcended that of the body and assumed its ascendancy, authority and dominion over the human world, this physical distinction hath ceased to be of consequence; and, as the sway and influence of the spirit have become apparent, women have come to be the full equals of men. Today, therefore, there is no respect or circumstance in which a person's sex provideth grounds for the exercise of either discrimination or favour.[883]

The facts mentioned above tell us that females have the right to claim their superiority because of those commendable features they have. If we males openly demonstrate their worth now, they will become admired by all and will be a stimulus for them to lead the path for food security for all. Because lovingly pursuing paths to peacefully liberate farmers of the cumulative effect of imitation, which is the root

---

of New Zealand.

[880] ID 39044397 © Darakchi | Dreamstime.com

[881] The Universal House of Justice, Family Life.

[882] 'Abdu'l-Bahá,'Abdu'l-Bahá in London, p. 104.

[883] 'Abdu'l Bahá, Additional Tablets, Extracts and Talks.

cause of prejudice. Because supporting and participating in discovering the immense potential of biodiversity to generate employment in the world. Their collaborative enthusiasm to making the youth competent in applying scientific methods to address the environmental crisis and any other problematic situation.

I understand the above is the path to embracing the lesser peace! We should try to learn from them to reach an equal spiritual level.

Humanity's well-being will be impendent under the same curriculum for both genders to realize completeness in their lives. Part of the difference is attributable to the educational contents of separate schooling for girls and boys and parental guidelines received at home. It also requires focusing on developing in males the potentialities of those faculties in which females are superior. The author believes that perfections within the human station are limitless for both genders, because, in *Some Answered Questions*, the Master, 'Abdu'l Bahá, taught:

> Man, having reached the human station, can progress only in perfections and not in station, for there is no higher station to which he can find passage than that of a perfect man. He can progress solely within the human station, as human perfections are infinite. Thus, however learned a man may be, it is always possible to imagine one even more learned.
>
> And as the perfections of man are infinite, he can also advance in these perfections after his ascension from this world.[884]

In, The Promulgation of Universal Peace, by 'Abdu'l-Bahá, we find:

> [T]here must be an equality of rights between men and women. Women shall receive an equal privilege of education. This will enable them to qualify and progress in all degrees of occupation and accomplishment. For the world of humanity possesses two wings: man and woman. If one wing remains incapable and defective, it will restrict the power of the other, and full flight will be impossible. Therefore, the completeness and perfection of the human world are dependent upon the equal development of these two wings.[885]
>
> Daughters and sons must follow the same curriculum of study, thereby promoting unity of the sexes. When all mankind shall receive the same opportunity of education and the equality of men and women be realized, the foundations of war will be utterly destroyed. Without equality this will be impossible because all differences and distinction are conducive to discord and strife. Equality between men and women is conducive to the abolition of warfare for the reason that women will never be willing to sanction it. Mothers will not give their sons as sacrifices upon the battlefield after twenty years of anxiety and loving devotion in rearing them from infancy, no matter what cause they are called upon to defend. There is no doubt that when women obtain equality of rights, war will entirely cease among mankind.[886]

---

[884] 'Abdu'l Bahá, Some Answered Questions.

[885] 'Abdu'l Bahá, The Promulgation of Universal Peace.

[886] Idem.

11.5 The Science's Value-free Ideal, Stripped of Emotions and Feelings

Definitions of quantitative and qualitative research

In *Understanding Quantitative Analysis in Chemistry,* by Anne Marie Helmenstine, Ph.D. Updated on July 07, 2019

> Quantitative analysis refers to the determination of how much of a given component is present in a sample. The quantity may be expressed in terms of mass, concentration, or relative abundance of one or all components of a sample. Here are a few sample results of quantitative analysis:
>
> Ore contains 42.88% silver by mass.
>
> The chemical reaction yielded 3.22 moles of product. The solution is 0.102 M NaCl.
>
> Quantitative Versus Qualitative Analysis
>
> Qualitative analysis tells 'what' is in a sample, while quantitative analysis is used to tell 'how much' is in a sample. The two types of analysis are often used together and are considered examples of analytical chemistry.
>
> As an alternative to quantitative research, *qualitative* research is also employed in social science research and is contrasted with quantitative research as such:
>
> Insider rather than outsider
>
> Person-centered rather than variable-centered Holistic rather than particularistic
>
> Depth rather than breadth.[887]

In *Social Sciences Research: Qualitative vs. Quantitative Research*, we learn:

> Quantitative research "is the systematic examination of social phenomena, using statistical models and mathematical theories to develop, accumulate, and refine the scientific knowledge base" ("Quantitative Research," 2008). Quantitative research also provides "generalizable" findings, and according to Marlow (1993), is "characterized by hypothesis testing, using large samples, standardized measures, a deductive approach, and rigorously structured data collection instruments" (cited in "Quantitative Research").[888]
>
> Quantitative methods emphasize objective measurements and the statistical, mathematical, or numerical analysis of data collected through polls, questionnaires, and surveys, or by manipulating pre-existing statistical data using computational techniques. Quantitative research focuses on gathering numerical data and generalizing it across groups of people or to explain a particular phenomenon.[889]

In, Investigación Cualitativa y Cuantitativa La Investigación Cualitativa (Síntesis Conceptual, by Miguel Martínez M., we read:

---

[887] Anne Marie Helmenstine, Ph.D., Understanding Quantitative Analysis in Chemistry.

[888] Focus group at Saint Petersburg College, Social Sciences Research: Qualitative vs. Quantitative Research.

[889] Babbie, Earl R. The Practice of Social Research. 12th ed.

"Mendoza (2006)[890] describes the quantitative method, saying: that quantitative methods emerged in the eighteenth and nineteenth centuries as elements within capitalism to analyze social and economic conflicts as a complex whole ... .[891]

Hurtado and Toro affirm that "the research subject is conceived as a person capable of stripping himself of his emotions and feelings. Study the object from an outside "perspective, without getting involved. Attribute objectivity in research. Therefore, their relationship is independent of each other" (Hurtado & Toro, 1998)."[892]

To start, let me say that I consider it impossible to divest ourselves from emotions and feelings to be a researcher. Love and feelings are inherent to human nature. Clearly, those practicing the conventional quantitative analysis of reality do not want to show off their desire to establish justice in the world to other scientists. Bahá'u'lláh revealed:

"O Son of Spirit! *The best beloved of all things in My sight is Justice*; turn not away therefrom if thou *desirest* Me, and neglect it not that I may confide in thee. By its aid thou shalt see with thine own eyes and not through the eyes of others, and shalt know of thine own knowledge and not through the knowledge of thy neighbor. Ponder this in thy *heart*; how it behooveth thee to be. Verily *justice* is My gift to thee and the sign of My *loving-kindness*. Set it then before thine eyes."[893]

They that are just and fair-minded in their judgment occupy a sublime station and hold an exalted rank. The light of piety and uprightness shineth resplendent from these souls. We earnestly hope that the peoples and countries of the world may not be deprived of the splendors of these two luminaries.[894]

Of course, there are plenty of scientists deviating from the conception of a scientist "as a person capable of stripping himself of his emotions and feelings." I have plenty of friends that are medical doctors, public health professionals, physicists, mathematicians, engineers, sociologists, journalists, social workers, and nutritionists: working to liberate the oppressed, developing fair trade relations, eliminating prejudices, and providing access to health to the nobodies.

A survey of scientists who are members of the American Association for the Advancement of Science, conducted by the Pew Research Center for the People & the Press in May and June 2009, finds that members of this group are, on the whole, much less religious than the general public. Indeed, the survey shows that scientists are roughly half as likely as the general public to believe in God or a higher power. According to the poll, just over half of scientists (51%) believe in some form of deity or higher power; specifically, 33% of scientists say they believe in God, while 18% believe in a universal spirit or higher power, according to a survey of the general public conducted by the Pew Research Center in July 2006.

---

[890] Rudy Mendoza Palacios, Investigación cualitativa y cuantitativa. Diferencias y limitaciones.

[891] Theodore M. Porter, History of statistics.

[892] Miguel Martínez M., Investigacion Cualitativa y Cuantitativa. La Investigación Cualitativa (Síntesis Conceptual).

[893] Bahá'u'lláh, The Hidden Words. No. 2 from the Arabic. Emphasis added.

[894] Bahá'u'lláh, Tablets of Bahá'u'lláh revealed after the Kitáb-i-Aqdas.

Specifically, more than eight-in-ten Americans (83%) say they believe in God and 12% believe in a universal spirit or higher power. Finally, the poll of scientists finds that four-in-ten scientists (41%) say they do not believe in God or a higher power, while the poll of the public finds that only 4% of Americans share this view.[895]

In `Abdu'l-Bahá's Tablet to Dr. Forel we learn:

> By materialists, whose belief with regard to Divinity hath been explained, is not meant philosophers in general, but rather that group of materialists of narrow vision who worship that which is sensed, who depend upon the five senses only, and whose criterion of knowledge is limited to that which can be perceived by the senses. All that can be sensed is to them real, whilst whatever falleth not under the power of the senses is either unreal or doubtful. The existence of the Deity they regard as wholly doubtful.[896]

In, Clarifying spiritual values among organizational development personnel, by Akbar Husain and Aqeel Khan, we learn:

Early assumptions about values Naturalism

It is the belief that the "universe is self-sufficient, without supernatural cause or control" (Honer and Hunt, 1987, p.225). Naturalists assume that human beings and the universe can be understood without restoring to spiritual explanations and that "the explanation of the world given by the sciences is the only satisfactory explanation of reality" (Honer and Hunt, 1987, p.225). This assumption led many behavioral scientists to conclude that all moral values are ephemeral and of human origin.

Ethical relativism

This is the belief that "there are no universally valid principles, since all moral principles are valid relative to cultural and individual values" (Percesepe, 1991, p.572).

Thus, "whatever a culture or society holds to be right is therefore right or at least, right for them" (Solomon, 1990, p.235). Values are considered as relevant to professionals and organizations. Ethical relativism led to conclude that, if values are relative, then organizations should lay emphasis on the values of the personnel.

Ethical hedonism

This is the belief that "we always ought to seek our own pleasure and that the highest good for us is the most pleasure together with the least pain" (Honer and Hunt, 1987, p.222). According to some behavioral scientists (Hillyer, 194; Lundin, 1985; Watson 1924/1983), human beings are basically hedonistic and reward seeking. This is the reason for contradicting the assumptions of ethical relativism by endorsing hedonistic ethical values. Relying on this assumption, organizations should encourage their personnel to "throw off the shackles" of religion and be more accepting of their hedonistic tendencies.

Positivism

It holds that "knowledge is limited to observable facts and their interactions" (Honer and Hunt, 1987, p.226), and that the scientific theories can be "shown to be true on the basis of evidence" (Bechtel, 1988, p.18). Positivists assume that it is possible for scientists to be objective, impartial observers and that their empirical observations will eventually lead to a

---

[895] Pew Research Center, Scientists and Belief.

[896] `Abdu'l-Bahá, `Abdu'l-Bahá's Tablet to Dr. Forel

complete understanding of reality. They have advocated that only scientific thinking and logical assertions were to be cognitively meaningful (Tolmin and Leary, 1992) values (understood in ethical terms) were regarded as intellectually meaningless (O' Donahue, 1989; Putnam, 1993).[897]

Hilary Putnam in, Objectivity and the Science/Ethics Distinction, The fact/value dichotomy: background, teaches:

> The Logical Positivists argued for a sharp fact-value dichotomy in a very simple way: scientific statements (outside of logic and Pure mathematics), they said, are "empirically verifiable" and value judgements are "unverifiable". … If a sentence that does not, in and of itself, by its very meaning, have a "method of verification is meaningless, then most of theoretical science turns out to be meaningless!
>
> A second feature of the view that "ethical sentences are cognitively meaningless because they have no method of verification" is that even if it had been correct, what it would have drawn would not have been a fact-value dichotomy. For, according to the positivists themselves, metaphysical sentences are cognitively meaningless for the same reason as ethical sentences: they are "unverifiable in principle."(So are poetic sentences, among others.)
>
> The Positivist position is well summarized by Vivian Walsh (Walsh, 1987): "Consider the 'putative' proposition 'murder is wrong'. What empirical findings, the positivists would ask, tend to confirm or disconfirm this? If saying that murder is wrong is merely a misleading way of reporting what a given society believes, this is a perfectly good sociological fact, and the proposition is a respectable empirical one. But the person making a moral judgement will not accept this analysis. Positivists then wielded their absolute analytic/synthetic distinction: if 'murder is wrong' is not a synthetic (empirically testable) proposition it must be an analytic proposition, like (they believed) those of logic and mathematics in effect, a tautology. The person who wished to make the moral judgement would not accept this, and was told that the disputed utterance was a 'pseudo- proposition' like those of poets, theologians and metaphysicians."[898]

Alejandro Cassini in, Simulation models and probabilities: a Bayesian defense of the value-free ideal, says:

> Some philosophers of science have recently argued that the epistemic assessment of complex simulation models, such as climate models, cannot be free of the influence of social values. In their view, the assignment of probabilities to the different hypotheses or predictions that result from simulations presupposes some methodological decisions that rest on value judgments. In this article, I criticize this claim and put forward a Bayesian response to the arguments from inductive risk according to which the influence of social values on the calculation of probabilities is negligible. I conclude that the epistemic opacity of complex

---

[897] Akbar Husain and Aqeel Khan, Clarifying spiritual values among organizational development personnel.

[898] Hilary Putnam, Objectivity and the Science/Ethics Distinction, The fact/value dichotomy: background.

simulations, such as climate models, does not preclude the application of Bayesian methods.[899]

Wenceslao Gonzalez in, Value Ladenness and the Value-free Ideal in Scientific Research, says:

> Regarding scientific research, value ladennes and the value-free ideal represent two poles. At the beginning of the 20th century the influential view was science as "value-free" (Wertfrei), whereas in the first decade of the 21st century the dominant perspective is science as "value-laden". After considering the historical setting on values in science, the analysis here deals with the characteristics and relations between axiology of research and ethics of science. This involves taking into account the option in favor of holism of values and the alternative in terms of fractional orientations, either in "internal" terms or in "external" ones. Thus, the presence of values in basic science and applied science is considered. After that, economics as a relevant case-study is addressed through the distinction between positive economics and normative economics. There is also a coda with final remarks on the topics analyzed.[900]

In Introduction: *The Discipline and Practice of Qualitative Research* by Denzin, Norman. K. and Yvonna S. Lincoln, we learn:

> The word qualitative implies an emphasis on the qualities of entities and on processes and meanings that are not experimentally examined or measured [if measured at all] in terms of quantity, amount, intensity, or frequency. Qualitative researchers stress the socially constructed nature of reality, the intimate relationship between the researcher and what is studied, and the situational constraints that shape inquiry. Such researchers emphasize the value-laden nature of inquiry. They seek answers to questions that stress how social experience is created and given meaning. In contrast, quantitative studies emphasize the measurement and analysis of causal relationships between variables, not processes. Qualitative forms of inquiry are considered by many social and behavioral scientists to be as much a perspective on how to approach investigating a research problem as it is a method.[901]

In Qualitative Physics in a Metaphysical Perspective, by Aleksandr Kulieshov, we find:

> Abstract. The article deals with the problem concerning the possibility of qualitative physics paradigm development and its close connection with metaphysics. The idea of qualitative physics is based on the principles of Aristotelian physics and is opposed to quantitative modern physics (classical and non-classical). It is stated that the essential difference between the two physical paradigms lies in the ways of describing physical objects. Qualitative physics presuppose the qualitative description of physical objects independent of their quantitative description. In normal nowadays physics, on the contrary, physical objects are regarded to be fully determined through quantitative (numerical and structural-analytical) relationships with other objects. The statements of modern physics are considered reasonable if they can be self-consistently expressed by the apparatus of mathematics. The article shows that this way of describing and explaining physical reality is incomplete. There is ground to assert that the quantitative relations of physical objects do not encompass everything that exists in the

---

[899] Cassini A. Simulation models and probabilities: a Bayesian defense of the value-free ideal.

[900] Gonzalez, Wenceslao. (2013). Value Ladenness and the Value-free Ideal in Scientific Research.

[901] Norman. K. Denzin and Yvonna S. Lincoln. "Introduction: The Discipline and Practice of Qualitative Research."

relations of physical objects. It is argued that there are qualitative aspects of physical reality that are not defined quantitatively and may become the content of special qualitative physics. The conclusion is made that such qualitative physics in its principles and language must be close to traditional metaphysics and can appear to be an application of metaphysics to the field of physical reality.[902]

The author believes that conventional methods have made considerable advances in some fields but also are conducive to poverty and the degradation of human beings and the environment:

First: The Universal House of Justice in *The Promise of World Peace* says:

The time has come when those who preach the dogmas of materialism, whether of the east or the west, whether of capitalism or socialism, must give account of the *moral stewardship* they have presumed to exercise. Where is the "new world" promised by these ideologies? Where is the international peace to whose ideals they proclaim their devotion? Where are the breakthroughs into new realms of cultural achievement produced by the aggrandizement of this race, of that nation or of a particular class? Why is the vast majority of the world's peoples sinking ever deeper into hunger and wretchedness when wealth on a scale undreamed of by the Pharaohs, the Caesars, or even the imperialist powers of the nineteenth century is at the disposal of the present arbiters of human affairs?[903]

Second: Legalize pornography with access to almost all phones and computers as part of the right to free speech. In *The Scientist*, a magazine for life science professionals, we read: "Scientific examination of the subject has found that as the use of porn increases, the rate of sex crimes goes down."[904]

Todd Love, Christian Laier, Matthias Brand, Linda Hatch, and Raju Hajela, in *Neuroscience of Internet Pornography Addiction: A Review and Update*, we find the following statement:

"[W]e reviewed available neuroscientific literature on Internet pornography addiction and connect the results to the addiction model. The review leads to the conclusion that Internet pornography addiction fits into the addiction framework and shares similar basic mechanisms with substance addiction. Together with studies on Internet addiction and Internet Gaming Disorder we see strong evidence for considering addictive Internet behaviors as behavioral addiction."[905]

Third: When we examine scientists lobbying for the alcohol industry, whose business causes depression to all our faculties of protection, including our right to exercise our free will, we find:

"The alcohol industry works to undermine, alter and block evidence-based policy measures that protect health and well-being because these policy measures would threaten the profits of Big Alcohol.

Independent scientific evidence clearly shows: The harm of alcohol use outweighs by far the benefits from "moderate use." However, the alcohol industry works to perpetuate myths

---

[902] Aleksandr Kulieshov, Qualitative Physics in a Metaphysical Perspective.

[903] The Universal House of Justice, The Promise of World Peace. Emphasis added.

[904] Milton Diamond, Porn: Good for us?

[905] Todd Love, et al., Neuroscience of Internet Pornography Addiction: A Review and Update.

about their products and alcohol's effects, spread doubt about scientific evidence, and pressure decision-makers to refrain from regulating Big Alcohol."[906]

Fourth: Recreational and medicinal marijuana has recently been legalized in some states, in spite of scientific evidence such as Eric Groce in, The Health Effects of Cannabis and Cannabinoids: The Current State of Evidence and Recommendations for Research: "There is substantial evidence of a statistical association between cannabis use and The development of schizophrenia or other psychoses, with the highest risk among the most frequent users."[907] In this case, we can perceive that science has no respect among the groups interested in its legalization. Wall Street seems to be happy with the results:

"The more than 100,000 cannabis jobs created in 2021 represent a 33% increase in just one year. Despite the challenges of the ongoing Covid-19 pandemic, 2021 was the fifth consecutive year that the cannabis industry showed an annual job growth rate of 27% or higher." [T]he industry added another 107,059 new cannabis jobs in 2021.

"To put that in perspective, America's entire financial sector added 145,000 jobs last year," … "The construction industry, coast to coast, added 165,000 jobs."[908]

Hashish is a paste from oil extracted from the trichomes of the same cannabis plant. The outcome is a psychotic individual. In, Gone to pot – a review of the association between cannabis and psychosis, we find:

While psychosis refers to a heterogeneous group of disorders defined as consisting of positive symptoms (delusions[909], hallucinations, and thought- alienation phenomena), negative symptoms (alogia[910], avolition[911], anhedonia[912], asociality[913], and affective flattening[914]), and disorganization/cognitive symptoms (deficits in attention, working memory, problem-solving, and executive function); psychosis-like experiences are characterized by a

---

[906] Movendi International. Big Alcohol Exposed.

[907] Eric Groce, The Health Effects of Cannabis and Cannabinoids: The Current State of Evidence and Recommendations for Research.

[908] A.J. Herrington, New Cannabis Jobs Report Reveals Marijuana Industry's Explosive Employment Growth.

[909] A persistent false psychotic belief regarding the self or persons or objects outside the self that is maintained despite indisputable evidence to the contrary. Merriam-Webster Dictionary.

[910] Inability to speak: difficulty in speaking: reduced fluency of speech. Merriam Webster Dictionary.

[911] A lack of interest or engagement in goal-directed behavior. Merriam-Webster Dictionary.

[912] A psychological condition characterized by the inability to experience pleasure in normally pleasurable acts. Merriam-Webster Dictionary.

[913] Asociality refers to the lack of motivation to engage in social interaction or a preference for solitary activities. Wikipedia

[914] A loss or lack of emotional expressiveness. It is sometimes called blunted or restricted effect. Medical-dictionary.thefreedictionary.com/Affective+flattening

loss of reality-testing and include derealization[915], depersonalization[916], dissociation[917], hallucination[918], paranoia[919], impairment in concentration, and perceptual alterations, which are transient and self-limited.[920]

Other systematic reviews supporting the psychosis result of Cannabis addiction: Cannabis Use and Earlier Onset of Psychosis A Systematic Meta-analysis, by Matthew Large, et al., 2011; Meta-analysis of the Association Between the Level of Cannabis Use and Risk of Psychosis, by Arianna Marconi, et al., 2016; and Cannabis use in adolescence and risk of psychosis: Are there factors that moderate this relationship? A systematic review and meta-analysis, by Sarah Kanana Kiburi, et al., 2021.

If you are concerned about your friends, you can share the quote from The Kitáb-i-Aqdas, He orders: Alcohol consumeth the mind and causeth man to commit acts of absurdity, but this *opium*, this foul fruit of the infernal tree, and this wicked *hashish extinguish the mind, freeze the spirit, petrify the soul,* waste the body and leave man frustrated and lost.[921] If it ruins the power of reason, every spiritual virtue is frozen, and the love for truth is petrified, please tell me what is left of the human kingdom of an addicted to opium or hashish? Is there any sense of shame or responsibility left?

Fifth: An obvious purpose for any business is to have long-term clients. In the case of liquors, harmful drugs, and porn industries, it is intentional to increase the addictive power of their products, this is one example: "Overall, the potency of illicit cannabis plant material has consistently risen over time since 1995 from approximately 4% in 1995 to approximately 12% in 2014. On the other hand, the CBD content has fallen on average from approximately 0.28% in 2001 to <0.15% in 2014, resulting in a change in the ratio of T.H.C. to CBD from 14 times in 1995 to approximately 80 times in 2014."[922]

---

[915] A feeling of altered reality (such as that occurring in schizophrenia or in some drug reactions) in which one's surroundings appear unreal or unfamiliar. Merriam Webster Dictionary

[916] A psychopathological syndrome characterized by loss of identity and feelings of unreality and strangeness about one's own behavior. Merriam-Webster Dictionary.

[917] The separation of whole segments of the personality (as in dissociative identity disorder) or of discrete mental processes (as in schizophrenia) from the mainstream of consciousness or of behavior. Merriam-Webster Dictionary.

[918] A sensory perception (such as a visual image or a sound) that occurs in the absence of an actual external stimulus and usually arises from neurological disturbance (such as that associated with delirium tremens, schizophrenia, Parkinson's disease, or narcolepsy) or in response to drugs (such as LSD or phencyclidine). Merriam-Webster Dictionary.

[919] A tendency on the part of an individual or group toward excessive or irrational suspiciousness and distrustfulness of others. Merriam Webster Dictionary.

[920] Rajiv Radhakrishnan, Samuel T. Wilkinson, and Deepak Cyril D'Souza. Gone to pot – a review of the association between cannabis and psychosis.

[921] Bahá'u'lláh, The Kitáb-i-Aqdas. Emphasis added.

[922] THC binds to CB1 receptors in the brain and causes feelings of euphoria or 'a high'. CBD binds weakly to the CB1 receptor and only when THC is present. CBD does not produce euphoria or 'a high'. Web, August 2022 <https://www.cannasouth.co.nz/2021/cbdthc-ratio/>.

The World Health Organization has the following assessment:

> Psychoactive drugs are substances that, when taken in or administered into one's system, affect mental processes, e.g. perception, consciousness, cognition or mood and emotions. Psychoactive drugs belong to a broader category of psychoactive substances that include also alcohol and nicotine. "Psychoactive" does not necessarily imply dependence-producing, and in common parlance, the term is often left unstated, as in "drug use", "substance use" or "substance abuse".
>
> About 270 million people (or about 5.5% of global population aged 15-64) had used psychoactive drugs in the previous year and about 35 million people are estimated to be affected by drug use disorders (harmful pattern of drug use or drug dependence). It is estimated that about 0.5 million death annually are attributable to drug use with about 350 000 male and 150 000 female deaths. Opioid-related deaths, largely due to synthetic opioids, have recently changed the mortality trends in some high-income countries. More than 42 million years of healthy life loss (DALY) were attributable to drug use in 2017; that is about 1.3% of the global burden of disease. It is estimated that worldwide there are almost 11 million people who inject drugs, of whom 1.4 million live with HIV and 5.6 million - with hepatitis C."[923]

The World Health Organization expresses its concern:

> Drug use disorders, particularly when untreated, increase morbidity and mortality risks for individuals, can trigger substantial suffering and lead to impairment in personal, family, social, educational, occupational or other important areas of functioning. Drug use disorders are associated with significant costs to society due to lost productivity, premature mortality, increased health care expenditure, and costs related to criminal justice, social welfare, and other social consequences. [924]

The value-free science's ideal support the violation of the most fundamental human right when supporting addictions. Any harmful dependence affects our capacity to discern and inhibits the individual's self-control to exercise his(or her) volition to make sound choices, becoming a massive human rights violation. When I say "massive," it is not just the number of individuals with an addiction disorder but the number of human rights at stake when there is no free will. The family's human rights of the addicted are also on the line. Consider, for example, the following conception of science:

> The idea seems to be that science should be free from not merely ideological or religious values but also ideological or religious beliefs. ... With this clarification in mind, I propose that we define the value-free view of science more precisely in this way: The value-free view of science is the standpoint that science should be autonomous, neutral, impartial, non-responsible, and non-normative.[925]

---

[923] World Health Organization, Drugs (psychoactive).

[924] Idem.

[925] Mikael Stenmark, qtd. in LeRon Shults (ed.), The Evolution of Rationality, p. 51.

In supporting his perspective about an impartial science, Stenmark says: "Science should be impartial in the sense that it should not presuppose the truth of any particular political vision, religion or ideology in the validation of scientific theories."[926]

Wislawa Szymborska, Nobel Prize, in my view, has an excellent message for scientists supporting harmful addictions and their financiers, producers, merchants, and pushers:

In Praise of Self-Deprecation

The buzzard has nothing to fault himself with.
Scruples are alien to the black panther.
Piranhas do not doubt the rightness of their actions.
The rattlesnake approves of himself without reservations.

The self-critical jackal does not exist.
The locust, alligator, trichina, horsefly
live as they live and are glad of it.

The killer whale's heart weighs one hundred kilos
but in other respects it is light.

There is nothing more animal-like
than a clear conscience
on the third planet of the Sun.[927]

Is the autonomous, neutral, impartial, non-responsible, and non-normative value-free science supporting slavery? In Latin *addictus* means: "A debt slave; a person who has been bound as a slave to his creditor."[928]

Research has shown that addiction is a disease that chemically alters your brain, making you "a slave" to a substance or activity. The Latin definition gets support from the ancient myth of Addictus. The myth tells the story of a slave who is set free from his master but became so used to his chains, that he wandered the land with his chains still attached even though he could have removed them at any time. This story of course can be seen as a metaphor for the modern definition of addiction as we know it today. Because an addict becomes tolerant to the drugs they use, *they become a slave that doesn't recognize their own freedom.*[929]

Ian Kluge says in, *Ethics Based on Science Alone?*

We are left with the question of whether science has a role in ethical debate. If science insists on retaining the scientific method as currently formulated, then it is clear that science's role in ethics will be very limited. At most it will be able to supply facts to an ethical debate, but at no point will it be able to actually render an ethical decision on its own ground. It simply doesn't deal with such issues as good, evil, justice, intent, purity of motive, spiritual well-being all of which are the substance of ethics. To do so, science would have to abandon

---

[926] ibid. p. 51.

[927] Wislawa Szymborska, Nobel Prize. In Praise of Self Deprecation.

[928] Web. December 2022 <https://www.wordsense.eu/addictus/>.

[929] Northpoint Recovery, A Slave for Addiction: The Origins of the Word. Web. December 2022 <https://www.northpointrecovery.com/blog/slave-addiction-origins-word/>. Emphasis added.

materialism and re- invent the entire scientific method. Given its success in its own areas, there is little reason why it should do so.[930]

In *Science, Policy, and the Value-Free Ideal*, by Heather E. Douglas, we learn:

> The value-free ideal is a bad ideal for science. It is not restrictive enough on the proper role for cognitive values in science and it is too restrictive on the needed role for social and ethical values. The moral responsibility to consider the consequences of error requires the use of values, including social and ethical values, in scientific reasoning. Yet the inclusion of social and ethical values in scientific reasoning seems to threaten scientific objectivity. Our notion of objectivity should be reworked and clarified in light of the arguments of the previous two chapters. We need an understanding of objectivity.
>
> Reliance on the value-free ideal has produced something of a mess. Scientists have thought that any consideration of ethical or social values, particularly in the assessment of evidence, would undermine scientific integrity and authority. Yet one cannot adequately assess the sufficiency of evidence without such values, especially in cases where science has such a profound impact on society. Thus, a crucial source of disagreement among scientists has remained hidden, unexamined and unacknowledged.
>
> When considering the importance of science in policymaking, common wisdom contends that keeping science as far as possible from social and political concerns would be the best way to ensure science's reliability. This intuition is captured in the value-free ideal for science—that social, ethical, and political values should have no influence over the reasoning of scientists, and that scientists should proceed in their work with as little concern as possible for such values. Contrary to this intuition, I will argue in this book that the value-free ideal must be rejected precisely because of the importance of science in policymaking.[931]

[G]ender socialization and a greater tendency for female scientists to be aware of sexism in their fields makes them more likely to reject certain aspects of the value-free ideal.[932] Dan Hicks, On the Ideal of Autonomous Science, expresses: I suggest that alternative ideals for science might be developed by drawing on egalitarian liberal and communitarian political philosophy.[933]

Sarah S. Richardson in, Science, Policy, and the Value-Free Ideal, and: Philosophy of Science after Feminism (review), teaches:

> Heather E. Douglas's Science, Policy, and the Value-Free Ideal (2009) and Janet A. Kourany's in *Philosophy of Science after Feminism* (2010), advance a new plank in these discussions. Douglas and Kourany assert that, while philosophers of science have rejected the inaccurate image of science as ideally "value-free," they have not yet found a satisfactory alternative conception of the role of values in scientific practice. They set themselves the task of constructing a positive normative account of the role of social values and responsibilities in

---

[930] Ian Kluge, Ethics Based on Science Alone?

[931] Heather E. Douglas, "INTRODUCTION: Science Wars and Policy Wars." Science, Policy, and the Value-Free Ideal.

[932] Daniel Steel, Chad Gonnerman, Aaron M. McCright, Itai Bavli; Gender and Scientists' Views about the Value-Free Ideal.

[933] Dan Hicks, On the Ideal of Autonomous Science.

scientific work.[934] Kourany's vision of philosophy of science's future as "socially engaged and socially responsible."[935]

I agree with the following statement in Social Sciences Research: Qualitative vs. Quantitative Research:

> Trochim (2006), however, warns that researchers should not become so caught up in the polarizing differences between qualitative and quantitative research. He writes, "All quantitative data is based upon qualitative judgments; and all qualitative data can be described and manipulated numerically".[936]

In, *Can science be value-free? The "gap" argument*, by Chris ChoGlueck, we learn:

> While some philosophers think the value-free ideal should be defended or refined, others argue that it is time to look for more socially responsible ideals and guidelines. I, for one, think it is important to find a replacement that better guides scientists toward ethical choices. It is better for scientists to become more mindful of their value judgments and the political stakes of their science.
>
> Accordingly, they are better suited to discern their responsibilities as scientists and to improve society with science.[937]

What has been the historical role of the forces of Religious Institutions towards science? In, *Some Answered Questions*, the Master taught:

> Among the popes there have indeed been some blessed souls who followed in the footsteps of Christ, particularly in the early centuries of the Christian era when earthly means were lacking and heaven-sent trials were severe. But when the means of temporal sovereignty were secured, and worldly honour and prosperity were obtained, the papal government entirely forgot Christ and occupied itself with earthly dominion and grandeur, with material comforts and luxuries. It put people to death, opposed the diffusion of learning, persecuted men of science, obstructed the light of knowledge, and gave the order to slay and to pillage. Thousands of people, men of science and learning and innocent souls, perished in the prisons of Rome. With such ways and deeds, how can the claim of the vicarship of Christ be accepted?[938]

In, *The Secret of Divine Civilization*, we learn:

> In the early ages of Islám the peoples of Europe acquired the sciences and arts of civilization from Islám as practiced by the inhabitants of Andalusia. A careful and thorough investigation of the historical record will establish the fact that the major part of the civilization of Europe is derived from Islám; for all the writings of Muslim scholars and divines and philosophers were gradually collected in Europe and were with the most painstaking care weighed and debated at academic gatherings and in the centers of learning, after which their valued

---

[934] Sarah S., Richardson, Review of Science, Policy, and the Value-Free Ideal, and: Philosophy of Science after Feminism.

[935] Matthew J. Brown, The source and status of values for socially responsible science.

[936] Focus group at Saint Petersburg College, Social Sciences Research: Qualitative vs. Quantitative Research.

[937] Chris ChoGlueck, Can science be value-free? The "gap" argument.

[938] 'Abdu'l-Bahá, Some Answered Questions.

contents would be put to use. Today, numerous copies of the works of Muslim scholars which are not to be found in Islamic countries, are available in the libraries of Europe.

Furthermore, the laws and principles current in all European countries are derived to a considerable degree and indeed virtually in their entirety from the works on jurisprudence and the legal decision of Muslim theologians. Were it not for the fear of unduly lengthening the present text, We would cite these borrowings one by one.

… For example Draper, the well-known French authority, a writer whose accuracy, ability and learning are attested by all European scholars, in one of his best- known works, The Intellectual Development of Europe, has written a detailed account in this connection, that is, with reference to the derivation by the peoples of Europe of the fundamentals of civilization and the bases of progress and well-being from Islám. His account is exhaustive, and a translation here would unduly lengthen out the present work and would indeed be irrelevant to Our purpose. If further details are desired the reader may refer to that text.

In essence, the author shows how the totality of Europe's civilization—its laws, principles, institutions, its sciences, philosophies, varied learning, its civilized manners and customs, its literature, art and industry, its organization, its discipline, its behavior, its commendable character traits, and even many of the words current in the French language, derives from the Arabs.[939]

In, *Religion and the Rise and Fall of Islamic Science*, by Eric Chaney, May 2016, we learn:

While the Islamic world stood at the vanguard of scientific and technological production during the medieval period, today it produces a disproportionately small share of world scientific output. This paper contributes to our understanding of this reversal by providing the first large-scale empirical investigation of the evolution of scientific output in the Islamic world over a millennium. The empirical patterns suggest that a surge in the political power of religious leaders in the mid/late eleventh century CE caused a decline in scientific production and the patterns cast doubt on the most prominent alternative explanations for the decline.

I hypothesize that these newly empowered religious leaders worked to limit the study of scientific topics because they believed that the unrestricted study of science led Muslims to both embrace rationalistic interpretations of Islam and to disregard their teachings. Thus, religious leaders altered the institutional framework in order to develop an education system that both discouraged scientific research and rewarded obedience to authority. I provide empirical evidence consistent with this motive and argue that the evidence suggests that religious elites, like any elite, will rent-seek unless otherwise constrained.[940]

I want to express my profound admiration for the coherence with science and religion, of the Jesuits' Christian order, especially the *Liberation Theology* of the Oppressed. When I visited Georgetown University in Washington D.C., I found a Masjid (Mosque) on its campus. Besides a Chapel, there are also spaces for a Jewish Gathering, and Dharma Meditation. The Jesuit Javeriana University in Cali, Colombia, opened its doors to a graduate program proposal in Social Management suggested by three bahá'ís: Dr. Farzam Arbab, Dr. Eloy Anello, and Mr. Gustavo Correa. I graduated in the first class, and because of my thesis (first book), In Synthesis Science is Love, I was appointed academic coordinator of the Social Management graduate

---

[939] 'Abdu'l-Bahá, The Secret of Divine Civilization.

[940] Eric Chaney, Religion and the Rise and Fall of Islamic Science.

program. Later on, Javeriana University published twice my second book, At the Crossroads a New Perspective, In Honor of the Women of the World. When suggesting a new concept of governance in that book, I concluded that women should play a more significant role than men in managing human affairs.[941] Indeed, not a small thing for a Catholic Order. Most likely, there are similar experiences in other religious communities.

In *Paris Talks*, the Master taught: "We may think of science as one wing and religion as the other; a bird needs two wings for flight, one alone would be useless. Any religion that contradicts science or that is opposed to it, is only ignorance – for ignorance is the opposite of knowledge."[942]

The conventional quantitative method refers exclusively to size. The author called it the Formative Method of Science instead of the "quantitative method" to keep its proper perspective. The form, arrangement, and design are much more important than the size to perceive the beauty of justice, harmony, and unity. When human beings organize the arrangement of things, manage the general properties of matter (form, size, mass, temperature, movement, phases of matter, and position in time and space). Love is the Formative Cause's essence; feelings and emotions are also part of it.

I suggest that uniqueness should also be considerd a general property of matter, we may invite a person with a unique talent to be part of a team, or choose a particular piece of art to beautify a room. What about purpose? "God, the True One, beareth Me witness, and every atom in existence is moved to testify that such means as lead to the elevation, the advancement, the education, the protection and the regeneration of the peoples of the earth have been clearly set forth by Us and are revealed in the Holy Books and Tablets by the Pen of Glory."[943] It is evident that identity is also a general property of matter.

In *The Single Science*, all scientific methods are complementary to read the same reality from different perspectives to build a strategy. When studying the Qualitative method, it is critical to realize that the conventional approach ignores spiritual virtues because they are not considered universal notions regarded as intellectually meaningless and moral values as temporary and of human origin. Instead, within *The Single Science*, spiritual values are perceived holistically connected to several other key concepts and all the other methods. The spirit is the Material Cause's essence. The specific properties of matter are also part of the qualitative research method.

---

[941] Participating in more significant numbers, until they reach the majority, at the administrative level of all human affairs, even at the international level, except in the Universal House of Justice as ordained by Bahá'u'lláh.

[942] 'Abdu'l-Bahá, Paris Talks.

[943] Bahá'u'lláh, Tablets of Bahá'u'lláh Revealed after the Kitáb-i-Aqdas, p. 130.

In the table below, you can perceive a proposed way to link values to the research methods suggested in this book.

Exploratory & Propositional methods	Formative method	Qualitative method	Unique Method	Explanatory method	Descriptive & Experimental methods
Compassion Stewardship	Equity & Love	Knowledge	Particular aspiration	Equality	Thoughtful
Wisdom	Solidarity	Purity	Perfections	Responsible	Happiness
Tenderness of Heart & Mercy	Friendship	Honesty	Pardon	Devotion	Radiant
Humbleness Self-control	Beauty & Harmony	Dignity	Fear of God	Obedience	Service
Free-will Patience	Courtesy	Faith	Submission to God's Will	Respect	Fruitful
Empathy Forbearance	Justice	Faithfulness	Self-accountability	Loyalty	Protagonist

Interconnected values to the ability to make sound life choices, such as mercy, compassion, wisdom, patience, perseverance, tenderness of heart, freedom, trust, empathy, and forbearance, are part of the exploratory and propositional method. I suggest that adolescents learn and practice these qualities before reaching the age of maturity at 15 years old.

Writing of religion as a social force, Bahá'u'lláh said:

> "Religion is the greatest of all means for the establishment of order in the world and for the peaceful contentment of all that dwell therein." Referring to the eclipse or corruption of religion, he wrote: "Should the lamp of religion be obscured, chaos and confusion will ensue, and the lights of fairness, of justice, of tranquillity and peace cease to shine." In an enumeration of such consequences the Bahá'í writings point out that the "perversion of human nature, the degradation of human conduct, the corruption and dissolution of human institutions, reveal themselves, under such circumstances, in their worst and most revolting aspects. Human character is debased, confidence is shaken, the nerves of discipline are relaxed, the voice of human conscience is stilled, the sense of decency and shame is obscured, conceptions of duty, of solidarity, of reciprocity and loyalty are distorted, and the very feeling of peacefulness, of joy and of hope is gradually extinguished."

> If, therefore, humanity has come to a point of paralyzing conflict it must look to itself, to its own negligence, to the siren voices to which it has listened, for the source of the misunderstandings and confusion perpetrated in the name of religion. Those who have held blindly and selfishly to their particular orthodoxies, who have imposed on their votaries erroneous and conflicting interpretations of the pronouncements of the Prophets of God, bear heavy responsibility for this confusion—a confusion compounded by the artificial barriers erected between faith and reason, science and religion. For from a fair-minded examination of the actual utterances of the Founders of the great religions, and of the social milieus in which they were obliged to carry out their missions, there is nothing to support the contentions and prejudices deranging the religious communities of mankind and therefore all human affairs.[944]

---

[944] The Universal House of Justice, The Promise of World Peace.

"And among the teachings of Bahá'u'lláh is the equality of women and men. The world of humanity has two wings—one is women and the other men. Not until both wings are equally developed can the bird fly."[945]

'Abdu'l-Bahá in, *The Promulgation of Universal Peace*, taught:

> Scientific knowledge is the highest attainment upon the human plane, for science is the discoverer of realities. It is of two kinds: material and spiritual. Material science is the investigation of natural phenomena; divine science is the discovery and realization of spiritual verities. The world of humanity must acquire both. A bird has two wings; it cannot fly with one. Material and spiritual science are the two wings of human uplift and attainment. Both are necessary—one the natural, the other supernatural; one material, the other divine. By the divine we mean the discovery of the mysteries of God, the comprehension of spiritual realities, the wisdom of God, comprehension of spiritual realities, the wisdom of God, inner significances of the heavenly religions and foundation of the law.[946]

Indeed, the reader of The Single Science deserves our respect if s(he) chooses to keep utilizing the current research methods based on the value-free idea of science, which also asks researchers to divest themselves of any emotions and feelings. In making a choice, consider if the conventional methods will ever think about those faculties that we males have in less abundance than females: *tenderness of heart, empathy, and mercy*. If we want the bird to fly, we need to learn early in life to listen to those emotions when females describe a child's situation, the environment, or the world, to make better decisions when consulting together. Does a human with emotional feelings deserve attention from a scientist who considers that s(he) lacks objectivity? The reader could instead further explore The Single Science proposal, where science and religion are joined and welded in reality, having the potential to unite humanity under the banner of justice, the best beloved of all things.

In Leo's view, the destiny of science's value-free ideal, divested of emotions and feelings, should be determined after considering what The Universal House of Justice in *The Promise of World Peace* message in 1985 recommended:

> Those who care for the future of the human race may well ponder this advice. "If long-cherished *ideals* and time-honoured institutions, if certain social assumptions and religious formulae have ceased to promote the welfare of the generality of mankind, if they no longer minister to the needs of a continually evolving humanity, let them be swept away and relegated to the limbo of obsolescent and forgotten doctrines. Why should these, in a world subject to the immutable law of change and decay, be exempt from the deterioration that must needs overtake every human institution? For legal standards, political and economic theories are solely designed to safeguard the interests of humanity as a whole, and not humanity to be crucified for the preservation of the integrity of any particular law or doctrine."[947]

---

[945] 'Abdu'l-Bahá, Tablets to The Hague. Emphasis added.

[946] 'Abdu'l-Bahá, The Promulgation of Universal Peace, No. 52. Emphasis added.

[947] The Universal House of Justice, The Promise of World Peace. Emphasis added.

11.6 Dealing with multiple crises

Because of the science value-free ideal, stripped of emotions and feelings, and considering religion in conflict with science, the author of The Single Science believes it will be impossible for the conventional methods of analysis to reach a proposal as the one mentioned-bellow.

Studying the following quote is crucial to address the current world's situation. The Universal House of Justice on 25 November 2020 in a message to the Bahá'ís of the World, expressed during the Covid-19 pandemic:

> Your resilience and your unwavering commitment to the well-being of those around you, persistent through all difficulties, have filled us with tremendous hope. But it is no wonder that, in some other quarters, hope has become a depleted resource. There is a mounting realization on the part of the world's people that the *decades ahead are set to bring with them challenges among the most daunting that the human family has ever had to face*.
>
> The current global health crisis is but one such challenge, the ultimate severity of whose cost, both to lives and livelihoods, is yet unknown; your efforts to succour and support one another as well as your sisters and brothers in society at large will certainly need to be sustained, and in places expanded.
>
> It is against this background of *furious storms* lashing humanity that the ark of the Cause is about to embark upon a series of Plans that will carry it into the third century of the Bahá'í Era and significantly strengthen the Bahá'í community's capacity for realizing the society-building powers of the Faith.[948]
>
> In Selections from the Writings of 'Abdu' l-Bahá, we find the world reconstruction: "Although our vision must be world-embracing, the initial stage of *economic reconstruction* is at the local level, beginning with agricultural reform. 'Abdu'l-Bahá said the solution begins with the village, and when the village is *reconstructed*, then will the cities be also".[949]

Turkey and Syria suffered recently the consequences of the 7.5 - 7.8 magnitude earthquake. Does "reconstruction" mean a remodel, rebuilding what they had before, or achieving a holistic reconception of foundations, columns, beams, structures, and materials for building houses and infrastructure to respond to the conditions in those countries, even profoundly affecting university courses teaching engineering and architecture? Would it have an impact on building law regulations?

If *furious storms* lashing humanity for decades ahead, is humanity's immediate destiny, what kind of stewardship can we offer, carefully and responsibly managing the kingdoms entrusted to our care?

The climate change crisis
Nitrification limits of water:

> Nitrogen is a critical limiting element for plant growth and production. It is a major component of chlorophyll, the most important pigment needed for photosynthesis, as well as

---

[948] The Universal House of Justice, To the Bahá'ís of the World. 25 November 2020. Emphasis added.

[949] 'Abdu' l-Bahá, Selections from the Writings of 'Abdu' l-Bahá. Emphasis added.

amino acids, the key building blocks of proteins. It is also found in other important biomolecules, such as ATP and nucleic acids. Even though it is one of the most abundant elements (predominately in the form of nitrogen gas (N2) in the Earth's atmosphere), plants can only utilize reduced forms of this element. Plants acquire these forms of "combined" nitrogen by: 1) the addition of ammonia and/or nitrate fertilizer (from the Haber-Bosch process) or manure to soil, 2) the release of these compounds during organic matter decomposition, 3) the conversion of atmospheric nitrogen into the compounds by natural processes, such as lightning, and 4) biological nitrogen fixation (Vance 2001).[950]

Nitrogen and phosphorus are nutrients that are natural parts of aquatic ecosystems. Nitrogen is also the most abundant element in the air we breathe. Nitrogen and phosphorus support the growth of algae and aquatic plants, which provide food and habitat for fish, shellfish and smaller organisms that live in water.

But when too much nitrogen and phosphorus enter the environment - usually from a wide range of human activities - the air and water can become polluted. Nutrient pollution has impacted many streams, rivers, lakes, bays and coastal waters for the past several decades, resulting in serious environmental and human health issues, and impacting the economy.

Too much nitrogen and phosphorus in the water causes algae to grow faster than ecosystems can handle. Significant increases in algae harm water quality, food resources and habitats, and decrease the oxygen that fish and other aquatic life need to survive. Large growths of algae are called algal blooms and they can severely reduce or eliminate oxygen in the water, leading to illnesses in fish and the death of large numbers of fish. Some algal blooms are harmful to humans because they produce elevated toxins and bacterial growth that can make people sick if they come into contact with polluted water, consume tainted fish or shellfish, or drink contaminated water.[951]

Nitrate is regarded as an undesirable substance in public water. Although it occurs naturally in water, elevated levels of nitrate in groundwater usually result from human activities, such as *over use of chemical fertilizers in agriculture and improper disposal of human and animal wastes*. High nitrate concentration in drinking water may cause serious problems in humans and animals. In order to protect against this effort, the US EPA established the maximum contamination level of nitrate in drinking water at 10 mg NO_3^-.[952]

Carbon sequestration

Soils are made in part of broken-down plant matter. This means they contain a lot of carbon that those plants took in from the atmosphere while they were alive. Especially in colder climates where decomposition is slow, soils can store—or "sequester"—this carbon for a very long time. If not for soil, this carbon would return to the atmosphere as carbon dioxide (CO_2), the main greenhouse gas causing climate change.

But converting natural ecosystems like forests and grasslands to farmland disturbs soil structure, releasing much of that stored carbon and contributing to climate change. Over the past 12,000 years, the growth of farmland has released about 110 billion metric tons of

---

[950] Stephen C. Wagner, Biological Nitrogen Fixation.

[951] EPA - United States Environmental Protection Agency, Nutrient Pollution.

[952] Mountain Empire Community College, Water/Wastewater courses. Lesson 8: Nitrification and Denitrification. Emphasis added.

carbon from the top layer of soil—roughly equivalent to 80 years' worth of present-day U.S. emissions. The question is: Can this trend be reversed at the global scale as part of a strategy to help fight climate change?[953]

Yes, it can be reversed; using cover crops is the answer to serve as shield for the moderation of temperature of the soil and providing food security to the microscopic animals in the soil. "Vegetative cover: A bare soil quickly absorbs heat, becomes hot during the hot season and becomes cold during the cold season. Vegetation acts as a thermal insulator and significantly affects the soil temperature. It does not allow the soil to become either too hot during the dry season or too cold during the rainy season."[954]

"Scientists have estimated that soils—mostly, agricultural ones—could sequester over a billion additional tons of carbon each year. This has led policymakers to increasingly look to soil-based carbon sequestration as a "negative emissions" technology—that is, one that removes CO_2 from the air and stores it somewhere it can't easily escape.[955]

Lowering CO2 and chemical nitrogen and phosphates fertilizers planting diversified crops and cover crops including legumes:

Because, "[t]he more microscopic animals exist in the soil, the better the plants will grow,"[956] is expected an increase in productivity that will increase the farmers' income. The author believes that the biodiversity above the ground of the planted crop and of the vegetative cover crop, have a meaningful correlation with the amount and biodiversity of the microscopic animals in the soil, and of course with the resulting productivity contributing to the farmers' income. Studying the vast experience of small farmers with the three sisters, we discover the symbiotic potential of plants of different *taxonomical families* becoming much more *efficient*. There are some legumes that may serve as cover crops, such as Pinto peanuts (Arachis pintoi), field peas, and clover providing nitrogen to the other sisters.

Biological nitrogen fixation (BNF), discovered by Beijerinck in 1901 (Beijerinck 1901), is carried out by a specialized group of prokaryotes. These organisms utilize the enzyme nitrogenase to catalyze the conversion of atmospheric nitrogen (N2) to ammonia (NH3). Plants can readily assimilate NH3 to produce the aforementioned nitrogenous biomolecules. These prokaryotes include aquatic organisms, such as cyanobacteria, free-living soil bacteria, such as Azotobacter, bacteria that form associative relationships with plants, such as Azospirillum, and most importantly, bacteria, *such as Rhizobium and Bradyrhizobium, that form symbioses with legumes and other plants* (Postgate 1982).[957]

Most of the legumes possess two types of microbial symbionts namely mycorrhizal fungi and nitrogen fixing bacteria thereby establishing triple association, capable of supplying N and P contents to the plants (Silveira & Cardoso, 2004). Dual inoculation with both microorganisms

---

[953] MIT Climate Portal, Soil-Based Carbon Sequestration.

[954] Brownmang onwuka and Brown Mang, Effects of soil temperature on some soil properties and plant growth.

[955] MIT Climate Portal, Soil-Based Carbon Sequestration..

[956] 'Abdu'l-Bahá, Additional Tablets, Extracts and Talks.

[957] Mountain Empire Community College, Water/Wastewater courses. Lesson 8: Nitrification and Denitrification. Emphasis added.

results in a tripartite mutualistic symbiosis and generally increases plant growth to a greater extent than inoculation with only one (Chalk et al., 2006). Inoculation alone or in combination of beneficial microorganisms including AMF, rhizobia, PGPR (Plant-growth promoting rhizobacteria) and PSB (Phosphate Solubilizing Bacteria) have been observed to increase production in green gram and chickpea, nitrogen fixation and nutrient uptake (Jain et al., 2007; Rahman et al., 2008; Jain et al., 2008; Ray & Valsalakumar, 2009; Pir et al., 2009; Akhtar & Siddiqui, 2009; Jain et al., 2009; Singh & Singh, 2010; Thenua et al., 2010). Murat et al. (2011) reported that AMF inoculation, alone or in combination with rhizobial inoculation, increased in yield, root colonization and phosphorus content of the seed and shoot.[958]

The extreme poverty crisis

In the section Increasing farmers' income the author expressed:

What happens at the macroeconomic level if the government of a country like The Ivory Coast decides to have a storehouse in each village, buying products at the farm gate from smallholder farmers to feed everyone who cannot afford enough food to keep him(her) in good health?

For example, the True Price[959] to pay is increasing the farm-gate price of cacao beans by 91%, explained as follows: 54% to increase income and pay their workers higher wages, 11% for replacing child labor conditioned on sending boys and girls to school, 13% to start building or planting barriers for controlling soil erosion, planting cover crops, and reforesting, 8% to eliminate adult and child forced labor, and 5% to contributing to social security.

A similar approach to deciding the true-price for the rest of the food staples should be on the table. How many problems will the Ivory Coast solve?

Is it fair to postpone for a **fifth stage** the other problem I mentioned early on: taking into account the *risks* involved in being a farmer, *risks* associated with the environment, pest control, insecticides, and price fluctuations? Compare farmers' *risks* to those assumed by the agribusiness transforming cereal into oats or flour, the merchant who sells the product, and the profits each receives. If one of the basic principles of economics states that *risk* and profit are directly related: more *risk* exposure should lead to higher gain and vice versa, why do most of the world's farmers receive a meager income? Who manages a greater complexity of the three actors (the farmer, the agribusiness, or the retailer)? Who works physically hardest?

"Perishability of the product: Most farm products are perishable in nature; but the period of their perishability varies from a few hours to a few months. Their perishability makes it almost impossible for producers to fix the reserve price for their farm grown products. The more perishable products require speedy handling and often-special refrigeration, which raises the cost of marketing."[960]

Because perishability of farm-grown products, farmers are frequently exposed to a power asymmetry between the farmer and the buyer as a decisive factor in how the relationships develop and disputes are steered and resolved. Establishing the storehouse in

---

[958] Tabassum Yaseen, et al., Influence of Arbuscular Mycorrhizal Fungi, Rhizobium Inoculation and Rock Phosphate on Growth and Quality of Lentil.

[959] https://trueprice.org/wp-content/uploads/2022/07/The-True-Price-of-Cocoa.-Progress-Tonys-Chocolonely-2018.pdf

[960] Insights IAS, Role of Agriculture in Indian Economy..

every village will solve the power asymmetry ensuring the payment to the farmer at the previously agreed true-price and food bulk sale because of its capacity to process the harvest to lower its perishability. It implies previously planning the amounts to purchase.

True Price found:

"84% of the total external costs of cultivation are social costs, 54% are due to underpayment of hired and family workers. The other largest external cost drivers are land use, child labor, forced labor, and lack of social security."

Income (54%): more than half of the external costs during cocoa cultivation result from the underpayment of hired workers and the underearning of family workers.

Underpayment of hired workers and underearning of smallholder farmers constitute by far the largest external cost in cocoa cultivation in Ivory Coast. Hired workers receive an average total wage of €1.6/day, which is only about 20% of the living wage. A farmer household earns on average €3.5/day from the activities on its cocoa farm, which results in a yearly income of around 40% of the living income. On an annual basis, the wage of workers is €477, while the legal minimum wage is €659. The annual living wage for an Ivorian worker, as calculated by True Price, is €2,869. This size of poverty is not only problematic for the livelihoods of workers and their families, but as well triggers other social issues, such as child and forced labour (Potts, et al., 2014).

It is hard for farmers to pay their workers higher wages, as they themselves do not earn a living income. Raising legal minimum wages, adjusting tax and subsidy structures, increasing farm productivity (e.g. by adopting Good Agricultural Practices) and raising minimum cocoa bean prices are a few possible routes in decreasing the external cost of income.[961]

In the Bahá'í Writings we find the following ordinance: "In every village there must be established a general storehouse which will have a number of revenues."[962]

"The storehouse has seven revenues: Tithes, taxes on animals, property without an heir, all lost objects found whose owners cannot be traced, one third of all treasure-trove, one third of the produce of all mines, and voluntary contributions."[963]

"The strain on today's farm economy is no accident; it's the result of policies designed to enrich corporations at the expense of farmers and ranchers. If the American family farmer is to survive, farm policy needs a massive shift in direction — one that delivers fair prices to farmers that allow them to make a living."[964] Proving the storehouse for buying the harvest at the farm gate at a true-price: will increase substantially the contribution of the agricultural sector to the GDP (Gross Domestic Product), and gradually increase the collection of taxes in the rural areas, which is usually very low.

The storehouse will ensure that every farmer gets the previously agreed true-price. The storehouse, if farmers decide, could also deduct and pay to the institution in charge the farmers' and their hired workers social security benefits. This information is necessary to determine the taxes to pay.

---

[961] IDH and True Price, The True Price of Cocoa from Ivory Coast.

[962] 'Abdu'l-Bahá, The Promulgation of Universal Peace.

[963] 'Abdu'l Bahá, Additional Tablets, Extracts and Talks. Emphasis Added.

[964] Farm Aid, Understanding the Economic Crisis Family Farms are Facing.

We will need to consult to reach agreements on how to proceed, if the above-mentioned ideas are worth considering.

IDH and True Price say about Child labor: "child workers remain common in the Ivorian cocoa industry, often performing hazardous tasks and missing out on education."[965]

The village's Local Spiritual Assembly, which will be called Local House of Justice in the future, will verify if the farmer sends her (his) children to school. Bahá'u'lláh revealed: "Everyone, whether man or woman, should hand over to a trusted person a portion of what he or she earneth through trade, agriculture or other occupation, for the training and education of children, to be spent for this purpose with the knowledge of the Trustees of the House of Justice."[966]

In the Bahá'í Writings we find the following ordinance: "In every village there must be established a general storehouse which will have a number of revenues."[967]

" A certain amount must be provided … for the village's *system of education*. "[968] Strategically, this allocation of resources will be of maximum priority to reaching the goal of empowering the youth living in rural areas with scientific methods encouraging the discovery of the potential of biodiversity to replace the construction industry as the employment's engine of the economy, addressing public health issues, and any problem within their village.

The goal is to create a scientific community with a presence in universities, the public sector, corporations, farmers, agroindustry, the health sector, and the youth focusing on specifying the challenges that have to be overcome and determining the desired lines of inquiry.

The environmental and healthcare system crises

"With the prolonged nature of the COVID-19 pandemic, across the globe, healthcare workers involved in COVID-19 care have hit their physical and mental limits. A healthcare system's collapse due to a pandemic caused by a novel infectious disease".[969]

In the Introduction of The Single Science, the author has shown the potential increase in productivity based on microscopic animals in the soil, conditioning its reachability on avoiding herbicides, insecticides, and fungicides. Water and air pollution will also be part of monitoring the public health resources allocation.

IDH and True Price say: Land use (13%): deforestation and other land degradation, caused by the establishment of new farm land and plantations, is the second largest externality."[970]

The author believes that the following proposal will generate millions of jobs: Farmers receiving a true-price for their harvest at the farm gate, must commit to keeping: their soils protected with a vegetative cover crop, building/planting barriers for controlling soil erosion,

---

[965] IDH and True Price, The True Price of Cocoa from Ivory Coast.

[966] Bahá'u'lláh, Tablets of Bahá'u'lláh.

[967] 'Abdu'l-Bahá, The Promulgation of Universal Peace.

[968] 'Abdu'l Bahá, Additional Tablets, Extracts and Talks. Emphasis Added.

[969] Hwang, S., Kwon, K.T., Lee, S.H. et al. Correlates of burnout among healthcare workers during the COVID-19 pandemic in South Korea.

[970] IDH and True Price, The True Price of Cocoa from Ivory Coast.

and supporting reforestation and orchard initiatives as part of the planned biodiversity above ground.

In the Bahá'í Writings we find the following ordinance: "In every village there must be established a general storehouse which will have a number of revenues."[971]

"A certain amount must be put aside from the general storehouse for the orphans of the village and a certain sum for the incapacitated. A certain amount must be provided from this storehouse for those who are needy and incapable of earning a livelihood… ."[972] And, a certain amount must be set aside for the administration of public health. If anything is left in the storehouse, that must be transferred to the general treasury of the nation for national expenditures.[973] "

"Then there must be considered such emergencies as follows: A certain farmer whose expenses run up to ten thousand dollars and whose income is only five thousand will receive necessary expenses from the storehouse. Five thousand dollars will be allotted to him so he will not be in need.

Then the orphans will be looked after, all of whose expenses will be taken care of. The cripples in the village—all their expenses will be looked after. The poor in the village—their necessary expenses will be defrayed. And other members who for valid reasons are incapacitated—the blind, the old, the deaf—their comfort must be looked after. In the village no one will remain in need or in want. All will live in the utmost comfort and welfare. Yet no schism will assail the general order of the body politic."[974]

If the storehouses in each village of the country buy the export commodities to be sold at a fair price, the government will have and increase amount of resources to feed the hungry and the orphans, the incapacitated and incapable of earning a livelihood. With this measure , the government's balance of payments will become positive. The author believes that the public sector has a right to keep some of the profits of this venture. Mariana Mazzucato in, *The Entrepreneurial State* says:

In finance, it is commonly accepted that there is a relationship between risk and return. However, in the innovation game, this has not been the case. Risk-taking has been a collective endeavour while the returns have been much less collectively distributed. Often, the only return that the state gets for its risky investments are the indirect benefits of higher tax receipts that result from the growth that is generated by those investments. Is that enough?

There is indeed lots of talk of partnership between the government and private sector, yet while the efforts are collective, the returns remain private.[975]

---

[971] 'Abdu'l-Bahá, The Promulgation of Universal Peace.

[972] 'Abdu'l Bahá, Additional Tablets, Extracts and Talks. Emphasis Added.

[973] 'Abdu'l Bahá, Additional Tablets, Extracts and Talks. Emphasis Added.

[974] 'Abdu'l-Bahá, The Promulgation of Universal Peace.

[975] Mariana Mazzucato, The Entrepreneurial State.

The food security crisis

The recent lessons of the covid pandemic and the war in Ukraine for their disruptive effects have shown how vulnerable food security is for any nation. Food prices have skyrocketed because of the pandemic and the dramatic impact on fertilizers and cereals production of the war started by Russia in Ukraine. "Akrur Barua, Sizzling food prices are leading to global heartburn. Food prices have surged to record highs since the start of the war. For many around the world, even regular meals may soon become too expensive." [976]

The world's smallholder farmers produce around a third of the world's food, according to detailed new research by the Food and Agriculture Organization of the United Nations (FAO).

Five of every six farms in the world consist of less than two hectares, operate only around 12 percent of all agricultural land, and produce roughly 35 percent of the world's food, according to a study published in World Development. [977]

It Is crucial to understand the difference between small farmers and smallholder farmers. The latter are also small farmers but possess less than two hectares (less than five acres). The following is also from the Food and Agricultural Organization of the United Nations:

> *Small farmers produce much of the developing world's food. Yet, they are generally much poorer than the rest of the population in these countries and are less food secure than even the urban poor*. Furthermore, although rapid urbanization is taking place in many developing countries, farming populations in 2030 will not be much smaller than they are today. For the foreseeable future, therefore, *dealing with poverty and hunger in much of the world means confronting the problems that small farmers and their families face in their daily struggle for survival*.
>
> *Investment priorities and policies* must take into account the immense diversity of opportunities and *problems facing small farmers*. The resources on which they draw, their *choice* of activities, indeed the entire structure of their lives, are linked inseparably to the biological, physical, economic, and cultural environment in which they find themselves and over which they only have *limited control*. While every farmer is unique, those who share similar conditions also often share *common problems and priorities* that transcend *administrative* or political borders. [978]

One of the Sustainable Development Goals of the United Nations is to end world hunger by 2030. However, first the pandemic and then the conflict in Ukraine may likely end up reversing years of progress in reducing global hunger. The upheaval in global food markets since 2020 also serves as a warning that the world must be better prepared for global disruptions in the future, such as climate change. Recent research reveals that over half of all shocks to crop production systems globally result from extreme weather events—with droughts causing the highest damage. Changing precipitation patterns and increasing temperatures have also put the food system under pressure. According to FAO estimates, global mean yields of maize have declined by 3.8% and wheat by 2.5% between 1980 to 2008 due to climate change. Failure to learn lessons from the last two years with regard to

---

[976] Akrur Barua, Sizzling food prices are leading to global heartburn.

[977] FAO, Small family farmers produce a third of the world's food.

[978] FAO, Dixon, Gulliver and Gibbon, Farming Systems and Poverty: Improving Farmers' Livelihoods In A Changing World. Emphasis added.

preparing for other challenges may therefore mean that the world gets hungrier and even more unequal. That will not only be an economic setback, but also a global tragedy.[979]

Is the answer to it trying to keep food prices low by oppressing farmers further? In a crisis of this magnitude lasting decades, giant banks and corporations may go bankrupt. Will invest in those who are -too big to fail- better? Would this proposal contribute to lowering the migratory crisis? How many hectares will be destine to feed the world instead of planting crops for increasing masses of humans addicted to tobacco, alcohol, cocaine, hashish or opioids? Is there another way to address these crises simultaneously and not fragmented? Would it contribute to peace in the world?

Refugees on a big boat in the middle of the sea that require help.[980]

Life Jackets and boats left on Greek beach by refugees.[981]

Children war refugees.[982]

---

[979] Akrur Barua, Sizzling food prices are leading to global heartburn.

[980] Photo 137117243 / Migrants Boat © Vampy1 | Dreamstime.com

[981] Photo 63060167 / Migrants Boat © Anjo Kan | Dreamstime.com

[982] Photo 60156198 / Migrants Boat © Dimaberkut | Dreamstime.com

12 The Primal Point of Nine Concentric Circles

Bahá'u'lláh in, The Book of Certitude, said: "Knowledge is one point, which the foolish have multiplied."[983]

How often hath it been observed that certain human minds, far from being a source of guidance, have become as fetters upon the feet of the wayfarers and prevented them from treading the straight Path! The lesser intellect being thus circumscribed, one must search after Him Who is the ultimate Source of knowledge and strive to recognize Him. And should one come to acknowledge that Source round Whom every mind doth revolve, then whatsoever He should ordain is the expression of the dictates of a consummate wisdom.[984]

"The prophets of God have founded the laws of divine civilization. They have been the root and fundamental source of all knowledge".[985]

For, just as in the realm of the spirit, the reality of the Báb has been hailed by the Author of the Bahá'í Revelation as "The Point round Whom the realities of the Prophets and Messengers revolve," so, on this visible plane, His sacred remains constitute the heart and center of what may be regarded as nine concentric circles, paralleling thereby, and adding further emphasis to the central position accorded by the Founder of our Faith to One "from Whom God hath caused to proceed the knowledge of all that was and shall be," "the Primal Point from which have been generated all created things."[986]

ცხცp

This search started in 1989, after a challenging conversation with three native representatives of Indigenous tribes from South America. After presenting to them the almost 80 textbooks for secondary education developed during the first 15 years of FUNDAEC, they told me that they perceive Nature differently than FUNDAEC's approach, which had opened the door in 1974 to examine the reality of peasants in Colombia and Latin America.

Because I have not yet fully addressed the promise I made to them, I have started a new search for: Indigenous Wisdom for an Environmental Curricular Proposal. I want to start a consultation, mostly a humble learning process, based on the following quote:

Particular attention, I feel, should, at this juncture, be directed to the various Indian tribes, the aboriginal inhabitants of the Latin republics, whom the Author of the Tablets of the Divine Plan has compared to the "ancient inhabitants of the Arabian Peninsula." "Attach great importance," is His admonition to the entire body of the believers in the United States and the Dominion of Canada, "to the indigenous population of America. For these souls may be likened unto the ancient inhabitants of the Arabian Peninsula, who, prior to the Mission of Muḥammad, were like unto savages. When the light of Muḥammad shone forth in their midst, however, they became so radiant as to illumine the world.

Likewise, these Indians, should they be educated and guided, there can be no doubt that they will become so illumined as to enlighten the whole world." The initial contact already

---

[983] Bahá'u'lláh, The Kitáb-i-Íqán The Book of Certitude.

[984] Bahá'u'lláh, The Tabernacle of Unity.

[985] 'Abdu'l-Bahá, The Promulgation of Universal Peace, No. 52. Emphasis added.

[986] Shoghi Effendi, Citadel of Faith. Nine Concentric Circles.

established, in the concluding years of the first Bahá'í century, in obedience to 'Abdu'l-Bahá's Mandate, with the Cherokee and Oneida Indians in North Carolina and Wisconsin, with the Patagonian, the Mexican and the Inca Indians, and the Mayans in Argentina, Mexico, Peru and Yucatan, respectively, should, as the Latin American Bahá'í communities gain in stature and strength, be consolidated and extended.[987]

---

[987] Shoghi Effendi, Citadel of Faith. Importance of the American Indians.

13 Works Cited

'Abdu'l-Bahá. *'Abdu'l-Bahá in London*. Web. October 2016: Ocean Research Library.

_____. 'Abdu'l-Bahá 'Abbás in Pulpit. Centenary News. Web. September 2020 <https://centenary.bahai.us/news/abdul-baha-abbas-pulpit>.

_____. *Additional Tablets, Extracts and Talks*, Extract from a Tablet of 'Abdu'l-Bahá. Bahá'í Reference Library, Web. March 2021 <https://www.bahai.org/library/authoritative-texts/abdul-baha/additional-tablets-extracts-talks/442883358/1#739873643>.

_____. *Commentary in Ottoman Turkish on the Qur'ánic Sura 95.* Web. January 2021 <https://ocean-server production.dev2.us/?key=oceanviewer&bookId=798cf8de8419c6a1fcb13727888a7557&paraId=para_6>.

_____. *Divine Philosophy.* Web. June 2012 <http://bahairesearch.com/english/Baha%27i/Authoritative_Baha%27i/Abdu%27l-Baha/Divine_Philosophy.aspx>.

_____. *Foundations of World Unity*. Bahá'í Reference Library, Web. October 2015 <http://reference.bahai.org/en/t/ab/>.

_____. *Paris Talks: Addresses Given by 'Abdu'l-Bahá in Paris in 1911* in Writings and Utterances of 'Abdu'l-Bahá. New Delhi, India: Bahá'í Publishing Trust, 2001. 695-795. Print.

_____. *Selections from the Writings of 'Abdu'l-Bahá*. Great Britain: W & J Mackay Limited, Chatham, 1978. Print. Translated by a Committee at the Bahá'í World Centre and by Marzieh Gail. Haifa: Bahá'í World Centre, 1978.

_____. *Some Answered Questions*. Bahá'í Reference Library, Web. May 2021 <https://www.bahai.org/library/authoritative-texts/abdul-baha/some-answered-questions/13#358668317>.

_____. Star of the West. <http://www.bahairesearch.com/english/Baha'i/Authoritative_Baha'i/Abdu'l-Baha/SOW_-_Star_of_the_West/Star%20of%20the%20West%20-%20%201.aspx>.

_____. *Tablet to Dr. Forel.* Bahá'í Reference Library, Web. October 2015 <http://www.bahai.org/library/authoritative-texts/downloads>.

_____. *Tablets of 'Abdu'l-Bahá 'Abbás vol. 1-3.* Bahá'í Publishing Committee, 1909.

_____. *Tablets of 'Abdu'l-Bahá.* Bahá'í Reference Library, Web. June 2012 <http://reference.bahai.org/en/t/ab/TAB/tab-64.html>.

_____. *Tablets to The Hague*. Bahá'í Reference Library, Web. October 2021 <https://www.bahai.org/library/authoritative-texts/abdul-baha/tablets-hague-abdul-baha/2#234667906>.

_____. *The Promulgation of Universal Peace* in Writings and Utterances of 'Abdu'l-Bahá. New Delhi, India: Bahá'í Publishing Trust, 2001. 797-1214. Print.

_____. *The Secret of Divine Civilization.* Wilmette, Ill.: Bahá'í Publishing Trust, 1990. Print.

_____ . *A Traveller's Narrative.* Web. March 2014: Ocean Research Library.

'Abdu'l-Bahá, Bahá'u'lláh, Shoghi Effendi. *Bahai Education.* Web. December 2020 Ocean: <https://ocean-server-production.dev2.us/?key=ocean->.

'Abdu'l-Bahá, Bahá'u'lláh, Shoghi Effendi. *Excellence in All Things.* On behalf of Shoghi Effendi, 5 April 1942 to an individual believer. Web. May 2021 <https://ocean-server-production.dev2.us/?key=ocean-viewer&bookId=927157c9db0ea1b77747e6f32bc9af24&paraId=para_143>.

'Abdu'l-Bahá, Bahá'u'lláh, Shoghi Effendi. *Living the Life.* From a letter 12 May 1925 written on behalf of Shoghi Effendi to an individual believer. Web. September 2020 <https://ocean-server- production.dev2.us/?key=oceanviewer&bookId=25abac9244c93d77588b3ece09af34bc&para>

Almeida, Alexandre, Alex L. Mitchell, Miguel Boland, Samuel C. Forster, Gregory B. Gloor, Aleksandra Tarkowska, Trevor D. Lawley & Robert D. Finn. *A new genomic blueprint of the human gut microbiota.* Published in Nature 11 February 2019. Web. January 2022 <https://www.nature.com/articles/s41586-019-0965-1>.

Arbab, Farzam. Gustavo Correa and Francia de Valcarcel. *Fundaec: Its Principles and its Activities.* Cali, Colombia, 1988. Published by CELATER. Web. June 2018 <http://www.fundaec.org/en/institution/celater_doc.htm>.

Aristotle. *A Treatise on Government.* Trans. by William Ellis, A.M. The Project Gutenberg EBook of Politics, by Aristotle, 2009. Web. June 2012 <http://www.gutenberg.org/files/6762/6762-h/6762-h.htm#2HCH0084>.

_____ . *AzQuotes* . Web. October 2020, AzQuotes: <https://www.azquotes.com/author/524-Aristotle

_____ . *The Metaphysics.* Web. Nov 2015. <http://classics.mit.edu/Aristotle/metaphysics.mb.txt>.

_____ . *Nicomachean Ethics.* Wikipedia. Web. October 2020 <https://en.wikipedia.org/wiki/Nicomachean_Ethics#cite_note-124>.

Babbie, Earl R. *The Practice of Social Research.* 12th ed. Belmont, CA: Wadsworth Cengage, 2010; Muijs, Daniel. Doing Quantitative Research in Education with SPSS. 2nd edition. London: SAGE Publications, 2010. Web. August 2022 <https://libguides.usc.edu/writingguide/quantitative>.

Badee, Hooshmand. (2015). *Bahá'í Teachings on Economics and Their Implications for the Bahá'í Community and the Wider Society.* Ph.D. dissertation for the University of Leeds, Faculty of Education and Theology. Web. September 2020 <https://bahai library.com/badee_economics_society>.

Bahá'u'lláh. *Days of Remembrance.* Web. March 5, 2021 <https://ocean-server-production.dev2.us/?key=ocean-viewer&bookId=7a93329d24bd94f6e43c6c3485f2b9d6&paraId=para_439>.

_____ . *Epistle to the Son of the Wolf.* Trans. Shoghi Effendi. Wilmette, Ill.: Bahá'í Publishing Trust, 1962. Print.

_____ . *Gems of Divine Mysteries.* Australia: Griffin Press, 2002. Print.

_____. *Gleanings from the Writings of Bahá'u'lláh*. Trans. Shoghi Effendi. Wilmette, Ill.: Bahá'í Publishing Trust, 1976. Print.

_____. *Prayers and Meditations by Bahá'u'lláh*. Trans. Shoghi Effendi. Web. January 2022 <https://www.bahai.org/library/authoritative- texts/search#q=colleges%20universities>.

_____. *Tablet to a Physician* (Lawh-i-Tibb) Web. March 2021 <https://bahai-library.com/uhj_lawh_tibb>.

_____. *Tablets of Bahá'u'lláh*. Great Britain: W & J Mackay Limited, Chatham, 1978. Print.

_____. *Tablets of Bahá'u'lláh Revealed After the Kitáb-i-Aqdas.* Bahá'í Reference Library, Web. October 2015 <http://reference.bahai.org>.

_____. *The Call of the Divine Beloved.* Bahá'í, 2019. Bahá'í Reference Library, Web. May 2021 <https://www.bahai.org/library/authoritative-texts/bahaullah/call-divine-beloved/1#959114648>.

_____. **The Hidden Words of Bahá'u'lláh.** Web. July 2021 <https://www.bahai.org/library/authoritative-texts/search#q=The%20Hidden%20Words>.

_____. *The Kitáb-i-Aqdas*: The Most Holy Book. Ann Arbor, Michigan: Edward Brothers, 1992. Print.

_____. *The Kitáb-i-Íqán The Book of Certitude.* 1989. (Shogui Effendi, Trans.) Bahá'í Reference Library. Web. <https://www.bahai.org/library/authoritative-texts/search#q=destiny%20all>.

_____. *The Proclamation of Bahá'u'lláh.* Web. October 2016: Ocean Research Library.

_____. *The Call of the Divine Beloved. The Seven Valleys*. Web. May 2021 <https://www.bahai.org/library/authoritative-texts/search#q=Just%20as%20the%20phenomenal%20sun%20shines>.

_____. *The Summons of the Lord of Hosts.* Haifa, Israel: Bahá'í World Center, 2002. Bahá'í Reference Library, Web. October 2020 <https://reference.bahai.org/en/t/b/SLH/slh-9.html>..utf8?query=Summons&action=highlight#gr192>.

_____. *The Tabernacle of Unity*. Bahá'í Reference Library, Web. May 2021 <https://www.bahai.org/library/authoritative-texts/bahaullah/tabernacle-unity/3#734353285>.

Bailey, Regina. Understanding Plant Tropisms. Web. April 2022 <https://www.thoughtco.com/plant-tropisms-4159843>.

Bakalar, Nicholas. *Earth May Be Home to a Trillion Species of Microbes.* The New York Times, May 23, 2016. Web, March 2021 <https://www.nytimes.com/2016/05/24/science/one-trillion-microbes-on- earth.html >.

Baker, Richard St Barbe. *Igi Qko: The Tree Farms at Sapobai, Nigeria circa 1927*; quoted in The Spirit of Agriculture edited by Paul Hanley. Oxford, England: George Ronald Publisher 2005, Print.

Baldwin, Ian T. *Plant volatiles*. Current Biology, Volume 20, Issue 9, 11 May 2010. Web, March 2021 <https://www.sciencedirect.com/science/article/pii/S0960982210002411>.

Bancroft, Dani. *Microbiome Myth Debunked: Counting Bacteria in Our Gut = A New Ratio.* Web. January 2022 <https://www.labiotech.eu/trends-news/microbiome-myth- debunked-

counting-bacteria-in-our-gut/>.

Barrett, Christopher B., Bachke, Maren E., Bellemare, Marc F., Michelson, Hope C., Narayanan, Sudha and Walker, Thomas F. *Smallholder Participation in Agricultural Value Chains: Comparative Evidence from Three Continents*. December 2010 revision. Web. January 2018 <barrett.dyson.cornell.edu/.../SmallholderMarketParticipationDec2010Submission.pdf>.

Barry, Ellen. *After Farmers Commit Suicide, Debts Fall on Families in India* in New York Times 22 February 2014. Web. March 2014 <http://www.nytimes.com/2014/02/23/world/asia/after-farmers-commit- suicide-debts-fall-on-families-in-india.html>.

Barua, Akrur. Sizzling food prices are leading to global heartburn. Web. February 2023 <https://www2.deloitte.com/uk/en/insights/economy/global-food-prices-inflation.html>.

Bertocchini, Federica. *A caterpillar that eats and digests plastic in record time*. Web. March 2021 <https://www.dw.com/en/a-caterpillar-that-eats-and-digests-plastic-in- record-time/a-38567055>.

Blavatsky, Helena Petrovna. *Law of Cycles. Biology Online*, Web. July 2017 <http://www.biology-online.org/>. Theosophy Wiki. Web. June 2021 <https://theosophy.wiki/en/Law_of_Cycles#:~:text=Law%20of%20Cycles.%20The%20Law%20of%20Cycles%20postulates,some%20event%20either%20cosmic%2C%20mundane%2C%20physical%20or%20metaphysical.>.

Bonanno, Alessandro and Lawrence Busch. *The international political economy of agriculture and food: An introduction.* 24 April 2015. Web. June 2021 <https://doi.org/10.4337/9781782548263.00006>.

Borghini, Andrea. *Plato and Aristotle on Women*: Selected Quotes. Web. December 2022 <https://www.thoughtco.com/plato-aristotle-on-women-selected-quotes- 2670553 >.

Born, Max, Nobel Prize-winning physicist, qtd. in Gerald Holton's *Thematic Origins of Scientific Thought*. Web. October 2015 <https://books.google.com/books?id=vAv5YmGWosoC&pg=PA7&dq=Born,+Max,+Thematic+Origins+of+Scientific+Thought+scouts+erect&hl=en&sa=X&ved=0ahUKEwjS3oihwqLRAhXGZCYKHcFdBcAQ6AEIHDAA#v=onepage&q=Born%2C%20Max%2C%20Thematic%20Origins%20of%20Scientific%20Thought%20scouts%20erect &f=false>.

Bradley, Ian M., *Iron Oxide Amended Biosand Filters for Virus Removal*. Web. June 2020 <https://www.ideals.illinois.edu/bitstream/handle/2142/26107/Bradly_Ian.pdf>.

Brown, Matthew J. *The source and status of values for socially responsible science*. Web. December 2022 < https://www.jstor.org/stable/41932652>.

Britannica. *Pheromone*. Web. March 2021 <https://www.britannica.com/science/pheromone>.

Brahmkshatriya, Priyanka P. and Pathik S. Brahmkshatriya, *Terpenes: Chemistry, Biological Role, and Therapeutic Applications*. 2013. In: Ramawat K., Mérillon JM. (eds) Natural Products. Springer, Berlin, Heidelberg. Web. March 2021 <https://link.springer.com/referenceworkentry/10.1007%2F978-3-642-22144- 6_120>.

Brockett, Charles D., Land Power and Poverty. Agrarian Transformation and Political Conflict in Central America. Westview Press, 1998, Print.

Brown, Nino. *Plant Domestication*. Web. March 1, 2021 <http://plantbreeding.coe.uga.edu/index.php?title=Plant_Domestication>.

Bukalo, Olena and Alexander Dityatev. *Synaptic Cell Adhesion Molecules*. Web. June 2020 <https://pubmed.ncbi.nlm.nih.gov/22351053/>.

Burkart, Karl. *Boy discovers microbe that eats plastic*. Web. March 2021 <http://www.mnn.com/green-tech/research-innovations/blogs/boy-discovers- microbe-that-eats-plastic>.

Business Dictionary. Web. July 2017 <http://www.businessdictionary.com/>.

Canning, Anna. *Low Coffee Prices – A Dire Call to Action. 45 Year Historical Chart* Web. April 2023 Web. February 2023 <a href='https://www.macrotrends.net/2535/coffee-prices-historical-chart-data'>Coffee Prices - 45 Year Historical Chart</a>.

Cassini A. Simulation models and probabilities: a Bayesian defense of the value-free ideal. SIMULATION. 2022;98(2):113-125. doi:10.1177/00375497211028815

Cavicchioli, Ricardo, William J. Ripple, [...]Nicole S. Webster. *Scientists' warning to humanity: microorganisms and climate change.* Nature Reviews Microbiology volume 17, pages569–586 (2019). November 2021 <https://www.nature.com/articles/s41579-019-0222- 5?fbclid=IwAR3245HEXWi0k_SAg26sBGfMZDnDLL_d5MCN1SNMFSyHRv2v-f8cSzz6UKo>.

Center for Sustainable Systems, University of Michigan. *U.S. Food System*. Web. March 2014 <http://css.snre.umich.edu/css_doc/CSS01-06.pdf>.

CDC - Centers for Disease Control and Prevention. Access to Clean Water, Sanitation, and Hygiene. Web. February 2022 <https://www.cdc.gov/healthywater/global/wash_statistics.html>.

Chaney, Eric. Religion and the Rise and Fall of Islamic Science. Web. August 2022 <https://scholar.harvard.edu/files/chaney/files/paper.pdf>.

Chase, Thornton. The Bahai Revelation. Published By The Bahai Publishing Society. Chicago, ILL., U.S.A. 1909.Web. January 2022 <https://view.officeapps.live.com/op/view.aspx?src=https%3A%2F%2Fbahai-library.com%2Fdocs%2Fc%2Fchase_bahai_revelation.doc&wdOrigin=BROWSELI NK>.

Chemistry of Iron, edited by J. Silver. Web. February 2019 <https://books.google.com/books?id=Pj3rCAAAQBAJ&pg=PA2&lpg=PA2&dq=number+of+different+molecules+with+iron?&source=bl&ots=itZyk-_phD&sig=ACfU3U2GcbDf83qE6QRLxcUe3iwUyOeeHg&hl=en&sa=X&ved=2ahUKEwjop8bE76XgAhVF4qwKHdfZCkkQ6AEwIHoECBcQAQ#v=onepage&q=number%20of%20different%20molecules%20with%20iron%3F&f=false>.

Chen, Martha Alter. *Rethinking the Informal Economy: Linkages with the Formal Economy and the Formal Regulatory Environment.* DESA Working Paper No. 46 July 2007.

ChoGlueck, Chris. *Can science be value-free?* The "gap" argument. Posted on May 7, 2019. Web. December 2022 <https://blogs.iu.edu/sciu/2019/05/07/value-free-sci-part- 1/>.

Collier, Paul, and Stefan Dercon. *African Agriculture in 50 Years: Smallholders in a Rapidly Changing World?*, Elsevier . Web. December 2013.
<http://www.sciencedirect.com/science/article/pii/S0305750X13002131>.

Compilation for the 2018 Counsellors' Conference. January 2023
<https://www.bahai.org/library/authoritative-texts/compilations/give-me-thy- grace-serve-thy-loved-ones/6#277294374>.

Conant, James Bryant. qtd. in *Science and Common Sense.* Web. October 2015 < http://www.gly.uga.edu/railsback/1122sciencedefns.html >.

CUESA. *How far do your fruit and vegetables travel?* Web. March 2021
<https://cuesa.org/learn/how-far-does-your-food-travel-get-your-plate>.

d'Estries, Michael. *Researchers Turn to 'Sentinel Trees' to Warn of Destructive Pests.* Updated February 19, 2021. Web. April 2021 <https://www.treehugger.com/global- sentinel-tree-network-aims-warn-destructive-pests-4865301>.

DANE, *estimaciones de población 1985 - 2005 y proyecciones de población 2005 - 2020 total municipal por área.* Web. May 2021.
<https://www.google.com/url?sa=t&rct=j&q=&esrc=s&source=web&cd=&cad=rja&uact=8&ved=2ahUKEwjRoLaF6LrwAhWEVc0KHTc4D_cQFjAAegQIBBAD&url=https%3A%2F%2Fwww.dane.gov.co%2Findex.php%2Festadisticas-por-tema%2Fdemografia-y-poblacion%2Fproyecciones-de-poblacion&usg=AOvVaw3oskzOzN4LUzExhlONSI2V>.

Davison, Sophie and Aleksandar Janca. *Personality disorder and criminal behaviour: what is the nature of the relationship?* Web. April 2021
<https://pubmed.ncbi.nlm.nih.gov/22156936/>.

Davidson, Thomas. *Aristotle and Ancient Educational Ideals.* Web, June 2017
<http://www.gutenberg.org/files/40552/40552-h/40552-h.htm>.

Dematteis, Philip. *Was Plato a Libertarian?* The Magazine of Free Minds and Free Markets.
February, 1979.

Denzin, Norman. K. and Yvonna S. Lincoln. *Introduction: The Discipline and Practice of Qualitative Research." In The Sage Handbook of Qualitative Research.* 3rd edition. (Thousand Oaks, CA: Sage, 2005), p. 10. Web. August 2022
<https://libguides.usc.edu/writingguide/qualitative>.

Dewar, Gwen, Ph.D. *Boosting iron absorption: A guide for the science-minded* . Web. January 2020 <https://www.parentingscience.com/iron-absorption.html>.

Diamond, Milton. *Porn: Good for us?* Published in The Scientist on Mar 1, 2010. Web December 2022 <https://www.the-scientist.com/uncategorized/porn-good-for- us-43469 >.

Dixon, John, Aidan Gulliver and David Gibbon. *Farming Systems and Poverty: improving farmers' livelihoods in a changing world.* Web. March 2014
<http://www.fao.org/farmingsystems/>.

Douglas, Heather E. "INTRODUCTION: Science Wars and Policy Wars." Science, Policy, and the Value-Free Ideal, University of Pittsburgh Press, 2009, pp. 1–22. JSTOR. Web. December 2022 <https://doi.org/10.2307/j.ctt6wrc78.5>.

Dunlop, Greg Reporting by the BBC, *Australian scientists use "soybean oil" to create graphene*. Web. March 2021 <https://www.bbc.com/news/world-australia- 38804802>.

Duque, Leonardo, *En la Encrucijada: Una Perspectiva Nueva: En Honor a las Mujeres del Mundo*. Cali: Editorial Cargraphics, 1999. Print.

Ejrnæs, Rasmus. (2018, September). *Uniquity: A general metric for biotic uniqueness of sites*. Biological Conservation , 225, 98-105. Web. February 19, 2021 <https://www.sciencedirect.com/science/article/abs/pii/S0006320717318293#> Eleven Coffees, *25 Top Coffee-Producing Countries in 2022*. Web. January 2023 <https://elevencoffees.com/top-coffee-producing-countries/>.

Elitzak, H. *Food Cost Review, 1950-97*. USDA, Agricultural Economic Report 780. 1999. qtd. in Center for Sustainable Systems. Web. March 2014 <http://css.snre.umich.edu/css_doc/CSS01-06.pdf>.

Encyclopedia Britannica. Web.July 2017 <https://www.britannica.com/topic/European-Community-European-economic-association>.

EPA - United States Environmental Protection Agency, Nutrient Pollution. Web. March 2023 <https://www.epa.gov/nutrientpollution/issue>.

Ersek, Kaitlyn. *5 Types Of Soil Microbes And What They Do For Plants.* Web. June 2021. <https://www.holganix.com/blog/5-types-of-soil-microbes-and-what-they-do- for-plants>.

Esslemont J. E., *Bahá'u'lláh and the New Era.* 1923. Wilmette, Ill.: Bahá'í Publishing Trust, 1978. Print.

Fairtrade Foundation, About Coffee. Web. March 2021 <https://www.fairtrade.org.uk/farmers-and-workers/coffee/about-coffee/>. Fakhrzadegan, Shahin, Hossein Gholami-Doon, Bagher Shamloo, and Solmaz Shokouhi-

Moghaddam. *The Relationship between Personality Disorders and the Type of Crime Committed and Substance Used among Prisoners*, 2017. Web. May 2021 <https://www.ncbi.nlm.nih.gov/pmc/articles/PMC5742412/>.

FAO (Food and Agriculture Organization of the United Nations). *Farming Systems and Poverty: Improving Farmers' Livelihoods In A Changing World* by Dixon, Gulliver and Gibbon. Emphasis added. Web. February 2023 <https://agris.fao.org/agris-search/search.do?recordID=XF2002402935>.

_____ . Small family farmers produce a third of the world's food. Web. March 2023 <https://www.fao.org/news/story/en/item/1395127/icode/>.

_____ . *Women hold the key to building a world free from hunger and poverty.* Web. February 2022 <https://www.fao.org/news/story/pt/item/460267/icode/>.

Farrell, John. *Why Teleology Isn't Dead.* 2016. Web.October 2020 <https://www.forbes.com/sites/johnfarrell/2016/06/08/why-teleology-isnt-dead/#6c6452196d69>.

Faust Gallery. *Nature's Role in American Indian Culture*, May 20 2019. Web. August 2021 <https://www.faustgallery.com/natures-role-in-american-indian-culture/#:~:text=Native%20Americans%20hold%20a%20deep%20reverence%20for%20nature.,aspects%20of%20their%20understanding%20and%20way%20of% 20life.>.

Ferris, Shaun, Peter Robbins, Rupert Best, Don Seville, Abbi Buxton, Jefferson Shriver, and Emily Wei. *Linking Smallholder Farmers to Markets and the Implications for Extension and Advisory Services*. Web. December 2022 <https://agrilinks.org/sites/default/files/resource/files/MEAS%20Discussion%20Paper%204%20-%20Linking%20Farmers%20To%20Markets%20-%20May%202014_0.pdf>.

Feynman, Richard P., Nobel prize winning physicist. *Religion is a culture of faith; science is a culture of doubt*. Web. October 2015 <http://www.gly.uga.edu/railsback/1122sciencedefns.html>.

_____. *The Pleasure of Finding Things Out* 1999. Web. October 2015 <http://www.gly.uga.edu/railsback/1122sciencedefns.html>.

Fieldhouse, Paul and Chris Jones Kavelin. (2001). *Food, Justice, and the Bahá'í Faith, in Examination of the Environmental Crisis.* Bahá'í Library Online. Web. September 2020 <https://bahai library.com/jones_environmental_crisis&chapter=all>.

Flajnik, Martin F. and Masanori Kasahara, *Origin and evolution of the adaptive immune system: genetic events and selective pressures*. Web. January 2023 <https://www.ncbi.nlm.nih.gov/pmc/articles/PMC3805090/>.

Farm Aid. *Understanding the Economic Crisis Family Farms are Facing*. Web. February 2023 <https://www.farmaid.org/blog/fact-sheet/understanding-economic-crisis-family-farms-are-facing/#:~:text=The%20strain%20on%20today%E2%80%99s%20farm%20economy%20is%20no,farmers%20that%20allow%20them%20to%20make%20a%20living.>.

Focus group at Saint Petersburg College. *Social Sciences Research: Qualitative vs. Quantitative Research.* Web. August 2022 <https://spcollege.libguides.com/c.php?g=254343&p=1695372>.

Food Security Information Network (FSIN). *2020 Global Report on Food Crises*. Web. March 2021 <https://www.google.com/url?sa=t&rct=j&q=&esrc=s&source=web&cd=&ved=2ahUKEwjct7Kdo7_vAhXWKM0KHZNdA_c4MhAWMA16BAgnEAM&url=https%3A%2F%2Fwww.fsinplatform.org%2Fsites%2Fdefault%2Ffiles%2Fresources%2Ffiles%2FGRFC_2020_ONLINE_200420_FINAL.pdf&usg=AOvVaw1LlycOJS31n3ailW_zw uk7>.

Ford, Mary Hanford. *The economic teaching of 'Abdu'l-Bahá*. 21 March 1917. Star of the West – 5. p. 13. Web. June 2021 <http://bahairesearch.com/>.

Fresco, Louise O, Floor Geerling-Eiff, Anne-Charlotte Hoes, Lan van Wassenaer, Krijn J Poppe, and Jack G.A.J van der Vorst. *Sustainable food systems: do agricultural economists have a role?* European Review of Agricultural Economics, jbab026. Web. July 2021, <https://doi.org/10.1093/erae/jbab026>.

Fresh Harvest, *Your direct partnership with local, organic farmers*. Web. March 2021 <https://freshharvestga.com/>.

Gail, Marzieh. (1987). *Summon Up Remembrance.* Oxford: Oxford: George Ronald, 1987. Web. February 25, 2021 <https://bahai- library.com/gail_summon_up_remembrance>.Gould, Stephen J., *Nonoverlapping Magisteria.* 1997. Web. December 2016 <http://www.stephenjaygould.org/library/gould_noma.html>.

Gandhimohan, M. V. *Mahatma Gandhi and the Bahá'ís: Striving towards a Nonviolent Civilization,* 2000. New Delhi: Baha'i Publishing Trust of India. Web. August 2020 <https://bahai-library.com/gandhimohan_gandhi_bahais_nonviolence>.

Gomes, Leandro, Alexandre Madeira and Luís Soares Barbosa. *Synchronous searching for DNA patterns.* Web. January 2021 <http://mlcsb2018.web.ua.pt/images/LivroAbsMLCSB18.pdf#page=31>.

Gonzalez, Carmen G. *Trade Liberalization, Food Security and the Environment: The Neoliberal Threat to Sustainable Rural Development.* Web. July 2021 <https://www.google.com/url?sa=t&rct=j&q=&esrc=s&source=web&cd=&ved=2ahUKEwimm6yO9ezxAhXYQc0KHRa8CkMQFjAIegQIDRAD&url=https%3A%2F%2Fdigitalcommons.law.seattleu.edu%2Fcgi%2Fviewcontent.cgi%3Farticle%3D1385%26context%3Dfaculty&usg=AOvVaw3wulDsxJVeRGHMkUYH20v8>.

Gonzalez, Wenceslao. (2013). *Value Ladenness and the Value-free Ideal in Scientific Research.* Handbook of the Philosophical Foundations of Business Ethics. 1503-1521.

Groce, Eric. *The Health Effects of Cannabis and Cannabinoids: The Current State of Evidence and Recommendations for Research*. February 2023 <https://www.ncbi.nlm.nih.gov/books/NBK425748/>.

Hacker, J. David. *New Estimate Raises Civil War Death Toll.* (G. Gugliotta, Ed.) The New York Times. (2012, April 2). Web. November 2020, <https://www.nytimes.com/2012/04/03/science/civil-war-toll-up-by-20-percent-in-new-estimate.html>.

Hajjar, Lisa, Eduardo de Leon Buendia, Patrick Fairbanks, Emma Kuskey, Sasha Misco, and Ada Quevedo. *Cultures of Resistance: The Struggle Against Domestic Violence in Arab Societies*. Handbook of Healthcare in the Arab World. Web. January 2020 <https://link.springer.com/referenceworkentry/10.1007%2F978-3-030-36811-1_201>.

Hamilton, Alan C., Medicinal plants, conservation and livelihoods. Biodiversity and Conservation 13: 1477–1517, 2004. Web. September 2021 <https://d1wqtxts1xzle7.cloudfront.net/30197352/a-00116-with-cover-page-v2.pdf?Expires=1668999626&Signature=NPQHMd1L4bkPs9GZne2q86XlB778KTXI0My3wxzNfUJwODMLMdjrCj9~RFJ50il7V9J3CGWM1IAR9Et08vFR5FteszL~8UvffFsReL8iAEXagFUOXSirduoZvrd7Tx8pTJtu3edTd7QptOVcO-gKAxcpq3fua22KhjM4rFd-o-TxhsHawPePOEqFXnPMhhnNC7QFV74raoivYR0p4L2dnhhXKcoD9iWiWrueGJdjcUzIRP7KmxQK0uj2~bLOhfSPfibQulHiQYUl04kYv1vud3iy6fWbSsEhVSGzkZyTIE1GSuYIv4aDocn65uGSY9Vq-CZmaqI5WZTJLKwWr1qLow__&Key-Pair-Id=APKAJLOHF5GGSLRBV4ZA>.

Han Seo, Dong, et al., *Single-step ambient-air synthesis of graphene from renewable precursors as electrochemical genosensor*. January 30 2017. Web. April 3, 2021 < https://www.nature.com/articles/ncomms14217>.

Hanle, Aline. *Love and Science: A New Dimension to Life*. 2011. @CATALYSTWORLD. Web. September 2020 <https://www.modernlifeblogs.com/2011/09/love-and-science-a-new-dimension-to-life/>.

Hatcher, Hatcher, William S. *The Concept of Spirituality*. Ottawa: Bahá'í Studies, volume 11. Association for Bahá'í Studies. Web. September 2020 <https://bahai-library.com/hatcher_bw18_spiritua

_____. *Logic and Logos*. Oxford: George Ronald, 1990. Print.

_____. *Love, Power, and Justice.* (A. f. America, Ed.) Ottawa: Journal of Bahá'í Studies, 9:3. Web. September 2020 <https://bahai-library.com/hatcher_love_power_justice>.

Hauser, Philip M. (1909-1994), Demographer and Census Expert, qtd. in Theodore Berland's *The Scientific Life.* Web. June 2012 <http://www.gly.uga.edu/railsback/1122sciencedefns.html>.

Heller, Martin C. and Keoleian Gregory A. *Life Cycle-Based Sustainability Indicators for Assessment of the U.S. Food System*, The University of Michigan Center for Sustainable Systems, 2000. qtd. in Center for Sustainable Systems. Web. March 2014 <http://css.snre.umich.edu/css_doc/CSS01-06.pdf>.

Helmenstine, Anne Marie Ph.D. *Understanding Quantitative Analysis in Chemistry*, Updated on July 07, 2019Web. August 2022 <https://www.thoughtco.com/definition-of-quantitative-analysis-604627>.

Heiskanen, Markku and Anni Lietonen, *Crime and Gender. A Study on how Men and Women are Represented in International Crime Statistics.* Helsinki 2016. Web. April 2021 <https://heuni.fi/documents/47074104/49491250/Crime_and_gender_taitto.pdf/b11659ba-06db-7f05-ca4d- c2b7c05f492a/Crime_and_gender_taitto.pdf?t=1607458023820>.

Herrick, Elizabeth. *The Bahá'í Dispensation*. 2, May 1923. Star of the West (Vol. 8). Web. June 2021 < http://bahairesearch.com/>.

Herrington, A.J. *New Cannabis Jobs Report Reveals Marijuana Industry's Explosive Employment Growth*. Web. September 2020 <https://www.forbes.com/sites/ajherrington/2022/02/23/new-cannabis-jobs- report-reveals-marijuana-industrys-explosive-employment- growth/?sh=7943b98023f2>.

Heussner, Ki Mae. ABC News. Web. April 2015 <http://abcnews.go.com/WN/Technology/stephen-hawking-religion-science-win/story?id=10830164http://abcnews.go.com/WN/Technology/stephen- hawking-religion-science-win/story?id=10830164>.

Hicks, Dan. *On the Ideal of Autonomous Science.* Philosophy of Science, (2011), 78(5), 1235-1248. doi:10.1086/662255. Web. December 2022 <https://www.cambridge.org/core/search?filters%5BauthorTerms%5D=Dan%20Hicks&eventCode=SE-AU>.

Hilo, Alexander et al. *A specific role of iron in promoting meristematic cell division during adventitious root formation.* Journal of Experimental Botany, Volume 68, Issue 15, 9 September 2017, Pages 4233–4247. Web. January 2020 <https://academic.oup.com/jxb/article/68/15/4233/4068695>.

Hirst, K. Kris. (2019). *Plant Domestication, a unique opportunity to identify the genetic basis of adaptation.* Web. June 2020, <https://www.thoughtco.com/plant- domestication-table-dates-places-170638>.

_____ (2019). *Plant Domestication. Dates and Locations of Human Farming Advances.* ThoughtCo . Web. September 2020, from • By K. Kris Hirst Plant Domestication. Dates and Locations <https://www.thoughtco.com/plant-domestication-table- dates-places-170638>.

Hawksworth, David L. and Robert Lücking, *Fungal Diversity Revisited: 2.2 to 3.8 Million Species.* Microbiol Spectr. 2017 Jul;5(4). Web. April 2021 <https://pubmed.ncbi.nlm.nih.gov/28752818/>.

Hodgson, Erin W. *Beneficial Insects: Lacewings and Antilions*. Published by Utah State University Extension and Utah Plant Pest Diagnostic Laboratory ENT-124-08 July 2008. Web. March 2021 <https://www.google.com/url?sa=t&rct=j&q=&esrc=s&source=web&cd=&ved=2ahUKEwia5aT5kLbvAhUq1lkKHWEjBekQFjABegQIARAD&url=https%3A%2F%2Fdigitalcommons.usu.edu%2Fcgi%2Fviewcontent.cgi%3Farticle%3D1856%26context%3Dextension_curall&usg=AOvVaw06i9FvTaR6oK1gCM4DcmtC>.

Hofstede, Geert. *Culture's Consequences: International Differences in Work Related Values.* Beverly Hills CA: Sage Publications, 1980.

Hornby, Helen. comp. *Lights of Guidanc.* Web. March 2014: Ocean Research Library. *A Bahá'í.*

Huitt, William. *What Is a Human Being and Why Is Education Necessary.* May 2017. Web. August 2017 <http://www.edpsycinteractive.org/topics/intro/human.html>.

_____ . *Citizenship. Cosmic-Citizenship.* 2015. Web. August 2017 <http://www.cosmic-citizenship.org/>.

Husain, Akbar and Aqeel Khan. *Clarifying spiritual values among organizational development personnel*. Web. November 2022 <https://www.internationalscholarsjournals.com/articles/clarifying-spiritual- values-among-organizational-development-personnel.pdf>.

Hwang, S., Kwon, K.T., Lee, S.H. et al. *Correlates of burnout among healthcare workers during the COVID-19 pandemic in South Korea*. Web. February 2023 <https://www.nature.com/articles/s41598-023-30372-x>

Independent, *A female revolution? How necessity may have been the mother of agricultural invention*. Web, February 2009 <https://www.independent.co.uk/news/world/world-history/a-female- revolution-how-necessity-may-have-been-the-mother-of-agricultural-invention- 1604969.html>.

Insights IAS. *Role of Agriculture in Indian Economy.* Web. February 2023 <https://www.insightsonindia.com/agriculture/role-of-agriculture-in-indian-economy/>.

International Labor Organization (ILO). *Agriculture; plantations; other rural sectors.* Web. January 2015 <http://ilo.org/global/industries-and-sectors/agriculture- plantations-other-rural-sectors/lang--en/index.htm>.

_____ . ILO Statement to the 56th Commission on the Status of Women. *Adoption of international labour standards key to supporting rural women.* New York March 2012. Web. July 2017 <http://www.ilo.org/newyork/at-the-un/commission-on- the-status-of-women/WCMS_209379/lang--en/index.htm>.

_____ . *Global Farm Worker Issues*. October 2003 Volume 9 Number 4.Web. February 2014 <http://migration.ucdavis.edu/rmn/more.php?id=785_0_5_0>.

Jammer, Max. *Einstein and Religion.* Princeton University Press, 1999. Web. December 2016 <file:///C:/Users/Leonardo%20Duque/Dropbox/Downloads/s6681(1).pdf>.

Jamshidi-Kia, Fatemeh, Zahra Lorigooini, and Hossein Amini-Khoe. *Medicinal plants: Past history and future perspective.* Journal of Herbmed Pharmacol. 2018. Web. September 2021 <http://herbmedpharmacol.com/Article/jhp-1198>.

Johnson, Ian. *Leaving the Land: China's Great Uprooting: Moving 250 Million Into Cities.* New York Times June 2013. Web. March 2014. <http://www.nytimes.com/2013/06/16/world/asia/chinas-great-uprooting- moving-250-million-into-cities.html?pagewanted=all>.

Joseph, Anu. Women against arrack. Organizing for change: India. Web. March 2021 <https://pubmed.ncbi.nlm.nih.gov/12290000/>.

Joseph, Enamuthu, *Global Nutrition and Development.* Center for Human Resources, State University of New York at Plattsburgh. 1996.

Kemerling, Garth. *A Dictionary of Philosophical Terms and Names.* Web. 18 June 2012 <http://www.philosophypages.com/dy/index.htm>.

Kessler, H. G. *The Diary of a Cosmopolitan*, London: Weidenfeld and Nicolson, 1971, p.157; quoted in *Einstein and Religion* by Max Jammer (Princeton University Press, 1999) pp. 39-40.Web. April 2015 <http://einsteinandreligion.com/religioncomments.html>.

Kluge, Ian. *Ethics Based on Science Alone?* Published in Studies in Bahá'í Philosophy, vol. 4, pages 1-23. Idyllwild, CA: Charles Schlacks, 2015. Web. December 2022 <https://bahai-library.com/author/Ian+Kluge>.

_____ . *Nietzsche and the Bahá'í Writings: A First Look.* Haj Mehdi Arjmand Colloquium, *2017. 18*, pp. 351-424. Wilmette, Illinois: Lights of Irfan, 18, pages 351-424. Web. February 24, 2021 <https://bahai-library.com/kluge_nietzsche_bahai_reprint>.

_____ . *Reason and the Bahá'í Writings.* Web. May 2020 <https://bahai-library.com/kluge_reason_writings_2013>.

_____ . *Reason and the Bahá'í Writings: The Use and Misuse of Logic and Persuasion.* Seattle: ABS Conference. Web. September 2020 <https://bahai-library.com/kluge_reason_bahai_writings>.

_____ . *Relativism and the Bahá'í Writings* (Vols. Volume 9, pages 179-238). Wilmette, Illinois: Lights of Irfan. Irfan Colloquia. Web. September 2020 <https://bahai-library.com/kluge_relativism_bahai_writings>.

_____ . *Some Answered Questions: A Philosophical Perspective* (Vols. Volume 10, pages 149-274). Wilmette, Illinois: Lights of Irfan. Web. September 2020 <https://bahai-library.com/kluge_saq_philosophical_perspective>.

_____ . *The Aristotelian Substratum of the Baha'i Writings.* Web. July 2017 <https://www.bahaiphilosophy.com/the-aristotelian-substratum-1-.html>.

_____ . *The Bahá'í Philosophy of Human Nature.* The Journal of Bahá'í Studies 27.1-2 2017. Web. March 2021 <https://bahai- library.com/kluge_philosophy_human_nature>.

_____. *The Bahá'í Writings and the Buddhist Doctrine of Emptiness: An Initial Survey*. (W. H. Colloquium, Producer) Web. September 2020, from Bahá'í Library Online: <https://bahai-library.com/kluge_buddhist_doctrine_emptiness>.

Konikow, L. *Groundwater depletion in the United States (1900-2008).* U.S. Geological Survey (USGS) Scientific Investigations Report. 2013. qtd. in Center for Sustainable Systems. Web. March 2014 <http://css.snre.umich.edu/css_doc/CSS01-06.pdf>.

Kulieshov, Aleksandr. *Qualitative Physics in a Metaphysical Perspective*. Web. November 2022 <https://pathofscience.org/index.php/ps/article/view/607>.

Kwelagobe, Mpule K. *Commentary - Investing in Rural Women: Closing the Gender Gap in African Agriculture.* Web. February 2014 <http://globalfoodforthought.typepad.com/global-food-for- thought/2013/10/commentary-investing-in-rural-women-closing-the-gender-gap-in-african-agriculture-1.html>.

Latimer, George O. *Economic Justice*. Pilgrimage in November 1919. Web. January 2022 <https://oceanlibrary.com/fountain-of wisdom_bahaullah>.

Liodakis, George. *Political Economy, Capitalism and Sustainable Development*. Department of Sciences, Technical University of Crete, 73100 Chania, Greece. Sustainability 2010, 2(8), 2601-2616. Web. July 2021 <https://doi.org/10.3390/su2082601>.

Lodish H., Berk A., Zipursky S.L., et al. *Molecular Cell Biology Chapter 22, Integrating Cells into Tissues.* 4th edition. New York: W. H. Freeman; 2000. Web. March 2021 <https://www.ncbi.nlm.nih.gov/books/NBK21717/>.

Loehle, Craig. *On Human Origins: A Bahá'í Perspective.* Published in the Journal of Bahá'í Studies Vol. 2, number 4, 1990. Web. September 2020 <https://pdfs.semanticscholar.org/1ebf/ba4ab74d1c60bc5a2e23d58380d1fed97 7f4.pdf>.

Lunt, Alfred E. *The supreme affliction. A study in Bahá'í economics and socialization.* p.1204, July 1932. Star of the West – 10. p. 97

Martínez M., Miguel. *Investigacion Cualitativa y Cuantitativa. La Investigación Cualitativa (Síntesis Conceptual)*. Translated by the author. Web. August 2022 <https://kevinchristianpuchuyucra.blogspot.com/p/invetigacion.html>.

Martorell, Reynaldo. *Undernutrition During Pregnancy and Early Childhood: Consequences for Cognitive and Behavioral Development*. Elsevier Science B.V. 1.997.

Maultsby, Beth E. *High Conflict Family Law Matters and Personality Disorders*. Web. August 2021 <https://www.gbfamilylaw.com/wp-content/uploads/2014/03/Beth-High- Conflict-Family-Law-Matters.pdf>.

May, Monica. *Surprising science: Not all our cells have the same DNA*. Sanford Burnham Prebys, published August 15, 2018. Web. January 2021 < https://www.sbpdiscovery.org/news/beaker-blog/surprising-science-not-all-our- cells-have-same-dna>.

Mayo Clinic. *Narcissistic personality disorder.* Web. August 2020 <https://www.mayoclinic.org/diseases-conditions/narcissistic-personality-disorder/symptoms-causes/syc-20366662>.

_____ . Personality Disorders. Web. August 2020 <https://www.mayoclinic.org/diseases-conditions/personality-disorders/symptoms-causes/syc-20354463>.

Mazzucato, Mariana The Value of Everything. Making and taking in the global economy. Web. December 2022 <https://marianamazzucato.com/books/the-value-of- everything>.

McCarthy, Eugene M. *Online Biology Dictionary*. Web. July 2017 <http://www.macroevolution.net/biology-dictionary.html>.

McKeon, Richard. *The Basic Works of Aristotle*. New York: Random House, 1941. Print. McLean, Jack. *The Art of Rhetoric in the Writings of Shoghi Effendi*, 2007. Wilmette, Illinois: Published in Lights of Irfan, Volume 8, pages 203-256.

McMurry John. *Organic Chemistry with Biological Applications.* Cengage Learning, Apr 14, 2014 - Science - 672 pages. Web. <https://books.google.com/books?id=KDIeCgAAQBAJ&pg=PP4&dq=Organic+Chemistry+with+Biological+Applications++John+McMurry&hl=en&sa=X&ved=0CCcQ6AEwAGoVChMIybrm2ubCxwIViXc-Ch2pBgqq#v=onepage&q=Organic%20Chemistry%20with%20Biological%20Applications%20%20John%20McMurry&f=false>.

McDonough, William and Michael Braungart. *Cradle to Cradle: Remaking the Way We Make Things.* 2.002.

McGregor, Deborah. *Traditional Ecological Knowledge.* Ideas: the Arts and Science Review, vol. 3, no. 1, spring 2006. Faculty of Arts & Science, University of Toronto. Web. August 2021 <https://www.silvafor.org/assets/silva/PDF/DebMcGregor.pdf>.

Mendoza Palacios, Rudy. *Investigación cualitativa y cuantitativa. Diferencias y limitaciones*. Web. August 2022 <https://recursos.salonesvirtuales.com/assets/bloques/investigacionDIFERENY_LIMITACIONES.pdf>.

Merriam-Webster Dictionary. Web. October 2012.<www.Merriam-Webster.com>.

Messina, Nena P. *Shocking Statistics and Facts about Alcohol-Related Crimes,* 2020, September 4. Web. February 17, 2021, from Addiction Resource: <https://addictionresource.com/alcohol/effects/alcohol-related-crimes/>.

Ministry of Foreign Affairs of Denmark. August 2010. *Gender and Value Chain Development.* Web. January 2018 <http://www.netpublikationer.dk/um/10511/html/entire_publication.htm#Section4.1 >.

MIT Climate Portal. *Soil-Based Carbon Sequestration*. Web. February 2023 <https://climate.mit.edu/explainers/soil-based-carbon-sequestration>.

Molnar, Charles and Jane Gair. *Concepts of Biology* – 1st Canadian Edition. Web. May 2021 <https://opentextbc.ca/biology/chapter/5-2-the-light-dependent-reactions-of-photosynthesis/>.

Mora, Camilo, Derek P. Tittensor, SinaAdl, Alastair G. B. Simpson and Boris Worm**.** *How Many Species on Earth and in the Ocean?* Web. April 2014 <http://www.plosbiology.org/article/info%3Adoi%2F10.1371%2Fjournal.pbio.10 01127>

Morgenstern, Julian. *A Jewish Interpretation of the Book of Genesis.* Web. October 2015 <https://books.google.com/books?id=SJ4sAAAAYAAJ&pg=PA74&lpg=PA74&dq=%22Eye+for+eye,+and+tooth+for+tooth%22+nomad&source=bl&ots=xjGsfD34-p&sig=eEnN9poHgfsqLiVq-cscorzoOQo&hl=en&sa=X&ved=0CB0Q6AEwAGoVChMIjuqzjqrryAIVRzkmCh14HAYL#v=onepage&q=%22Eye%20for%20eye%2C%20and%20tooth%20for%20tooth%22%20nomad&f=false>.

Mountain Empire Community College, *Water/Wastewater courses. Lesson 8: Nitrification and Denitrification.* Emphasis added. Web. March 2023 <https://water.mecc.edu/courses/Env149/lesson8_print.htm>.

Mouritsen, Henrik. *Long-distance navigation and magnetoreception in migratory animals.* Web. June 2020 <https://www.researchgate.net/publication/325605112_Long-distance_navigation_and_magnetoreception_in_migratory_animal>.

Movendi International. *Big Alcohol Exposed.* Web. August 2022 <https://iogt.org/the-issues/advocacy/exposing-big-alcohol/>.

Mycorrhizal Online LLC. Web. June 2021 <https://www.mycorrhizalonline.com/about>.

Nagamangala, Chidananda Kanchiswamy, *Chemical diversity of microbial volatiles and their potential for plant growth and productivity.* Web. March 2021 <https://www.ncbi.nlm.nih.gov/pmc/articles/PMC4358370/>.

National Human Genome Research Institute (NHGRI). *Lysosome* . Web. April 2021 <https://www.genome.gov/genetics-glossary/Lysosome>.

National Oceanic and Atmospheric Administration (NOAA). "NOAA Scientists: Midwest drought brings fourth smallest Gulf of Mexico 'Dead Zone' since 1985. 2008. qtd. in Center for Sustainable Systems. Web. March 2014 <http://css.snre.umich.edu/css_doc/CSS01-06.pdf>.

National Pesticide Information Center, Oregon State University. *Beneficial Insects.* Web. April 2021 <http://npic.orst.edu/envir/beneficial/index.html>.

Natural Resources Conservation Service. United States Department of Agriculture. *Soil Formation.* Web. April 2021 <https://www.nrcs.usda.gov/wps/portal/nrcs/detail/wa/soils/?cid=nrcs144p2_0 36333>.

Nawaz, Farzana. *Are Women Less Corrupt than Men? And Other Gender/Corruption Questions*, 2011. Transparency International . Web. August 2020 <https://blog.transparency.org/2011/10/07/are-women-less-corrupt-than-men- and-other-gendercorruption-questions/index.html>.

Northpoint Recovery, A Slave for Addiction: The Origins of the Word. Web. December 2022 <https://www.northpointrecovery.com/blog/slave-addiction-origins-word/>.

O'Connor, James. *Political economy of ecology of socialism and capitalism.* 25 Feb 2009. Web. June 2021 <https://www.tandfonline.com/doi/abs/10.1080/10455758909358386>.

Ohio State Test - *Physical Science: Practice & Study Guide / Science Courses, Gravitational Force: Definition, Equation & Examples.* Web. January 2020 <https://study.com/academy/lesson/gravitational-force-definition-equation-examples.html>.

onwuka, Brownmang and Brown Mang. *Effects of soil temperature on some soil properties and plant growth*. Web. February 2023 <https://medcraveonline.com/APAR/effects-of-soil-temperature-on-some-soil-properties-and-plant-growth.html>.

Ortega, Rocel Amor, Alexander Mahnert, Christian Berg, Henry Müller, Gabriele Berg. *The plant is crucial: specific composition and function of the phyllosphere microbiome of indoor ornamentals.* FEMS Microbiology Ecology, December 2016. Volumen 92 Issue 12. Web. February 19, 2021 <https://academic.oup.com/femsec/article/92/12/fiw173/2570375>.

Paradies, Mark. *Definition of a Root Cause,* July 19, 2019. TapRooT® Web. March 2020 <https://www.taproot.com/definition-of-a-root-cause/>.

Paran, Ilan and Esther van der Knaap. *Genetic and molecular regulation of fruit and plant domestication traits in tomato and pepper*. Journal of Experimental Botany, Volume 58, Issue 14, November 2007, Pages 3841–3852. Web. February 2022 <https://doi.org/10.1093/jxb/erm257>.

PBS News Hour Science. *The Science of Snowflakes, and Why No Two Are Alike*, 2011. Web. October 2020 <https://www.pbs.org/newshour/science/the-science-of-snowflakes>.

Pew Research Center. *The Gender Gap in Religion Around the World. Women are generally more religious than men, particularly among Christian.* Web. March 2021 <https://www.pewforum.org/2016/03/22/the-gender-gap-in-religion-around-the-world/>.

_____ . *Scientists and Belief*. Web. March 2023 <https://www.pewresearch.org/religion/2009/11/05/scientists-and-belief/>.

Pitawanakwat, Lillian. *The Medicine Wheel Teachings.* Web. August 2022 <https://ecampusontario.pressbooks.pub/movementtowardsreconciliation/chapter/the-medicine-wheel-teachings/>.

Plato. *Azquotes*. Web. October 2020 <https://www.azquotes.com/author/37843-Plato?p=4>.

_____ . *Crito: or, the Duty of a Citizen.* Project Gutenberg's EBook of Apology, Crito, and Phaedo of Socrates, by Plato 2004.Web. June 2012 <http://www.gutenberg.org/files/13726/13726-h/13726-h.htm - crito_or_the_duty_of_a_citizen>.

_____ . *Sophist.* Trans. Benjamin Jowett. The Project Gutenberg EBook of Sophist, by Plato, 2008.Web. June 2012 <http://www.gutenberg.org/files/1735/1735-h/1735-h.htm>.

_____ . *Symposium*. Trans. Benjamin Jowett. The Project Gutenberg EBook of Symposium, by Plato, 2008.Web. June 2012 <http://www.gutenberg.org/files/1600/1600-h/1600-h.htm>.

_____ . *Theaetetus.* Trans. Benjamin Jowett. The Project Gutenberg EBook of Theaetetus, by Plato 2008.Web. June 2012 <http://www.gutenberg.org/files/1726/1726-h/1726-h.htm>.

Plotinus. *The Six Enneads*. Trans. by Stephen Mackenna and B. S. Page.Web. June 2012 <http://classics.mit.edu/Plotinus/enneads.3.third.html>.

Popper, Karl. *Truth and the growth of knowledge*. 1962. Web. March 2015 <http://books.google.com.co/books?hl=en&lr=&id=YzvKJ-

2nJn4C&oi=fnd&pg=PA285&dq=%22Popper,+Karl%22+1962+falsifiability&ots=U
VEzPyo3Ao&sig=r2DUgYQAM7WlJ- 36rgqi_9nsOMA#v=onepage&q=refutability&f=false>.

Popular Mechanics. *17,000 miles more than a conventional touring tire.* Web. March 2021 <http://www.popularmechanics.com/cars/hybrid-electric/a7593/the-science- behind-yokohamas-orange-oil-tires-8146348/ 33% more mileage>.

Porter, Theodore M., *History of statistics*, From Wikipedia, the free encyclopedia; probability and statistics. Web. August 2022 <https://en.wikipedia.org/wiki/History_of_statistics>.

Prezi, *Harlequin Ladybird.* Web. March 2021 <https://prezi.com/50g5-pxjdjry/harlequin-ladybird/>.

Price, Michael. *Strong women did a lot of the heavy lifting in ancient farming societies*, 2017. Web. January 2021 <https://www.sciencemag.org/news/2017/11/strong- women-did-lot-heavy-lifting-ancient-farming-societies>.

Putnam, Hilary. *Objectivity and the Science/Ethics Distinction, The fact/value dichotomy: background*. Web. November 2022 <https://citeseerx.ist.psu.edu/viewdoc/download?doi=10.1.1.642.7638&rep=rep 1&type=pdf>.

Radhakrishnan, Rajiv, Samuel T. Wilkinson, and Deepak Cyril D'Souza. *Gone to pot – a review of the association between cannabis and psychosis.* Web. December 2022 <https://www.frontiersin.org/articles/10.3389/fpsyt.2014.00054/full>.

Ramírez Rojas, Manuel Alvaro, Diego Andrés Guevara F. y Ana María Korena G. *Mercado de Trabajo y Condiciones del Empleo en Colombia: Los efectos de la globalización*, 2003. Web. May 2021 <https://www.google.com/url?sa=t&rct=j&q=&esrc=s&source=web&cd=&ved=2ahUKEwjHyZjDsKzwAhWIW80KHdeJCEsQFjABegQIAxAD&url=http%3A%2F%2Fwww.fuac.edu.co%2Frecursos_web%2Fobservatorio%2Fpublicaciones%2FMercado_de_Trabajo_y_condiciones_del_empleo_en_Colombia.pdf&usg=AOvVaw0dIfuH2g3rjuLaUkfY9Abd>.

Regeneration International. *Why Regenerative Agriculture?* February 16, 2017. Web. March 2021 <https://regenerationinternational.org/why-regenerative-agriculture/>.

Rexford, Orcella. *Radiant Acquiescence.* Wilmette, Illinois: Baha'i Publishing Committee World Order, 1937. Web. September 2020 <https://bahai-library.com/rexford_radiant_acquiescence>.

Richardson, Sarah S. Review of Science, Policy, and the Value-Free Ideal, and: Philosophy of Science after Feminism. Feminist Formations, vol. 24 no. 2, 2012, p. 199-205. Web. November 2022 <https://muse.jhu.edu/article/484109>.

Ridley, Matt. *Genome: the autobiography of a species in 23 chapters*, p. 271.1999. Web. October 2015 <http://www.gly.uga.edu/railsback/1122sciencedefns.html>.

Riederer, Markus. *Biology of the plant cuticle*. Oxford: Blackwell Publishing 2006. Web. April 2021 <https://doi.org/10.1002/9781119312994.apr0229>.

Riposati, Andrea. (2020, September 13). *Why is everyone's DNA so unique? The part of DNA which makes us unique*. Web. from Dante Labs: <https://us.dantelabs.com/blogs/blog/why-is-everyones-dna-so-unique>.

Robin, León. *El Pensamiento Griego y los Orígenes del Espíritu Científico. La Evolución de la Humanidad.* Trans. by the author. Enciclopedia Uteha, Vol. 14. Unión Tipográfica Editorial Hispano Americana, México, 1.962. Print.

Saiedi, Nader. *Gate of the Heart.* Ocean 2.0 Reader is an interfaith library. Web. January 2021.

_____ . *Logos and Civilization: spirit, history, and order in the Writings of Bahá'u'lláh.* Maryland: University Press of Maryland, 1995. Print.

Sánchez-Salas, José Luis et al. *Inactivation of Bacterial Spores and Vegetative Bacterial Cells by Interaction with ZnO-Fe2O3 Nanoparticles and UV Radiation.* Web. June 2020 <https://www.researchgate.net/publication/326298784_Inactivation_of_Bacterial_Spores_and_Vegetative_Bacterial_Cells_by_Interaction_with_ZnO-Fe2O3_Nanoparticles_and_UV_Radiation>.

Satz, Mario. *Ecología y Mitología.* Trans. by the author. Revista Nueva Conciencia. Barcelona, España. Integral Ed., 1991. Print.

Schiffman, Richard. *What do plants and people have in common? More than you think.* The Christian Science Monitor, May 7, 2021. Web. August 2021 <https://www.csmonitor.com/Books/Book-Reviews/2021/0507/What-do-plants-and-people-have-in-common-More-than-you-think>.

Seed, Tracy. *The Universal Force of Love – Albert Einstein.* Web. May 2022 <https://tracyseed.com/love-energy/>.

Schnitzer, Stefan A., Klironomos JN, Hillerislambers J, Kinkel LL, Reich PB, Xiao K, Rillig MC, Sikes BA, Callaway RM, Mangan SA, van Nes EH, Scheffer M. *Soil microbes drive the classic plant diversity-productivity pattern.* Ecology. 2011 Feb;92(2):296-303. Web. October 2015 <http://www.ncbi.nlm.nih.gov/pubmed/21618909>.

Seligman, Martin. *Quotes and Sayings.* Web. March 2020 Inspiring Quotes:|<https://www.inspiringquotes.us/author/5934-martin-seligman>; Goodreads: <https://www.goodreads.com/work/quotes/28610>.

Semple, Ian C. *Obedience*, 1991, July 26. Web. September 2020 <http://bahaitalks.blogspot.com/2009/12/obedience.html>.

Shiva, Vandana. *The Violence of the Green Revolution: Third World Agriculture, Ecology, and Politics.* The University Press of Kentucky, 2016. Web. June 2021 <https://muse.jhu.edu/book/44425>.

Shoghi Effendi. *Bahá'í Administration.* Wilmette, Illinois: Bahá'í Publishing Trust, 1960. Print.

_____ . *Call to the Nations*. Web. April 2021 <https://ocean-server-production.dev2.us/?key=ocean-viewer&bookId=618558523022cd067c376d2fa43db912&paraId=para_12>.

_____ . *Citadel of Faith.* Web. August 2021 <https://www.bahai.org/library/authoritative-texts/search#q=Importance%20of%20the%20American%20Indians >.

_____ . *Directives from the Guardian.* Web. March 2014: Ocean Research Library.

_____ . *From a letter written on behalf of the Guardian to an individual believer, May 12, 1925, quoted in Living the Life.*

_____ . *Letters from the Guardian to Australia and New Zealand.* Haifa: 1942. Ocean 2.0 Reader is an interfaith library.

_____ . Summary Statement - 1947, Special UN Committee on Palestine. Web. January 2015: Ocean Research Library.

_____ . *The Advent of Divine Justice.* New Delhi, India: Bahá'í Publishing Trust, 1968.

_____ . *The Promised Day is Come*, 1980. Wilmette: Bahá'í Publishing Trust. Bahá'í Reference Library, Web. September 2020 <https://www.bahai.org/library/authoritative-texts/shoghi-effendi/promised- day-come/1#617979506>.

_____ . *The World Order of Bahá'u'lláh. Selected Letters*. Bahá'í Reference Library. Web. May 2021 <https://www.bahai.org/library/authoritative-texts/shoghi- effendi/world-order-bahaullah/4#450837692>.

_____ . *The World Religion of Bahá'u'lláh: A Summary of Its Aims, Teachings and History to the High Comissioner for Palestine*, 1933. Web. July 2021 <https://ocean- server-production.dev2.us/?key=ocean-viewer&bookId=8abd2d15d021ef8bf4d1f5e34a6e7340&paraId=para_10>.

Simard, Suzanne and Peter Wohlleben. *Intelligent Trees - The Documentary*. Web. March 2021 <https://www.intelligent-trees.com/>.

Skok, Stephen. *What Impact Did the Belgian Presence in Rwanda Have to Spark Further Conflict?* Web. February 27, 2021 <https://education.seattlepi.com/impact-did- belgian-presence-rwanda-spark-further-conflict-5558.html>.

Skwarecki, Beth. Friendly Viruses Protect Us Against Bacteria. Web. April 2021 <https://www.sciencemag.org/news/2013/05/friendly-viruses-protect-us- against-bacteria>.

South African History Online (SAHO). (2016, August 2). *Rwanda*. Web. September 2020, from South African History Online (SAHO): <https://www.sahistory.org.za/place/rwanda>.

Star of the West. Web. July 2017: Ocean Research Library.

Steel, Alix. *Why Colombia Coffee Farmers Are 'Desperate'*. Bloomberg News provides economic, financial, and political market news. May 24th, 2018. Web. December 2019 <https://www.bloomberg.com/news/videos/2018-05-24/why-colombia- coffee-farmers-are-desperate-video>.

Steel, Daniel, Chad Gonnerman, Aaron M. McCright, and Itai Bavli; *Gender and Scientists' Views about the Value-Free Ideal. Perspectives on Science* 2018; 26 (6): 619–657. Web. December 2022 <https://doi.org/10.1162/posc_a_00292>.

Stenmark, Mikael. *Rationality and Different Conceptions of Science* quoted in Wentzel van Huyssteen on Rationality in Science and Theology. A Discussion Note on F. LeRon Shults (ed.). The Evolution of Rationality. Interdisciplinary Essays in Honor of J. Wentzel van Huyssteen. Grand Rapids, MI. /Cambridge, U.K.: Eerdmans,2006.ArsDisputandiVolume 9 (2009). Web. December 2016 <https://books.google.com/books?id=r3y4FdVCQ1sC&pg=PA51&lpg=PA51&dq=%E2%80%9CScience+should+be+nonresponsible+in+the+sense%22&source=bl&ots=AXA0f3sijJ&sig=sEwiliwS0tLikMPUvae2Zig_1AY&hl=en&sa=X&ved=0ahUKEwj

217Pp1fbQAhXmg1QKHT68AiQQ6AEIHDAA#v=onepage&q=%E2%80%9CScience%20should%20be%20non-responsible%20in%20the%20sense%22&f=false>.

Stephenson, Deborah. *How to Get Iron in Plants.* Web. June 2020 <https://www.gardenguides.com/82464-iron-plants.html>.

Streiff, Jeffrey. *The Golden Rule.* Web. December 2016 <http://www.goldenruleart.com/>. St. Fleur, Nicholas. *Songbirds Can Hear Tornadoes Long Before They Form.* The Atlantic. Web. May 2021 <https://www.theatlantic.com/technology/archive/2014/12/birds-can-hear-tornadoes-coming-long-before-they-form/383898/>.

Stice, Kyle and Sale, Andrew. *Aruligo Pineapple Value Chain – Mapping Report.* November 2008. Web. January 2018 <www.pacificfarmers.com/wp.../06/Solomon-Islands- Pineapple-Mapping-report.pdf>.

Sun, Joseph C., Lopez-Verges, Sandra, Kim, Charles C., DeRisi, Joseph L. and Lewis L. Lanier. *NK Cells and Immune "Memory".* Web. May 2014 <http://www.jimmunol.org/content/186/4/1891.short>.

Swiss Platform for Sustainable Cocoa. Web. January 2023 <https://www.kakaoplattform.ch/about-cocoa/cocoa-facts-and-figures>.

Szymborska, Wislawa. Nobel Prize. *In Praise of Self Deprecation.* Web. March 2022 <https://carolyncrantz.tumblr.com/post/154766189599/in-praise-of-self- deprecation>.

Ta'eed, Lata. *Sex, Gender, and New Age Stereotyping.* London: Association for Baha'i Studies English-Speaking Europe. 1994.

Taherzadeh, Adib. *The Covenant of Bahá'u'lláh.* Ocean by Chad Jones. Web. October 2020

_____. *The Revelation of Bahá'u'lláh* v 4. Oxford: George Ronald, Publisher, 1988. Print.

Tang. Ruifei. *An Analysis of Traditional Ecological Knowledge's Status and its Conservation Options.* Web. August 2021 <https://researcharchive.vuw.ac.nz/xmlui/bitstream/handle/10063/2785/thesis.pdf?sequence=>.

The Báb. *Selections from the Writings of the Báb.* Chatham: W & J Mackay Limited, 1976 Web. January 2023 <https://www.bahai.org/library/authoritative- texts/the-bab/selections-writings-bab/1#103864442>.

The Global Alliance for the Rights of Nature. *What is Rights of Nature?* Web. June 2021 <https://www.therightsofnature.org/what-is-rights-of-nature/>.

The Universal House of Justice. *A Chaste and Holy Life.* A Compilation Prepared by the Research Department of the Universal House of Justice, From a letter 6 February 1973 written by the Universal House of Justice to all National Spiritual Assemblies. Web. February 2022 <https://www.bahai.org/library/authoritative- texts/compilations/chaste-holy-life/4#500269670>.

_____. "*Alcoholics Anonymous*" (From a letter dated March 19, 1973, to a National Spiritual Assembly) <https://bahai-library.com/uhj_alcoholics_anonymoushtml>.

_____. *Consultation: A Compilation.* (Wilmette: Bahá'í Publishing Trust), p. 22 [Ed. - sel. 45]. Shoghi Effendi, Universal House of Justice, Research Department of the Universal House of Justice, "Community Functioning, Issues Concerning: Fostering the Development of

_____. Bahá'í Communities". Web. June 2021 <https://bahai-library.com/uhj_issues_community_functioning>.

_____. *Challenges for Bahá'í Youth in a Western Way of Life.* 2013. Haifa.

_____. *Family Life*, A Compilation of Extracts from the Bahá'í Writings and from Letters Written by and on Behalf of Shoghi Effendi and the Universal House of Justice. Prepared by the Research Department of the Universal House of Justice. March 2008. Bahá'í Reference Library, Web. May 2021 <https://www.bahai.org/library/authoritative-texts/search#q=la%20mujer>.

_____. *Issues Related to the Study of the Bahá'í Faith.* The Universal House of Justice, Department of the Secretariat, 8 Feb. 1998. Bahá'í Reference Library, Web. June 2021 <https://www.bahai.org/library/authoritative-texts/compilations/issues- related-study-bahai-faith/11#373231889>.

_____. *Letter to an individual* October 11/1978. The Universal House of Justice, Department of the Secretariat. Emphasis added. Bahá'í Reference Library, Web. June 2021 <https://www.bahai.org/library/authoritative-texts/search#q=Each%20individual%20is%20unique%20and%20has%20a%20unique%20path%20to%20tread%20in%20his%20lifetime>.

_____. *To the Bahá'ís of the World.* 25 November 2020 <https://www.bahai.org/library/authoritative-texts/the-universal-house-of-justice/messages/20201125_001/1#300076430>

_____. *Non-Involvement in Partisan Politics.* Web. March 2021 <https://bahai-library.com/uhj_non-involvement_partisan_politics>.

_____. *One Common Faith.* Bahá'í Reference Library. Web. June 2021 <https://www.bahai.org/library/other-literature/official-statements- commentaries/one-common-faith/one-common-faith.xhtml?3305dd4f >.

_____. *Social Action*. A Compilation Prepared by the Research Department of the Universal House of Justice. August 2020. Bahá'í Reference Library, Web. May 202 <www.bahai.org/library>.

_____. *The Promise of World Peace.* 1985. New Delhi, India: Bahá'í Publishing Trust, 1992. Print.

_____. *Women*. Compiled by the Research Department of the Universal House of Justice. January 1986. Revised July 1990. Bahá'í Reference Library, Web. May 2021 <https://www.bahai.org/library/authoritative-texts/downloads>.

ThoughtCo. *Van der Waals Forces: Properties and Components.* January 2020 <https://www.thoughtco.com/definition-of-van-der-waals-forces-604681>.

Toepfer, Georg. *Teleology and its constitutive role for biology as the Science of organized systems in nature*, 2012. (Vols. 43, Issue 1). Studies in History and Philosophy of Biological and Biomedical Sciences.

Toynbee, Arnold. *Estudio de la Historia.* Compendio. Madrid: Alianza Editorial. 1981. Print.

Traoré, Doussou. *Cocoa and Coffee Value Chains in West and Central Africa: Constraints and Options for Revenue-Raising Diversification*. Food and Agriculture Organization of the

United Nations. February 2009. Web. January 2018 <www.fao.org/fileadmin/templates/est/.../FAO_AAACP_Paper_Series_No_3_1_.pdf>.

True Price. Web. December 2022 <https://trueprice.org/>.

UN News. *UN report: one-third of world's food wasted annually, at great economic, environmental cost*. Web. March 2021 <https://news.un.org/en/story/2013/09/448652>.

United Nations. *Ending Poverty*. Web. February 2022 <https://www.un.org/en/global-issues/ending-poverty>.

US Bahá'í News Service, "Star of the West". (n.d.). Download from Ocean interfaith library, by Chad Jones. Web. January 2022 <https://oceanlibrary.com/library>.

U.S. Environmental Protection Agency (EPA). *Pesticide Industry Sales and Usage: 2006 and 2007 Market Estimates*. 2011. qtd. in Center for Sustainable Systems. Web. March 2014 <http://css.snre.umich.edu/css_doc/CSS01-06.pdf>.

USDA, Economic Research Service. *Characteristics of principal farm operator households, by gross farm sales, 2011.* 2012. qtd. in Center for Sustainable Systems. Web. March 2014 <http://css.snre.umich.edu/css_doc/CSS01-06.pdf>.

_____ . *Food Dollar Series*. 2013. Web. March 2014 <http://css.snre.umich.edu/css_doc/CSS01-06.pdf>.

USDA, National Resources Conservation Service. *2007 National Resources Inventory*. 2009. qtd. in Center for Sustainable Systems. Web. March 2014 <http://css.snre.umich.edu/css_doc/CSS01-06.pdf>.

Vardiman, Larry. *Evolution and the Snowflake*. Institute for Creation Research, 1986. Web. April 2021 <https://www.icr.org/article/evolution-snowflake>.

Virtual Medical Center. *Anatomy and Physiology of the Nasal Cavity (Inner Nose) and Mucosa*. Web. April 2021 <https://www.myvmc.com/medical-centres/lungs- breathing/anatomy-and-physiology-of-the-nasal-cavity-inner-nose-and- mucosa/>.

Wagner, Stephen C. *Biological Nitrogen Fixation*. Web. March 2023 <https://www.nature.com/scitable/knowledge/library/biological-nitrogen-fixation-23570419/>.

Walsh, Roger. *El Compromiso con el Planeta.* Revista Nueva Conciencia. p. 82. Trans. by the author.

_____ . *Staying Alive: the psychology of human survival.* Boulder, Colorado: Shambhala Publications Inc., 1984. Print.

Weisberger, Mindy. *Billions of Viruses Are Falling to Earth Right Now (But That Isn't Why You Have the Flu)*. February 07, 2018. Web. April 2021 < https://www.livescience.com/61689-viruses-fall-from-sky.html>.

Wikipedia. *Anthropocentrism*. Web. December 2016 <http://en.wikipedia.org/wiki/Anthropocentrism>.

_____ . *Innate Immune System.* <https://en.wikipedia.org/wiki/Innate_immune_system>. WHO, World

Health Organization, Diabetes. Web. December 2022 <https://www.who.int/health-topics/diabetes#tab=tab_1>.

———. Drugs (psychoactive). Web, December 2022 <https://www.who.int/health-topics/drugs-psychoactive#tab=tab_2>.

———. Global status report on alcohol and health, 2018. Web. March 2021 <https://www.google.com/url?sa=t&rct=j&q=&esrc=s&source=web&cd=&ved=2ahUKEwjR6crOhqvvAhU6VzABHYNuApoQFjABegQIAxAD&url=https%3A%2F%2Fapps.who.int%2Firis%2Fbitstream%2Fhandle%2F10665%2F274603%2F978924156 5639-eng.pdf&usg=AOvVaw1SduA7alO9hM0JDfEMsy7c>.

Wu, Rui, et al. *Inter-Species Grafting Caused Extensive and Heritable Alterations of DNA Methylation in Solanaceae Plants*, 2013. Web. March 2021 <https://www.ncbi.nlm.nih.gov/pmc/articles/PMC3628911/>.

Yang, Yu, Jun Yang, Wei-Min Wu, Jiao Zhao, Yiling Song, Longcheng Gao, Ruifu Yang, and Lei Jiang. *Biodegradation and Mineralization of Polystyrene by Plastic-Eating Mealworms: Part 1. Chemical and Physical Characterization and Isotopic Test*. Environ. Sci. Technol. 2015, 49, 20, 12080–12086, Web. April 2021 <https://pubs.acs.org/doi/abs/10.1021/acs.est.5b02661>.

Yaseen, Tabassum, Kawsar Ali, Fazal Munsif, Abdur Rab, Masood Ahmad, Muhammad Israr, and Aziz Khan Barai. *Influence of Arbuscular Mycorrhizal Fungi, Rhizobium Inoculation and Rock Phosphate on Growth and Quality of Lentil.* Web. March 2023 <https://pakbs.org/pjbot/PDFs/48(5)/42.pdf>.

York, Susan Morris. *Understanding Neutrophils: Function, Counts, and More.* Web. March 2021 <https://www.healthline.com/health/neutrophils>.

Zeder, Melinda A. *Core questions in domestication research.* Proceedings of the National Academy of Sciences (PNAS), 2015. Web. September 2020 <https://www.pnas.org/content/112/11/3191>.

Zeigler, David. *The Science of destiny: what is the meaning of life, anyway?* Altadena, California: Skeptics Society & Skeptic Magazine, 2015. Web. November 2020 <https://go.gale.com/ps/anonymous?id=GALE%7CA439185733&sid=googleScholar&v=2.1&it=r&linkaccess=abs&issn=10639330&p=AONE&sw=w>.

Zimmer, Carl. *Welcome to the virosphere, the unimaginably vast world of virus diversity.* The New York Times, Mar 24, 2020. Web. May 2020 <https://www.nytimes.com/2020/03/24/science/viruses-coranavirus- biology.html >.

ISBN 978-1-7320081-8-2 Second Edition

www.ingramcontent.com/pod-product-compliance
Lightning Source LLC
Chambersburg PA
CBHW061109070526
44583CB00027B/3240